CAMBRIDGE LIBRARY COLLECTION

Books of enduring scholarly value

Physical Sciences

From ancient times, humans have tried to understand the workings of
the world around them. The roots of modern physical science go back to
the very earliest mechanical devices such as levers and rollers, the mixing
of paints and dyes, and the importance of the heavenly bodies in early
religious observance and navigation. The physical sciences as we know them
today began to emerge as independent academic subjects during the early
modern period, in the work of Newton and other 'natural philosophers',
and numerous sub-disciplines developed during the centuries that followed.
This part of the Cambridge Library Collection is devoted to landmark
publications in this area which will be of interest to historians of science
concerned with individual scientists, particular discoveries, and advances in
scientific method, or with the establishment and development of scientific
institutions around the world.

Autobiography of Sir George Biddell Airy

Sir George Biddell Airy (1801–92) was a prominent mathematician and
astronomer. He was an honorary fellow of Trinity College, Cambridge, fellow
of the Royal Society and Astronomer Royal from 1835 until 1881. His many
achievements include important work on planetary orbits, the calculation of
the mean density of the earth and the establishment of the prime meridian
at Greenwich. He was also consulted by the government on a wide range of
issues and projects, serving on the weights and measures commission, the
tidal harbours commission and the railway gauge commission as well as
acting as an advisor for the repair of Big Ben and the laying of the Atlantic
cable. His autobiography, edited by his son Wilfred, comprises ten chapters
and is drawn from the astronomer's own records of the scientific work he
carried out at Greenwich Observatory along with his printed reports and
private and business correspondence.

Cambridge University Press has long been a pioneer in the reissuing of out-of-print titles from its own backlist, producing digital reprints of books that are still sought after by scholars and students but could not be reprinted economically using traditional technology. The Cambridge Library Collection extends this activity to a wider range of books which are still of importance to researchers and professionals, either for the source material they contain, or as landmarks in the history of their academic discipline.

Drawing from the world-renowned collections in the Cambridge University Library, and guided by the advice of experts in each subject area, Cambridge University Press is using state-of-the-art scanning machines in its own Printing House to capture the content of each book selected for inclusion. The files are processed to give a consistently clear, crisp image, and the books finished to the high quality standard for which the Press is recognised around the world. The latest print-on-demand technology ensures that the books will remain available indefinitely, and that orders for single or multiple copies can quickly be supplied.

The Cambridge Library Collection will bring back to life books of enduring scholarly value (including out-of-copyright works originally issued by other publishers) across a wide range of disciplines in the humanities and social sciences and in science and technology.

Autobiography of Sir George Biddell Airy

GEORGE BIDDELL AIRY
EDITED BY WILFRED AIRY

CAMBRIDGE
UNIVERSITY PRESS

CAMBRIDGE UNIVERSITY PRESS

Cambridge, New York, Melbourne, Madrid, Cape Town, Singapore,
São Paolo, Delhi, Dubai, Tokyo

Published in the United States of America by Cambridge University Press, New York

www.cambridge.org
Information on this title: www.cambridge.org/9781108008945

© in this compilation Cambridge University Press 2010

This edition first published 1896
This digitally printed version 2010

ISBN 978-1-108-00894-5 Paperback

AUTOBIOGRAPHY

OF

SIR GEORGE BIDDELL AIRY, K.C.B.

𝕷𝖔𝖓𝖉𝖔𝖓: C. J. CLAY AND SONS,
CAMBRIDGE UNIVERSITY PRESS WAREHOUSE,
AVE MARIA LANE.
𝕲𝖑𝖆𝖘𝖌𝖔𝖂: 263, ARGYLE STREET.

𝕷𝖊𝖎𝖕𝖟𝖎𝖌: F. A. BROCKHAUS.
𝕹𝖊𝖂 𝖄𝖔𝖗𝖐: THE MACMILLAN CO.

AUTOBIOGRAPHY

OF

SIR GEORGE BIDDELL AIRY, K.C.B.,

M.A., LL.D., D.C.L., F.R.S., F.R.A.S.,

HONORARY FELLOW OF TRINITY COLLEGE, CAMBRIDGE,

ASTRONOMER ROYAL FROM 1836 TO 1881.

EDITED BY

WILFRID AIRY, B.A., M. INST. C.E.

CAMBRIDGE:

AT THE UNIVERSITY PRESS.

1896

𝕮𝖆𝖒𝖇𝖗𝖎𝖉𝖌𝖊:

PRINTED BY J. AND C. F. CLAY,
AT THE UNIVERSITY PRESS.

PREFACE.

THE life of Airy was essentially that of a hard-working, business man, and differed from that of other hard-working people only in the quality and variety of his work. It was not an exciting life, but it was full of interest, and his work brought him into close relations with many scientific men, and with many men high in the State. His real business life commenced after he became Astronomer Royal, and from that time forward, during the 46 years that he remained in office, he was so entirely wrapped up in the duties of his post that the history of the Observatory is the history of his life. For writing his business life there is abundant material, for he preserved all his correspondence, and the chief sources of information are as follows :

(1) His Autobiography.
(2) His Annual Reports to the Board of Visitors.
(3) His printed Papers entitled " Papers by G. B. Airy."
(4) His miscellaneous private correspondence.
(5) His letters to his wife.
(6) His business correspondence.

(1) His Autobiography, after the time that he became Astronomer Royal, is, as might be expected, mainly a record of the scientific work carried on at the Greenwich Observatory: but by no means exclusively so. About the time when he took charge of the Observatory there was an immense development of astronomical enterprise: observatories were springing up in all directions, and the Astronomer

Royal was expected to advise upon all of the British and Colonial Observatories. It was necessary also for him to keep in touch with the Continental Observatories and their work, and this he did very diligently and successfully, both by correspondence and personal intercourse with the foreign astronomers. There was also much work on important subjects more or less connected with his official duties—such as geodetical survey work, the establishment of time-balls at different places, longitude determinations, observation of eclipses, and the determination of the density of the Earth. Lastly, there was a great deal of time and work given to questions not very immediately connected with his office, but on which the Government asked his assistance in the capacity of general scientific adviser: such were the Correction of the Compass in iron ships, the Railway Gauge Commission, the Commission for the Restoration of the Standards of Length and Weight, the Maine Boundary, Lighthouses, the Westminster Clock, the London University, and many other questions.

Besides those above-mentioned there were a great many subjects which he took up out of sheer interest in the investigations. For it may fairly be said that every subject of a distinctly practical nature, which could be advanced by mathematical knowledge, had an interest for him: and his incessant industry enabled him to find time for many of them. Amongst such subjects were Tides and Tidal Observations, Clockwork, and the Strains in Beams and Bridges. A certain portion of his time was also given to Lectures, generally on current astronomical questions, for he held it as his duty to popularize the science as far as lay in his power. And he attended the meetings of the Royal Astronomical Society with great regularity, and took a very active part in the discussions and business of the Society. He also did much work for the Royal Society, and (up to a certain date) for the British Association.

All of the foregoing matters are recorded pretty fully in his Autobiography up to the year 1861. After that date the Autobiography is given in a much more abbreviated form, and might rather be regarded as a collection of notes for his Biography. His private history is given very fully for the first part of his life, but is very lightly touched upon during his residence at Greenwich. A great part of the Autobiography is in a somewhat disjointed state, and appears to have been formed by extracts from a number of different sources, such as Official Journals, Official Correspondence, and Reports. In editing the Autobiography it has been thought advisable to omit a large number of short notes relating to the routine work of the Observatory, to technical and scientific correspondence, to Papers communicated to various Societies and official business connected with them, and to miscellaneous matters of minor importance. These in the aggregate occupied a great deal of time and attention. But, from their detached nature, they would have but little general interest. At various places will be found short Memoirs and other matter by the Editor.

(2) All of his Annual Reports to the Board of Visitors are attached to his Autobiography and were evidently intended to be read with it and to form part of it. These Reports are so carefully compiled and are so copious that they form a very complete history of the Greenwich Observatory and of the work carried on there during the time that he was Astronomer Royal. The first Report contained only four pages, but with the constantly increasing amount and range of work the Reports constantly increased in volume till the later Reports contained 21 pages. Extracts from these Reports relating to matters of novelty and importance, and illustrating the principles which guided him in his conduct of the Observatory, have been incorporated with the Autobiography.

(3) The printed "Papers by G. B. Airy" are bound in

14 large quarto volumes. There are 518 of these Papers, on a great variety of subjects: a list of them is appended to this history, as also is a list of the books that he wrote, and one or two of the Papers which were separately printed. They form a very important part of his life's work, and are frequently referred to in the present history. They are almost all to be found in the Transactions of Societies or in newspapers, and extend over a period of 63 years (1822 to 1885). The progress made in certain branches of science during this long period can very fairly be traced by these Papers.

(4) His private correspondence was large, and like his other papers it was carefully arranged. No business letters of any kind are included under this head. In this correspondence letters are occasionally found either dealing with matters of importance or in some way characteristic, and these have been inserted in this biography. As already stated the Autobiography left by Airy is confined almost entirely to science and business, and touches very lightly on private matters or correspondence.

(5) The letters to his wife are very numerous. They were written during his occasional absences from home on business or for relaxation. On these occasions he rarely let a day pass without writing to his wife, and sometimes he wrote twice on the same day. They are full of energy and interest and many extracts from them are inserted in this history. A great deal of the personal history is taken from them.

(6) All correspondence in any way connected with business during the time that he was Astronomer Royal is to be found at the Royal Observatory. It is all bound and arranged in the most perfect order, and any letter throughout this time can be found with the greatest ease. It is very bulky, and much of it is, in a historical sense, very interesting. It was no doubt mainly from this correspondence

that the Autobiography, which so far as related to the Greenwich part of it was almost entirely a business history, was compiled.

The history of the early part of his life was written in great detail and contained a large quantity of family matter which was evidently not intended for publication. This part of the Autobiography has been compressed. The history of the latter part of his life was not written by himself at all, and has been compiled from his Journal and other sources. In both these cases, and occasionally in short paragraphs throughout the narrative, it has been found convenient to write the history in the third person.

2, THE CIRCUS,
    GREENWICH.

NOTE.

The Syndics of the Cambridge University Press desire to express their thanks to Messrs Macmillan & Co. for their courteous permission to use in this work the steel engraving of Sir George Biddell Airy published in *Nature* on October 31, 1878.

TABLE OF CONTENTS.

CHAPTER I.

Personal Sketch of George Biddell Airy.

THE history of Airy's life, and especially the history of his life's work, is given in the chapters that follow. But it is felt that the present Memoir would be incomplete without a reference to those personal characteristics upon which the work of his life hinged and which can only be very faintly gathered from his Autobiography.

He was of medium stature and not powerfully built: as he advanced in years he stooped a good deal. His hands were large-boned and well-formed. His constitution was remarkably sound. At no period in his life does he seem to have taken the least interest in athletic sports or competitions, but he was a very active pedestrian and could endure a great deal of fatigue. He was by no means wanting in physical courage, and on various occasions, especially in boating expeditions, he ran considerable risks. In debate and controversy he had great self-reliance, and was absolutely fearless. His eye-sight was peculiar, and required correction by spectacles the lenses of which were ground to peculiar curves according to formulæ which he himself investigated: with these spectacles he saw extremely well, and he commonly carried three pairs, adapted to different distances: he took great interest in the changes that took place in his eye-sight, and wrote several Papers on the subject. In his later years he became somewhat deaf, but not to the extent of serious personal inconvenience.

The ruling feature of his character was undoubtedly Order. From the time that he went up to Cambridge to the end of his life his system of order was strictly maintained. He wrote his autobiography up to date soon after he had taken his degree, and made his first will as soon as he had any money to leave. His accounts were perfectly kept by double entry throughout his life, and he valued extremely the order of book-keeping: this facility of keeping accounts was very useful to him. He seems not to have destroyed a document of any kind whatever: counterfoils of old cheque-books, notes for tradesmen, circulars, bills, and correspondence of all sorts were carefully preserved in the most complete order from the time that he went to Cambridge; and a huge mass they formed. To a high appreciation of order he attributed in a great degree his command of mathematics, and sometimes spoke of mathematics as nothing more than a system of order carried to a considerable extent. In everything he was methodical and orderly, and he had the greatest dread of disorder creeping into the routine work of the Observatory, even in the smallest matters. As an example, he spent a whole afternoon in writing the word "Empty" on large cards, to be nailed upon a great number of empty packing boxes, because he noticed a little confusion arising from their getting mixed with other boxes containing different articles; and an assistant could not be spared for this work without with-drawing him from his appointed duties. His arrangement of the Observatory correspondence was excellent and elaborate: probably no papers are more easy of reference than those arranged on his system. His strict habits of order made him insist very much upon detail in his business with others, and the rigid discipline arising out of his system of order made his rule irksome to such of his subordinates as did not conform readily to it: but the efficiency of the Observatory unquestionably depended mainly upon it. As his powers failed with age the ruling passion for order assumed a greater prominence; and in his last days he seemed to be more anxious to put letters which he received into their proper place for reference than even to master their contents.

His nature was eminently practical, and any subject which had a distinctly practical object, and could be advanced by mathematical investigation, possessed interest for him. And his dislike of mere theoretical problems and investigations was proportionately great. He was continually at war with some of the resident Cambridge mathematicians on this subject. Year after year he criticised the Senate House Papers and the Smith's Prize Papers question by question very severely: and conducted an interesting and acrimonious private correspondence with Professor Cayley on the same subject. His great mathematical powers and his command of mathematics are sufficiently evidenced by the numerous mathematical treatises of the highest order which he published, a list of which is appended to this biography. But a very important feature of his investigations was the thoroughness of them. He was never satisfied with leaving a result as a barren mathematical expression. He would reduce it, if possible, to a practical and numerical form, at any cost of labour: and would use any approximations which would conduce to this result, rather than leave the result in an un-fruitful condition. He never shirked arithmetical work: the longest and most laborious reductions had no terrors for him, and he was remarkably skilful with the various mathematical expedients for shortening and facilitating arithmetical work of a complex character. This power of handling arithmetic was of great value to him in the Observatory reductions and in the Observatory work generally. He regarded it as a duty to finish off his work, whatever it was, and the writer well remembers his comment on the mathematics of one of his old friends, to the effect that "he was too fond of leaving a result in the form of three complex equations with three unknown quantities." To one who had known, in some degree, of the enormous quantity of arithmetical work which he had turned out, and the unsparing manner in which he had devoted himself to it, there was something very pathetic in his discovery, towards the close of his long life, "that the figures would not add up."

His energy and business capacity were remarkable. He was made for work and could not long be happy without it. Whatever subject he was engaged upon, he kept his object clearly in view, and made straight for it, aiming far more at clearness and directness than at elegance of periods or symmetry of arrangement. He wrote his letters with great ease and rapidity: and having written them he very rarely had occasion to re-write them, though he often added insertions and interlineations, even in the most important official letters. Without this it would have been impossible for him to have turned out the enormous quantity of correspondence that he did. He never dictated letters, and only availed himself of clerical assistance in matters of the most ordinary routine. In his excursions, as in his work, he was always energetic, and could not endure inaction. Whatever there was of interest in the places that he visited he examined thoroughly and without delay, and then passed on. And he thus accomplished a great deal in a short vacation. His letters written to his wife, while he was on his excursions, are very numerous and characteristic, and afford ample proofs of his incessant energy and activity both of body and mind. They are not brilliantly written, for it was not in his nature to write for effect, and he would never give himself the trouble to study the composition of his letters, but they are straight-forward, clear, and concise, and he was never at a loss for suitable language to express his ideas. He had a wonderful capacity for enjoyment: the subjects that chiefly interested him were scenery, architecture, and antiquities, but everything novel or curious had an interest for him. He made several journeys to the Continent, but by far the greater number of his excursions were made in England and Scotland, and there were few parts of the country which he had not visited. He was very fond of the Lake District of

Cumberland, and visited it very frequently, and each time that he went there the same set of views had an eternal freshness for him, and he wrote long descriptions of the scenery and effects with the same raptures as if he had seen it for the first time. Many of his letters were written from Playford, a village in a beautiful part of Suffolk, a few miles from Ipswich. Here he had a small property, and generally stayed there for a short time once or twice a year. He was extremely fond of this country, and was never tired of repeating his walks by the well-known lanes and footpaths. And, as in Cumberland, the Suffolk country had an eternal freshness and novelty for him. Wherever he went he was indefatigable in keeping up his acquaintance with his numerous friends and his letters abound in social reminiscences.

His memory was singularly retentive. It was much remarked at school in his early days, and in the course of his life he had stored up in his memory an incredible quantity of poetry, ballads, and miscellaneous facts and information of all sorts, which was all constantly ready and at his service. It is almost needless to add that his memory was equally accurate and extensive in matters connected with science or business.

His independence of character was no doubt due to and inseparable from his great powers. The value of his scientific work greatly depended upon his self-reliance and independence of thought. And in the heavy work of remodelling the Observatory it was a very valuable quality. This same self-reliance made him in his latter years apt to draw conclusions too confidently and hastily on subjects which he had taken up more as a pastime than as work. But whatever he touched he dealt with ably and in the most fearless truthseeking manner, and left original and vigorous opinions.

He had a remarkably well-balanced mind, and a simplicity of nature that appeared invulnerable. No amount of hero-worship seemed to have the least effect upon him. And from a very early time he was exposed to a great deal of it. His mind was incessantly engaged on investigations of Nature, and this seems to have been with him, as has been the case with others, a preserving influence. This simplicity of character he retained throughout his life. At the same time he was sensible and shrewd in his money matters and attentive to his personal interests. And his practical good sense in the general affairs of life, combined with his calm and steady consideration of points submitted to him, made his advice very valuable. This was especially recognized by his own and his wife's relations, who consulted him on many occasions and placed the fullest confidence in his absolute sense of justice as well as in his wise counsel. He was extremely liberal in proportion to his means, and gave away money to a large extent to all who had any claim upon him. But he was not in any sense reckless, and kept a most cautious eye on his expenses. He was not indifferent to the honours which he received in the scientific world, but he does not appear to have sought them in any way, and he certainly did not trouble himself about them.

His courtesy was unfailing: no amount of trouble could shake it. Whether it was the Secretary of the Admiralty, or a servant girl wanting her fortune told: whether a begging-letter for money, or miscellaneous invitations: all had their answer in the most clear and courteous language. But he would not grant personal interviews when he could avoid it: they took up too much of his time. His head was so clear that he never seemed to want for the clearest and most direct language in expressing his meaning, and his letters are models of terseness.

In all his views and opinions he was strongly liberal. At Cambridge at an early date he was one of the 83 members of the Senate who supported the application to permit the granting of medical degrees without requiring an expression of assent to the religious doctrines of the Church of England. And in 1868 he declined to sign a petition against the abolition of religious declarations required of persons admitted to Fellowships or proceeding to the degree of M.A. And he was opposed to every kind of narrowness and exclusiveness. When he was appointed to the post of Astronomer Royal, he stipulated that he should not be asked to vote in any political election. But all his views were in the liberal direction. He was a great reader of theology and church history, and as regarded forms of worship and the interpretation of the Scriptures, he treated them with great respect, but from the point of view of a freethinking layman. In the Preface to his "Notes on the Earlier Hebrew Scriptures" he says, "In regard to the general tone of these notes, I will first remark that I have nothing to say on the subject of verbal inspiration. With those who entertain that doctrine, I can have nothing in common. Nor do I recognize, in the professedly historical accounts, any other inspiration which can exempt them from the severest criticism that would be applicable to so-called profane accounts, written under the same general circumstances, and in the same countries." And his treatment of the subject in the "Notes" shews how entirely he took a rationalistic view of the whole question. He also strongly sided with Bishop Colenso in his fearless criticism of the Pentateuch, though he dissented from some of his conclusions. But he was deeply imbued with the spirit of religion and reflected much upon it. His whole correspondence conveys the impression of the most sterling integrity and high-mindedness, without a trace of affectation. In no letter does there appear a shadow of wavering on matters of principle, whether in public or private matters, and he was very clear and positive in his convictions.

The great secret of his long and successful official career was that he was a good servant and thoroughly understood his position. He never set himself in opposition to his masters, the Admiralty. He never hesitated to ask the Admiralty for what he thought right, whether in the way of money grants for various objects, or for occasional permission to give his services to scientific matters not immediately connected with the Observatory. Sometimes the Admiralty refused his requests, and he felt this very keenly, but he was far too busy and energetic to trouble himself about such little slights, and cheerfully accepted the situation. What was refused by one Administration was frequently granted by another; and in the meantime he was always ready to give his most zealous assistance in any matter that was officially brought before him. This cheerful readiness to help, combined with his great ability and punctuality in business matters, made him a very valuable servant, and speaking generally he had the confidence of the Admiralty in a remarkable degree. In many of his Reports to the Board of Visitors he speaks gratefully of the liberality of the Admiralty in forwarding scientific progress and research. In matters too which are perhaps of minor importance from the high stand-point of science, but which are invaluable in the conduct of an important business office, such for example as estimates and official correspondence, he was orderly and punctual in the highest degree. And, what is by no means unimportant, he possessed an excellent official style in correspondence, combined with great clearness of expression. His entire honesty of purpose, and the high respect in which he was held both at home and abroad, gave great weight to his recommendations.

With regard to his habits while he resided at the Observatory, his custom was to work in his official room from 9 to about 2.30, though in summer he was frequently at work before breakfast. He then took a brisk walk, and dined at about 3.30. This early hour had been prescribed and insisted upon by his physician, Dr Haviland of Cambridge, in whom he had great confidence. He ate heartily, though simply and moderately, and slept for about an hour after dinner. He then had tea, and from about 7 to 10 he worked in the same room with his family. He would never retire to a private room, and regarded the society of his family as highly beneficial in "taking the edge off his work." His powers of abstraction were remarkable: nothing seemed to disturb him; neither music, singing, nor miscellaneous conversation. He would then play a game or two at cards, read a few pages of a classical or historical book, and retire at 11. On Sundays he attended morning service at church, and in the evening read a few prayers very carefully and impressively to his whole household. He was very hospitable, and delighted to receive his friends in a simple and natural way at his house. In this he was most admirably aided by his wife, whose grace and skill made everything pleasant to their guests. But he avoided dinner-parties as much as possible—they interfered too much with his work—and with the exception of scientific and official dinners he seldom dined away from home. His tastes were entirely domestic, and he was very happy in his family. With his natural love of work, and with the incessant calls upon him, he would soon have broken down, had it not been for his system of regular relaxation. Two or three times a year he took a holiday: generally a short run of a week or ten days in the spring, a trip of a month or thereabouts in the early autumn, and about three weeks at Playford in the winter. These trips were always conducted in the most active manner, either in constant motion from place to place, or in daily active excursions. This system he maintained with great regularity, and from the exceeding interest and enjoyment that he took in these trips his mind was so much refreshed and steadied that he always kept himself equal to his work.

Airy seems to have had a strong bent in the direction of astronomy from his youth, and it is curious to note how well furnished he was, by the time that he became Astronomer Royal, both with astronomy in all its branches, and with the kindred sciences so necessary for the practical working and improvement of it. At the time that he went to Cambridge Physical Astronomy was greatly studied there and formed a

most important part of the University course. He eagerly
availed himself of this, and mastered the Physical Astronomy
in the most thorough manner, as was evidenced by his Papers
collected in his "Mathematical Tracts," his investigation of
the Long Inequality of the Earth and Venus, and many
other works. As Plumian Professor he had charge of the
small Observatory at Cambridge, where he did a great deal
of the observing and reduction work himself, and became
thoroughly versed in the practical working of an Observatory.
The result of this was immediately seen in the improved
methods which he introduced at Greenwich, and which were
speedily imitated at other Observatories. Optics and the
Undulatory Theory of Light had been very favourite subjects
with him, and he had written and lectured frequently upon
them. In the construction of the new and powerful telescopes
and other optical instruments required from time to time
this knowledge was very essential, for in its instrumental
equipment the Greenwich Observatory was entirely re-
modelled during his tenure of office. And in many of the
matters referred to him, as for instance that of the Lighthouses, a thorough knowledge of Optics was most valuable.
He had made a great study of the theory and construction
of clocks, and this knowledge was invaluable to him at
Greenwich in the establishment of new and more accurate
astronomical clocks, and especially in the improvement of
chronometers. He had carefully studied the theory of
pendulums, and had learned how to use them in his experiments in the Cornish mines. This knowledge he afterwards utilized very effectively at the Harton Pit in comparing
the density of the Earth's crust with its mean density ; and
it was very useful to him in connection with geodetic surveys
and experiments on which he was consulted. And his
mechanical knowledge was useful in almost everything.

The subjects (outside those required for his professional
work) in which he took most interest were Poetry, History,
Theology, Antiquities, Architecture, and Engineering. He

was well acquainted with standard English poetry, and had committed large quantities to memory, which he frequently referred to as a most valuable acquisition and an ever-present relief and comfort to his mind. History and Theology he had studied as opportunity offered, and without being widely read in them he was much at home with them, and his powerful memory made the most of what he did read. Antiquities and Architecture were very favourite subjects with him. He had visited most of the camps and castles in the United Kingdom and was never tired of tracing their connection with ancient military events: and he wrote several Papers on this subject, especially those relating to the Roman Invasions of Britain. Ecclesiastical Architecture he was very fond of: he had visited nearly all the cathedrals and principal churches in England, and many on the Continent, and was most enthusiastic on their different styles and merits: his letters abound in critical remarks on them. He was extremely well versed in mechanics, and in the principles and theory of construction, and took the greatest interest in large engineering works. This led to much communication with Stephenson, Brunel, and other Engineers, who consulted him freely on the subject of great works on which they were engaged: in particular he rendered much assistance in connection with the construction of the Britannia Bridge over the Menai Straits. There were various other subjects which he read with much interest (Geology in particular), but he made no study of Natural History, and knew very little about it beyond detached facts. His industry was untiring, and in going over his books one by one it was very noticeable how large a number of them were feathered with his paper " marks," shewing how carefully he had read them and referred to them. His nature was essentially cheerful, and literature of a witty and humourous character had a great charm for him. He was very fond of music and knew a great number of songs; and he was well acquainted with the theory of music: but he was no performer. He did not sketch freehand but made excellent drawings with his Camera Lucida.

At the time when he took his degree (1823) and for many years afterwards there was very great activity of scientific investigation and astronomical enterprise in England. And, as in the times of Flamsteed and Halley, the earnest zeal of men of science occasionally led to much controversy and bitterness amongst them. Airy was by no means exempt from such controversies. He was a man of keen sensitiveness, though it was combined with great steadiness of temper, and he never hesitated to attack theories and methods that he considered to be scientifically wrong. This led to differences with Ivory, Challis, South, Cayley, Archibald Smith, and others; but however much he might differ from them he was always personally courteous, and the disputes generally went no farther than as regarded the special matter in question. Almost all these controversial discussions were carried on openly, and were published in the Athenæum, the Philosophical Magazine, or elsewhere; for he printed nearly everything that he wrote, and was very careful in the selection of the most suitable channels for publication. He regarded it as a duty to popularize as much as possible the work done at the Observatory, and to take the public into his confidence. And this he effected by articles communicated to newspapers, lectures, numerous Papers written for scientific societies, reports, debates, and critiques.

His strong constitution and his regular habits, both of work and exercise, are sufficient explanation of the good health which in general he enjoyed. Not but what he had sharp touches of illness from time to time. At one period he suffered a good deal from an attack of eczema, and at another from a varicose vein in his leg, and he was occasionally troubled with severe colds. But he bore these ailments with great patience and threw them off in course of time. He was

happy in his marriage and in his family, and such troubles
and distresses as were inevitable he accepted calmly and
quietly. In his death, as in his life, he was fortunate : he had
no long or painful illness, and he was spared the calamity of
aberration of intellect, the saddest of all visitations.

CHAPTER II.

GEORGE BIDDELL AIRY was born at Alnwick in Northumberland on July 27th 1801.   His father was William Airy
of Luddington in Lincolnshire, the descendant of a long line
of Airys who have been traced back with a very high degree
of probability to a family of that name which was settled at
Kentmere in Westmorland in the 14th century.   A branch of
this family migrated to Pontefract in Yorkshire, where they
seem to have prospered for many years, but they were involved
in the consequences of the Civil Wars, and one member of
the family retired to Ousefleet in Yorkshire.   His grandson
removed to Luddington in Lincolnshire, where his descendants
for several generations pursued the calling of small farmers.
George Biddell Airy's mother, Ann Airy, was the daughter of
George Biddell, a well-to-do farmer in Suffolk.

William Airy, the father of George Biddell Airy, was a
man of great activity and strength, and of prudent and steady
character.   When a young man he became foreman on a
farm in the neighbourhood of Luddington, and laid by his
earnings in summer in order to educate himself in winter.
For a person in his rank, his education was unusually good,
in matters of science and in English literature.   But at the
age of 24 he grew tired of country labour, and obtained a
post in the Excise.   After serving in various Collections he
was appointed Collector of the Northumberland Collection

on the 15th August 1800, and during his service there his eldest son George Biddell Airy was born. The time over which his service as Officer and Supervisor extended was that in which smuggling rose to a very high pitch, and in which the position of Excise Officer was sometimes dangerous. He was remarkable for his activity and boldness in contests with smugglers, and made many seizures. Ann Airy, the mother of George Biddell Airy, was a woman of great natural abilities both speculative and practical, kind as a neighbour and as head of a family, and was deeply loved and respected. The family consisted of George Biddell, Elizabeth, William, and Arthur who died young.

William Airy was appointed to Hereford Collection on 22nd October 1802, and removed thither shortly after. He stayed at Hereford till he was appointed to Essex Collection on 28th February 1810, and during this time George Biddell was educated at elementary schools in writing, arithmetic, and a little Latin. He records of himself that he was not a favourite with the schoolboys, for he had very little animal vivacity and seldom joined in active play with his school-fellows. But in the proceedings of the school he was successful, and was a favourite with his master.

On the appointment of William Airy to Essex Collection, the family removed to Colchester on April 5th 1810. Here George Biddell was first sent to a large school in Sir Isaac's Walk, then kept by Mr Byatt Walker, and was soon noted for his correctness in orthography, geography, and arithmetic. He evidently made rapid progress, for on one occasion Mr Walker said openly in the schoolroom how remarkable it was that a boy 10 years old should be the first in the school. At this school he stayed till the end of 1813 and thoroughly learned arithmetic (from Walkingame's book), book-keeping by double entry (on which knowledge throughout his life he set a special value), the use of the sliding rule (which knowledge also was specially useful to him in after life), mensuration and algebra (from Bonnycastle's books). He also studied grammar in all its branches, and geography, and acquired some knowledge of English literature, beginning with that admirable book The Speaker, but it does not appear that Latin and Greek were attended to at this school. He records that at this time he learned an infinity of snatches of songs, small romances, &c., which his powerful memory retained most accurately throughout his life. He was no hand at active play: but was notorious for his skill in constructing guns for shooting peas and arrows, and other mechanical contrivances. At home he relates that he picked up a wonderful quantity of learning from his father's books. He read and remembered much poetry from such standard authors as Milton, Pope, Gay, Gray, Swift, &c., which was destined to prove in after life an invaluable relaxation for his mind. But he also studied deeply an excellent Cyclopædia called a Dictionary of Arts and Sciences in three volumes folio, and learned from it much about ship-building, navigation, fortification, and many other subjects.

During this period his valuable friendship with his uncle Arthur Biddell commenced. Arthur Biddell was a prosperous farmer and valuer at Playford near Ipswich. He was a well-informed and able man, of powerful and original mind, extremely kind and good-natured, and greatly respected throughout the county. In the Autobiography of George Biddell Airy he states as follows:

"I do not remember precisely when it was that I first visited my uncle Arthur Biddell. I think it was in a winter: certainly as early as the winter of 1812—13. Here I found a friend whose society I could enjoy, and I entirely appreciated and enjoyed the practical, mechanical, and at the same time speculative and enquiring talents of Arthur Biddell. He had a library which, for a person in middle life, may be called excellent, and his historical and antiquarian knowledge was not small. After spending one winter holiday with him, it easily came to pass that I spent the next summer holiday with him: and at the next winter holiday, finding that there

was no precise arrangement for my movements, I secretly
wrote him a letter begging him to come with a gig to fetch
me home with him : he complied with my request, giving no
hint to my father or mother of my letter: and from that
time, one-third of every year was regularly spent with him
till I went to College.   How great was the influence of this
on my character and education I cannot tell.   It was with
him that I became acquainted with the Messrs Ransome,
W. Cubitt the civil engineer (afterwards Sir W. Cubitt),
Bernard Barton, Thomas Clarkson (the slave-trade abolitionist), and other persons whose acquaintance I have valued
highly.   It was also with him that I became acquainted with
the works of the best modern poets, Scott, Byron, Campbell,
Hogg, and others: as also with the Waverley Novels and
other works of merit."

In 1813 William Airy lost his appointment of Collector
of Excise and was in consequence very much straitened
in his circumstances.   But there was no relaxation in the
education of his children, and at the beginning of 1814
George Biddell was sent to the endowed Grammar School
at Colchester, then kept by the Rev. E. Crosse, and remained
there till the summer of 1819, when he went to College.
The Autobiography proceeds as follows :

" I became here a respectable scholar in Latin and Greek,
to the extent of accurate translation, and composition of
prose Latin: in regard to Latin verses I was I think more
defective than most scholars who take the same pains, but
I am not much ashamed of this, for I entirely despise the
system of instruction in verse composition.

" My father on some occasion had to go to London and
brought back for me a pair of 12-inch globes.   They were
invaluable to me.   The first stars which I learnt from the
celestial globe were α Lyræ, α Aquilæ, α Cygni : and to this
time I involuntarily regard these stars as the birth-stars of
my astronomical knowledge.   Having somewhere seen a
description of a Gunter's quadrant, I perceived that I could

construct one by means of the globe: my father procured for me a board of the proper shape with paper pasted on it, and on this I traced the lines of the quadrant.

"My command of geometry was tolerably complete, and one way in which I frequently amused myself was by making paper models (most carefully drawn in outline) which were buttoned together without any cement or sewing. Thus I made models, not only of regular solids, regularly irregular solids, cones cut in all directions so as to shew the conic sections, and the like, but also of six-gun batteries, intrenchments and fortresses of various kinds &c.

"From various books I had learnt the construction of the steam-engine: the older forms from the Dictionary of Arts and Sciences; newer forms from modern books. The newest form however (with the sliding steam valve) I learnt from a 6-horse engine at Bawtrey's brewery (in which Mr Keeling the father of my schoolfellow had acquired a partnership). I frequently went to look at this engine, and on one occasion had the extreme felicity of examining some of its parts when it was opened for repair.

"In the mean time my education was advancing at Playford. The first record, I believe, which I have of my attention to mechanics there is the plan of a threshing-machine which I drew. But I was acquiring valuable information of all kinds from the Encyclopædia Londinensis, a work which without being high in any respect is one of the most generally useful that I have seen. But I well remember one of the most important steps that I ever made. I had tried experiments with the object-glass of an opera-glass and was greatly astonished at the appearance of the images of objects seen through the glass under different conditions. By these things my thoughts were turned to accurate optics, and I read with care Rutherford's Lectures, which my uncle possessed. The acquisition of an accurate knowledge of the effect of optical constructions was one of the most charming attainments that I ever reached. Long before I went to

College I understood the action of the lenses of a telescope better than most opticians. I also read with great zeal Nicholson's Dictionary of Chemistry, and occasionally made chemical experiments of an inexpensive kind : indeed I grew so fond of this subject that there was some thought of apprenticing me to a chemist. I also attended to surveying and made a tolerable survey and map of my uncle's farm.

"At school I was going on successfully, and distinguished myself particularly by my memory. It was the custom for each boy once a week to repeat a number of lines of Latin or Greek poetry, the number depending very much on his own choice. I determined on repeating 100 every week, and I never once fell below that number and was sometimes much above it. It was no distress to me, and great enjoyment. At Michaelmas 1816 I repeated 2394 lines, probably without missing a word. I do not think that I was a favourite with Mr Crosse, but he certainly had a high opinion of my powers and expressed this to my father. My father entertained the idea of sending me to College, which Mr Crosse recommended : but he heard from some college man that the expense would be £200 a year, and he laid aside all thoughts of it.

"The farm of Playford Hall was in 1813 or 1814 hired by Thomas Clarkson, the slave-trade abolitionist. My uncle transacted much business for him (as a neighbour and friend) in the management of the farm &c. for a time, and they became very intimate. My uncle begged him to examine me in Classical knowledge, and he did so, I think, twice. He also gave some better information about the probable expenses &c. at College. The result was a strong recommendation by my uncle or through my uncle that I should be sent to Cambridge, and this was adopted by my father. I think it likely that this was in 1816.

"In December 1816, Dealtry's Fluxions was bought for me, and I read it and understood it well. I borrowed Hutton's Course of Mathematics of old Mr Ransome, who had come to

reside at Greenstead near Colchester, and read a good' deal
of it.

"About Ladyday 1817 I began to read mathematics with
Mr Rogers (formerly, I think, a Fellow of Sidney College, and
an indifferent mathematician of the Cambridge school), who
had succeeded a Mr Tweed as assistant to Mr Crosse in the
school. I went to his house twice a week, on holiday after-
noons. I do not remember how long I received lessons from
him, but I think to June 1818. This course was extremely
valuable to me, not on account of Mr Rogers's abilities (for I
understood many things better than he did) but for its training me both in Cambridge subjects and in the Cambridge
accurate methods of treating them. I went through Euclid
(as far as usually read), Wood's Algebra, Wood's Mechanics,
Vince's Hydrostatics, Wood's Optics, Trigonometry (in a
geometrical treatise and also in Woodhouse's algebraical
form), Fluxions to a good extent, Newton's Principia to the
end of the 9th section. This was a large quantity, but I
read it accurately and understood it perfectly, and could
write out any one of the propositions which I had read in
the most exact form. My connexion with Mr Rogers was
terminated by *his* giving me notice that he could not undertake to receive me any longer : in fact I was too much for
him. I generally read these books in a garret in our house
in George Lane, which was indefinitely appropriated to my
brother and myself. I find that I copied out Vince's Conic
Sections in February 1819. The first book that I copied was
the small geometrical treatise on Trigonometry, in May
1817 : to this I was urged by old Mr Ransome, upon my
complaining that I could not purchase the book : and it was
no bad lesson of independence to me."

During the same period 1817—1819 he was occupied
at school on translations into blank verse from the Æneid
and Iliad, and read through the whole of Sophocles very
carefully.

The Classical knowledge which he thus gained at school

and subsequently at Cambridge was sound, and he took great pleasure in it: throughout his life he made a practice of keeping one or other of the Classical Authors at hand for occasional relaxation. He terminated his schooling in June 1819. Shortly afterwards his father left Colchester and went to reside at Bury St Edmund's. The Autobiography proceeds as follows:

"Mr Clarkson was at one time inclined to recommend me to go to St Peter's College (which had been much enriched by a bequest from a Mr Gisborne). But on giving some account of me to his friend Mr James D. Hustler, tutor of Trinity College, Mr Hustler urged upon him that I was exactly the proper sort of person to go to Trinity College. And thus it was settled (mainly by Mr Clarkson) that I should be entered at Trinity College. I think that I was sent for purposely from Colchester to Playford, and on March 6th, 1819, I rode in company with Mr Clarkson from Playford to Sproughton near Ipswich to be examined by the Rev. Mr Rogers, incumbent of Sproughton, an old M.A. of Trinity College: and was examined, and my certificate duly sent to Mr Hustler: and I was entered on Mr Hustler's side as Sizar of Trinity College.

"In the summer of 1819 I spent some time at Playford. On July 27th, 1819 (my birthday, 18 years old), Mr Clarkson invited me to dinner, to meet Mr Charles Musgrave, Fellow of Trinity College, who was residing for a short time at Grundisburgh, taking the church duty there for Dr Ramsden, the Rector. It was arranged that I should go to Grundisburgh the next day (I think) to be examined in mathematics by Mr Musgrave. I went accordingly, and Mr Musgrave set before me a paper of questions in geometry, algebra, mechanics, optics, &c. ending with the first proposition of the Principia. I knew nothing more about my answers at the time: but I found long after that they excited so much admiration that they were transmitted to Cambridge (I forget whether to Mr Musgrave's brother, a Fellow

of Trinity College and afterwards Archbishop of York, or
to Mr Peacock, afterwards Dean of Ely) and were long
preserved.

"The list of the Classical subjects for the first year in
Trinity College was transmitted to me, as usual, by Mr
Hustler.   They were—The Hippolytus of Euripides, the
3rd Book of Thucydides, and the 2nd Philippic of Cicero.
These I read carefully and noted before going up.   Mr
Hustler's family lived in Bury; and I called on him and
saw him in October, introduced by Mr Clarkson.   On the
morning of October 18th, 1819, I went on the top of the coach
to Cambridge, knowing nobody there but Mr Hustler, but
having letters of introduction from Mr Charles Musgrave to
Professor Sedgwick, Mr Thomas Musgrave, and Mr George
Peacock, all Fellows of Trinity College.

"I was set down at the Hoop, saw Trinity College for the
first time, found Mr Hustler, was conducted by his servant to
the robe-maker's, where I was invested in the cap and blue
gown, and after some further waiting was installed into
lodgings in Bridge Street.   At 4 o'clock I went to the
College Hall and was introduced by Mr Hustler to several
undergraduates, generally clever men, and in the evening I
attended Chapel in my surplice (it being St Luke's day)
and witnessed that splendid service of which the occasional
exhibition well befits the place.

"As soon as possible, I called on Mr Peacock, Mr Musgrave, and Professor Sedgwick.   By all I was received with
great kindness: my examination papers had been sent to them,
and a considerable reputation preceded me.   Mr Peacock at
once desired that I would not consider Mr C. Musgrave's
letter as an ordinary introduction, but that I would refer to
him on all occasions.   And I did so for several years, and
always received from him the greatest assistance that he
could give.   I think that I did not become acquainted with
Mr Whewell till the next term, when I met him at a breakfast party at Mr Peacock's.   Mr Peacock at once warned me

to arrange for taking regular exercise, and prescribed a walk of two hours every day before dinner: a rule to which I attended regularly, and to which I ascribe the continuance of good general health.

"I shewed Mr Peacock a manuscript book which contained a number of original Propositions which I had investigated. These much increased my reputation (I really had sense enough to set no particular value on it) and I was soon known by sight to almost everybody in the University. A ridiculous little circumstance aided in this. The former rule of the University (strictly enforced) had been that all students should wear drab knee-breeches: and I, at Mr Clarkson's recommendation, was so fitted up. The struggle between the old dress and the trowsers customary in society was still going on but almost terminated, and I was one of the very few freshmen who retained the old habiliments. This made me in some measure distinguishable: however at the end of my first three terms I laid these aside.

"The College Lectures began on Oct. 22: Mr Evans at 9 on the Hippolytus, and Mr Peacock at 10 on Euclid (these being the Assistant Tutors on Mr Hustler's side): and then I felt myself established.

"I wrote in a day or two to my uncle Arthur Biddell, and I received from him a letter of the utmost kindness. He entered gravely on the consideration of my prospects, my wants, &c.: and offered at all times to furnish me with money, which he thought my father's parsimonious habits might make him unwilling to do. I never had occasion to avail myself of this offer: but it was made in a way which in no small degree strengthened the kindly feelings that had long existed between us.

"I carefully attended the lectures, taking notes as appeared necessary. In Mathematics there were geometrical problems, algebra, trigonometry (which latter subjects the lectures did not reach till the terms of 1820). Mr Peacock gave me a copy of Lacroix's Differential Calculus as translated by himself and Herschel and Babbage, and also a copy of their Examples. At this time, the use of Differential Calculus was just prevailing over that of Fluxions (which I had learnt). I betook myself to it with great industry. I also made myself master of the theories of rectangular coordinates and some of the differential processes applying to them, which only a few of the best of the university mathematicians then wholly possessed. In Classical subjects I read the Latin (Seneca's) and English Hippolytus, Racine's Phèdre (which my sister translated for me), and all other books to which I was referred, Aristotle, Longinus, Horace, Bentley, Dawes &c., made verse translations of the Greek Hippolytus, and was constantly on the watch to read what might be advantageous.

"Early in December Mr Hustler sent for me to say that one of the Company of Fishmongers, Mr R. Sharp, had given to Mr John H. Smyth, M.P. for Norwich, the presentation to a small exhibition of £20 a year, which Mr Smyth had placed in Mr Hustler's hands, and which Mr Hustler immediately conferred on me. This was my first step towards pecuniary independence. I retained this exhibition till I became a Fellow of the College.

"I stayed at Cambridge during part of the winter vacation, and to avoid expense I quitted my lodgings and went for a time into somebody's rooms in the Bishop's Hostel. (It is customary for the tutors to place students in rooms when their right owners are absent.) I took with me Thucydides and all relating to it, and read the book, upon which the next term's lectures were to be founded, very carefully. The latter part of the vacation I spent at Bury, where I began with the assistance of my sister to pick up a little French: as I perceived that it was absolutely necessary for enabling me to read modern mathematics.

"During a part of the time I employed myself in writing out a paper on the geometrical interpretation of the algebraical expression $\sqrt{-1}$. I think that the original suggestion

of perpendicular line came from some book (I do not remember clearly), and I worked it out in several instances
pretty well, especially in De Moivre's Theorem. I had
spoken of it in the preceding term to Mr Peacock and he
encouraged me to work it out. The date at the end is 1820,
January 21. When some time afterwards I spoke of it to
Mr Hustler, he disapproved of my employing my time on
such speculations. About the last day of January I returned
to Cambridge, taking up my abode in my former lodgings. I
shewed my paper on $\sqrt{-1}$ to Mr Peacock, who was much
pleased with it and shewed it to Mr Whewell and others.

"On February 1 I commenced two excellent customs.
The first was that I always had upon my table a quire of
large-sized scribbling-paper sewn together: and upon this
paper everything was entered: translations into Latin and
out of Greek, mathematical problems, memoranda of every
kind (the latter transferred when necessary to the subsequent
pages), and generally with the date of the day. This is a
most valuable custom. The other was this: as I perceived
that to write Latin prose well would be useful to me, I wrote
a translation of English into Latin every day. However
much pressed I might be with other business, I endeavoured
to write at least three or four words, but if possible I wrote a
good many sentences.

"I may fix upon this as the time when my daily habits
were settled in the form in which they continued for several
years. I rose in time for the chapel service at 7. It was the
College regulation that every student should attend Chapel
four mornings and four evenings (Sunday being one of each)
in every week: and in this I never failed. After chapel
service I came to my lodgings and breakfasted. At 9 I went
to College lectures, which lasted to 11. Most of my contemporaries, being intended for the Church, attended also
divinity lectures: but I never did. I then returned, put my
lecture notes in order, wrote my piece of Latin prose, and
then employed myself on the subject which I was reading for

the time: usually taking mathematics at this hour. At 2 or a little sooner I went out for a long walk, usually 4 or 5 miles into the country: sometimes if I found companions I rowed on the Cam (a practice acquired rather later). A little before 4 I returned, and at 4 went to College Hall. After dinner I lounged till evening chapel time, ½ past 5, and returning about 6 I then had tea. Then I read quietly, usually a classical subject, till 11; and I never, even in the times when I might seem most severely pressed, sat up later.

"From this time to the close of the annual examination (beginning of June) I remained at Cambridge, stopping there through the Easter Vacation. The subjects of the mathematical lectures were ordinary algebra and trigonometry: but Mr Peacock always had some private problems of a higher class for me, and saw me I believe every day. The subjects of the Classical lectures were, the termination of Hippolytus, the book of Thucydides and the oration of Cicero. In mathematics I read Whewell's Mechanics, then just published (the first innovation made in the Cambridge system of Physical Sciences for many years): and I find in my scribbling-paper notes, integrals, central forces, Finite Differences, steam-engine constructions and powers, plans of bridges, spherical trigonometry, optical calculations relating to the achromatism of eye-pieces and achromatic object-glasses with lenses separated, mechanical problems, Transit of Venus, various problems in geometrical astronomy (I think it was at this time that Mr Peacock had given me a copy of Woodhouse's Astronomy 1st Edition), the rainbow, plans for anemometer and for a wind-pumping machine, clearing lunars, &c., with a great number of geometrical problems. I remark that my ideas on the Differential Calculus had not acquired on some important points the severe accuracy which they acquired in a few months. In Classics I read the Persæ of Æschylus, Greek and Roman history very much (Mitford, Hooke, Ferguson) and the books of Thucydides introductory to that of the lecture subject (the 3rd): and attended to Chronology.

On the scribbling-paper are verse-translations from Euripides, careful prose-translations from Thucydides, maps, notes on points of grammar &c. I have also little MS. books with abundant notes on all these subjects: I usually made a little book when I pursued any subject in a regular way.

"On May 1st Mr Dobree, the head lecturer, sent for me to say that he appointed me head-lecturer's Sizar for the next year. The stipend of this office was £10, a sum upon which I set considerable value in my anxiety for pecuniary independence: but it was also gratifying to me as shewing the way in which I was regarded by the College authorities.

"On Wednesday, May 24th, 1820, the examination began. I was anxious about the result of the examination, but only in such a degree as to make my conduct perfectly steady and calm, and to prevent me from attempting any extraordinary exertion.

"When the Classes were published the first Class of the Freshman's Year (alphabetically arranged, as is the custom) stood thus: Airy, Boileau, Childers, Drinkwater, Field, Iliff, Malkin, Myers, Romilly, Strutt, Tate, Winning. It was soon known however that I was first of the Class. It was generally expected (and certainly by me) that, considering how great a preponderance the Classics were understood, in the known system of the College, to have in determining the order of merit, Field would be first. However the number of marks which Field obtained was about 1700, and that which I obtained about 1900. No other competitor, I believe, was near us."—In a letter to Airy from his College Tutor, Mr J. D. Hustler, there is the following passage: "It is a matter of extreme satisfaction to me that in the late examination you stood not only in the First Class but first of the first. I trust that your future exertions and success will be commensurate with this honourable beginning."

"Of the men whom I have named, Drinkwater (Bethune) was afterwards Legal Member of the Supreme Court of India, Field was afterwards Rector of Reepham, Romilly (afterwards Lord Romilly) became Solicitor-General, Strutt (afterwards Lord Belper) became M.P. for Derby and First Commissioner of Railways, Tate was afterwards master of Richmond Endowed School, Childers was the father of Childers who was subsequently First Lord of the Admiralty.

"I returned to Bury immediately. While there, some students (some of them men about to take their B.A. degree at the next January) applied to me to take them as pupils, but I declined. This year of my life enabled me to understand how I stood among men. I returned to Cambridge about July 11th. As a general rule, undergraduates are not allowed to reside in the University during the Long Vacation. I believe that before I left, after the examination, I had made out that I should be permitted to reside: or I wrote to Mr Hustler. I applied to Mr Hustler to be lodged in rooms in College: and was put, first into rooms in Bishop's Hostel, and subsequently into rooms in the Great Court.

"The first affair that I had in College was one of disappointment by no means deserving the importance which it assumed in my thoughts. I had been entered a Sizar, but as the list of Foundation Sizars was full, my dinners in Hall were paid for. Some vacancies had arisen: and as these were to be filled up in order of merit, I expected one: and in my desire for pecuniary independence I wished for it very earnestly. However, as in theory all of the first class were equal, and as there were some Sizars in it senior in entrance to me, they obtained places first: and I was not actually appointed till after the next scholarship examination (Easter 1821). However a special arrangement was made, allowing me (I forget whether others) to sit at the Foundation-Sizars' table whenever any of the number was absent: and in consequence I received practically nearly the full benefits.

"Mr Peacock, who was going out for the Vacation, allowed me access to his books. I had also (by the assistance of various Fellows, who all treated me with great kindness, almost to a degree of respect) command of the University

Library and Trinity Library: and spent this Long Vacation, like several others, very happily indeed.

"The only non-mathematical subjects of the next examination were The Gospel of St Luke, Paley's Evidences, and Paley's Moral and Political Philosophy. Thus my time was left more free to mathematics and to general classics than last year. I now began a custom which I maintained for some years. Generally I read mathematics in the morning, and classics for lectures in the afternoon: but invariably I began at 10 o'clock in the evening to read with the utmost severity some standard classics (unconnected with the lectures) and at 11 precisely I left off and went to bed. I continued my daily translations into Latin prose as before.

"On August 24th, 1820, Rosser, a man of my own year, engaged me as private tutor, paying at the usual rate (£14 for a part of the Vacation, and £14 for a term): and immediately afterwards his friend Bedingfield did the same. This occupied two hours every day, and I felt that I was now completely earning my own living. I never received a penny from my friends after this time.

"I find on my scribbling-paper various words which shew that in reading Poisson I was struggling with French words. There are also Finite Differences and their Calculus, Figure of the Earth (force to the center), various Attractions (some evidently referring to Maclaurin's), Integrals, Conic Sections, Kepler's Problem, Analytical Geometry, D'Alembert's Theorem, Spherical Aberration, Rotations round three axes (apparently I had been reading Euler), Floating bodies, Evolute of Ellipse, Newton's treatment of the Moon's Variation. I attempted to extract something from Vince's Astronomy on the physical explanation of Precession: but in despair of understanding it, and having made out an explanation for myself by the motion round three axes, I put together a little treatise (Sept. 10, 1820) which with some corrections and additions was afterwards printed in my Mathematical Tracts. On Sept. 14th I bought Woodhouse's

Physical Astronomy, and this was quite an epoch in my mathematical knowledge. First, I was compelled by the process of " changing the independent variable" to examine severely the logic of the Differential Calculus. Secondly, I was now able to enter on the Theory of Perturbations, which for several years had been the desired land to me.

"At the Fellowship Election of Oct. 1st, Sydney Walker (among other persons) was elected Fellow. He then quitted the rooms in which he had lived (almost the worst in the College), and I immediately took them. They suited me well and I lived very happily in them till I was elected Scholar. They are small rooms above the middle staircase on the south side of Neville's Court. (Mr Peacock's rooms were on the same staircase.) I had access to the leads on the roof of the building from one of my windows. This was before the New Court was built : my best window looked upon the garden of the College butler.

"I had brought to Cambridge the telescope which I had made at Colchester, and about this time I had a stand made by a carpenter at Cambridge : and I find repeated observations of Jupiter and Saturn made in this October term.

"Other mathematical subjects on my scribbling-paper are: Geometrical Astronomy, Barometers (for elevations), Maclaurin's Figure of the Earth, Lagrange's Theorem, Integrals, Differential Equations of the second order, Particular Solutions. In general mathematics I had much discussion with Atkinson (who was Senior Wrangler, January 1821), and in Physics with Rosser, who was a friend of Sir Richard Phillips, a vain objector to gravitation. In Classics I read Æschylus and Herodotus.

"On October 5th I received notice from the Head Lecturer to declaim in English with Winning. (This exercise consists in preparing a controversial essay, learning it by heart, and speaking it in Chapel after the Thursday evening's service.) On October 6th we agreed on the subject, " Is natural difference to be ascribed to moral or to physical causes?" I taking the latter side. I spoke the declamation (reciting it without missing a word) on October 25th. On October 26th I received notice of Latin declamation with Myers: subject agreed on, "Utrum civitati plus utilitatis an incommodi afferant leges quæ ad vitas privatorum hominum ordinandas pertinent"; I took the former. The declamation was recited on November 11, when a curious circumstance occurred. My declamation was rather long: it was the first Saturday of the term on which a declamation had been spoken: and it was the day on which arrived the news of the withdrawal of the Bill of Pains and Penalties against Queen Caroline. (This trial had been going on through the summer, but I knew little about it.) In consequence the impatience of the undergraduates was very great, and there was such an uproar of coughing &c. in the Chapel as probably was never known. The Master (Dr Wordsworth, appointed in the beginning of the summer on the death of Dr Mansell, and to whom I had been indirectly introduced by Mrs Clarkson) and Tutors and Deans tried in vain to stop the hubbub. However I went on steadily to the end, not at all frightened. On the Monday the Master sent for me to make a sort of apology in the name of the authorities, and letters to the Tutors were read at the Lectures, and on the whole the transaction was nowise disagreeable to me.

"On the Commemoration Day, December 15th, I received my Prize (Mitford's Greece) as First-Class man, after dinner in the College Hall. After a short vacation spent at Bury and Playford I returned to Cambridge, walking from Bury on Jan. 22nd, 1821. During the next term I find in Mathematics Partial Differential Equations, Tides, Sound, Calculus of Variations, Composition of rotary motions, Motion in resisting medium, Lhuillier's theorem, Brightness of an object as seen through a medium with any possible law of refraction (a good investigation), star-reductions, numerical calculations connected with them, equilibrium of chain under centripetal force (geometrically treated, as an improvement upon Whe-

well's algebraical method), investigation of the magnitude of
attractive forces of glass, &c., required to produce refraction.
I forget about Mathematical Lectures; but I have an impression that I regularly attended Mr Peacock's lectures, and
that he always set me some private problems.

"I attended Mr Evans's lectures on St Luke: and I find
many notes about the history of the Jews, Cerinthus and
various heresies, Paley's Moral Philosophy, Paley's Evidences,
and Biblical Maps: also speculations about ancient pronun-
ciations.

"For a week or more before the annual examination I was
perfectly lazy. The Classes of my year (Junior Sophs) were
not published till June 11. It was soon known that I was
first with 2000 marks, the next being Drinkwater with 1200
marks. After a short holiday at Bury and Playford I
returned to Cambridge on July 18th, 1821. My daily life
went on as usual. I find that in writing Latin I began
Cicero De Senectute (retranslating Melmoth's translation,
and comparing). Some time in the Long Vacation the names
of the Prizemen for Declamations were published: I was
disappointed that not one, English or Latin, was assigned
to me: but it was foolish, for my declamations were rather
trumpery.

"My former pupil, Rosser, came again on August 14th.
On August 29th Dr Blomfield (afterwards Bishop of London)
called, to engage me as Tutor to his brother George Beecher
Blomfield, and he commenced attendance on Sept. 1st. With
these two pupils I finished at the end of the Long Vacation:
for the next three terms I had one pupil, Gibson, a Newcastle man, recommended by Mr Peacock, I believe, as a
personal friend (Mr Peacock being of Durham).

"The only classical subject appointed for the next examination was the 5th, 6th and 7th Books of the Odyssey: the
mathematical subjects all the Applied Mathematics and Newton. There was to be however the Scholarship Examination
(Sizars being allowed to sit for Scholarships only in their

3rd year : and the Scholarship being a kind of little Fellowship necessary to qualify for being a candidate for the real
Fellowship).

"When the October term began Mr Hustler, who usually
gave lectures in mathematics to his third-year pupils, said to
me that it was not worth my while to attend his lectures, and
he or Mr Peacock suggested that Drinkwater, Myers, and I
should attend the Questionists' examinations. The Questionists are those who are to take the degree of B.A. in the
next January : and it was customary, not to give them
lectures, but three times a week to examine them by setting
mathematical questions, as the best method of preparing for
the B.A. examination. Accordingly it was arranged that we
should attend the said examinations : but when we went the
Questionists of that year refused to attend. They were
reported to be a weak year, and we to be a strong one : and
they were disposed to take offence at us on any occasion.
From some of the scholars of our year who sat at table with
scholars of that year I heard that they distinguished us as
'the impudent year,' 'the annus mirabilis,' &c. On this
occasion they pretended to believe that the plan of our
attendance at the Questionists' examinations had been suggested by an undergraduate, and no explanation was of the
least use. So the Tutors agreed not to press the matter on
them : and instead of it, Drinkwater, Myers, and I went
three times a week to Mr Peacock's rooms, and he set us
questions. I think that this system was also continued during
the next two terms (ending in June 1822) or part of them,
but I am not certain.

"In August 1821 I copied out a MS. on Optics, I think
from Mr Whewell : on August 24th one on the Figure of the
Earth and Tides; and at some other time one on the motion
of a body round two centers of force; both from Mr Whewell.
On my scribbling paper I find—A problem on the vibrations
of a gig as depending on the horse's step (like that of a
pendulum whose support is disturbed), Maclaurin's Attrac-

tions, Effect of separating the lenses of an achromatic object-glass (suggested by my old telescope), Barlow's theory of numbers, and division of the circle into 17 parts, partial differentials, theory of eye-pieces, epicycloids, Figure of the Earth, Time of body in arc of parabola, Problem of Sound, Tides, Refraction of Lens, including thickness, &c., Ivory's paper on Equations, Achromatism of microscope, Capillary Attraction, Motions of Fluids, Euler's principal axes, Spherical pendulum, Equation $b^2 \dfrac{d^2y}{dx^2} = \dfrac{d^2y}{dt^2}$, barometer, Lunar Theory well worked out, ordinary differential equations, Calculus of Variations, Interpolations like Laplace's for Comets, Kepler's theorem. In September I had my old telescope mounted on a short tripod stand, and made experiments on its adjustments. I was possessed of White's Ephe-meris, and I find observations of Jupiter and Saturn in October. I planned an engine for describing ellipses by the polar equation $\dfrac{A}{1 + e \cos \theta}$, and tried to make a micrometer with silk threads converging to a point. Mr Cubitt called on Oct. 4 and Nov. 1; he was engaged in erecting a tread-mill at Cambridge Gaol, and had some thoughts of sending plans for the Cambridge Observatory, the erection of which was then proposed. On Nov. 19 I find that I had received from Cubitt a Nautical Almanac, the first that I had. On Dec. 11 I made some experiments with Drinkwater: I think it was whirling a glass containing oil on water. In Classics I was chiefly engaged upon Thucydides and Homer. On October 6th I had a letter from Charles Musgrave, introducing Challis, who succeeded me in the Cambridge Observatory in 1836.

"At this time my poor afflicted father was suffering much from a severe form of rheumatism or pain in the legs which sometimes prevented him from going to bed for weeks together.

"On the Commemoration Day, Dec. 18th, I received my

prize as first-class man in Hall again. The next day I walked to Bury, and passed the winter vacation there and at Playford.

"I returned to Cambridge on Jan. 24th, 1822. On Feb. 12th I kept my first Act, with great compliments from the Moderator, and with a most unusually large attendance of auditors. These disputations on mathematics, in Latin, are now discontinued. On March 20th I kept a first Opponency against Sandys. About this time I received Buckle, a Trinity man of my own year, who was generally supposed to come next after Drinkwater, as pupil. On my sheets I find integrals and differential equations of every kind, astronomical corrections (of which I prepared a book), chances, Englefield's comets, investigation of the brightness within a rainbow, proof of Clairaut's theorem in one case, metacentres, change of independent variable applied to a complicated case, generating functions, principal axes. On Apr. 8th I intended to write an account of my eye : I was then tormented with a double image, I suppose from some disease of the stomach : and on May 28th I find by a drawing of the appearance of a lamp that the disease of my eye continued.

"On Feb. 11th I gave Mr Peacock a paper on the alteration of the focal length of a telescope as directed with or against the Earth's orbital motion (on the theory of emissions) which was written out for reading to the Cambridge Philosophical Society on Feb. 24th and 25th. [This Society I think was then about a year old.] On Feb. 1 my MS. on Precession, Solar Inequality, and Nutation, was made complete.

"The important examination for Scholarships was now approaching. As I have said, this one opportunity only was given to Sizars (Pensioners having always two opportunities and sometimes three), and it is necessary to be a Scholar in order to be competent to be a candidate for a Fellowship. On Apr. 10th I addressed my formal Latin letter to the Seniors. There were 13 vacancies and 37 candidates. The election took place on Apr. 18th, 1822. I was by much

the first (which I hardly expected) and was complimented by
the Master and others. Wrote the formal letter of thanks as
usual. I was now entitled to claim better rooms, and I took
the rooms on the ground floor on the East side of the Queen's
Gate of the Great Court. Even now I think of my quiet
residence in the little rooms above the staircase in Neville's
Court with great pleasure. I took possession of my new
rooms on May 27th.

"The Annual Examination began on May 30th. The
Classes were published on June 5th, when my name was separated from the rest by two lines. It was understood that the
second man was Drinkwater, and that my number of marks
was very nearly double of his. Having at this time been disappointed of a proposed walking excursion into Derbyshire with
a college friend, who failed me at the last moment, I walked
to Bury and spent a short holiday there and at Playford.

"I returned to Cambridge on July 12th, 1822. I was
steadily busy during this Long Vacation, but by no means
oppressively so: indeed my time passed very happily. The
Scholars' Table is the only one in College at which the
regular possessors of the table are sure never to see a stranger,
and thus a sort of family intimacy grows up among the Scholars. Moreover the Scholars feel themselves to be a privileged
class 'on the foundation,' and this feeling gives them a sort
of conceited happiness. It was the duty of Scholars by turns
to read Grace after the Fellows' dinner and supper, and at
this time (1848) I know it by heart. They also read the
Lessons in Chapel on week days: but as there was no daily
chapel-service during the summer vacation, I had not much
of this. In the intimacy of which I speak I became much
acquainted with Drinkwater, Buckle, Rothman, and Sutcliffe:
and we formed a knot at the table (first the Undergraduate
Scholars' table, and afterwards the Bachelor Scholars' table)
for several years. During this Vacation I had for pupils
Buckle and Gibson.

"I wrote my daily Latin as usual, beginning with the

retranslation of Cicero's Epistles, but I interrupted it from Sept. 27th to Feb. 8th. I believe it was in this Vacation, or in the October term, that I began every evening to read Thucydides very carefully, as my notes are marked 1822 and 1823. On August 27 I find that I was reading Ovid's Fasti.

"In Mathematics I find the equation $x + y = a$, $x^q + y^q = b$, Caustics, Calculus of Variations, Partial Differentials, Aberration of Light, Motions of Comets, various Optical constructions computed with spherical aberrations, Particular Solutions, Mechanics of Solid Bodies, Attractions of Shells, Chances, Ivory's attraction-theorem, Lunar Theory (algebraical), Degrees across meridian, theoretical refraction, Newton's 3rd Book, Investigation of the tides in a shallow equatoreal canal, from which I found that there would be low-water under the moon, metacentres, rotation of a solid body round three axes, Attractions of Spheroids of variable density, finite differences, and complete Figure of the Earth. There is also a good deal of investigation of a mathematical nature not connected with College studies, as musical chords, organ-pipes, sketch for a computing machine (suggested by the publications relating to Babbage's), sketch of machine for solving equations. In August there is a plan of a MS. on the Differential Calculus, which it appears I wrote then: one on the Figure of the Earth written about August 15th; one on Tides, Sept. 25th; one on Newton's Principia with algebraical additions, Nov. 1st. On Sept. 6th and 10th there are Lunar Distances observed with Rothman's Sextant and completely worked out; for these I prepared a printed skeleton form, I believe my first. On December 13th there are references to books on Geology (Conybeare and Phillips, and Parkinson) which I was beginning to study. On July 27th, being the day on which I completed my 21st year, I carefully did nothing.

"Another subject partly occupied my thoughts, which, though not (with reference to practical science) very wise, yet gave me some Cambridge celebrity. In July 1819 I had (as before mentioned) sketched a plan for constructing reflecting telescopes with silvered glass, and had shewn it afterwards to Mr Peacock. I now completed the theory of this construction by correcting the aberrations, spherical as well as chromatic. On July 13th, 1822, I drew up a paper about it for Mr Peacock. He approved it much, and in some way communicated it to Mr (afterwards Sir John) Herschel. I was soon after introduced to Herschel at a breakfast with Mr Peacock: and he approved of the scheme generally. On August 5th I drew up a complete mathematical paper for the Cambridge Philosophical Society, which I entrusted to Mr Peacock. The aberrations, both spherical and chromatic, are here worked out very well. On Nov. 25th it was read at the meeting of the Philosophical Society, and was afterwards printed in their Transactions: this was my first printed Memoir. Before this time however I had arranged to try the scheme practically. Mr Peacock had engaged to bear the expense, but I had no occasion to ask him. Partly (I think) through Drinkwater, I communicated with an optician named Bancks, in the Strand, who constructed the optical part. I subsequently tried my telescope, but it would not do. The fault, as I had not and have not the smallest doubt, depends in some way on the crystallization of the mercury silvering. It must have been about this time that I was introduced to Mr (afterwards Sir James) South, at a party at Mr Peacock's rooms. He advised me to write to Tulley, a well-known practical optician, who made me some new reflectors, &c. (so that I had two specimens, one Gregorian, the other Cassegrainian). However the thing failed practically, and I was too busy ever after to try it again.

"During the October term I had no pupils. I kept my second Act on Nov. 6 (opponents Hamilton, Rusby, Field), and an Opponency against Jeffries on Nov. 7. I attended the Questionists' Examinations. I seem to have lived a very comfortable idle life. The Commemoration Day was Dec. 18th, when I received a Prize, and the next day I walked to Bury. On Jan. 4th, 1823, I returned to Cambridge, and until the B.A. Examination I read novels and played cards more than at any other time in College.

"On Thursday, Jan. 9th, 1823, the preliminary classes, for arrangement of details of the B.A. Examination, were published. The first class, Airy, Drinkwater, Jeffries, Mason. As far as I remember, the rule was then, that on certain days the classes were grouped (in regard to identity of questions given to each group) thus: 1st, $\left\{ \begin{matrix} 2\text{nd} \\ 3\text{rd} \end{matrix} \right\}$, $\left\{ \begin{matrix} 4\text{th} \\ 5\text{th} \end{matrix} \right\}$, &c., and on certain other days thus: $\left\{ \begin{matrix} 1\text{st} \\ 2\text{nd} \end{matrix} \right\}$, $\left\{ \begin{matrix} 3\text{rd} \\ 4\text{th} \end{matrix} \right\}$, &c. On Saturday, Jan. 11th, I paid fees. On Monday, Jan. 13th, the proceedings of examination began by a breakfast in the Combination Room. After this, Gibson gave me breakfast every day, and Buckle gave me and some others a glass of wine after dinner. The hours were sharp, the season a cold one, and no fire was allowed in the Senate House where the Examination was carried on (my place was in the East gallery), and altogether it was a severe time.

" The course of Examination was as follows:

"Monday, Jan. 13th. 8 to 9, printed paper of questions by Mr Hind (moderator); half-past 9 to 11, questions given orally; 1 to 3, ditto; 6 to 9, paper of problems at Mr Higman's rooms.

"Tuesday, Jan. 14th. 8 to 9, Higman's paper; half-past 9 to 11, questions given orally; 1 to 3, ditto; 6 to 9, paper of problems in Sidney College Hall.

"Wednesday, Jan. 15th. Questions given orally 8 to 9 and 1 to 3, with paper of questions on Paley and Locke (one question only in each was answered).

"Thursday, Jan. 16th. We went in at 9 and 1, but there seems to have been little serious examination.

"Friday, Jan. 17th. On this day the brackets or classes as resulting from the examination were published. 1st bracket Airy, 2nd bracket Jeffries, 3rd bracket Drinkwater, Fisher, Foley, Mason, Myers.

" On Saturday, Jan. 18th, the degrees were conferred in the usual way. It had been arranged that my brother and sister should come to see me take my degree of B.A., and I had asked Gibson to conduct them to the Senate House Gallery: but Mr Hawkes (a Trinity Fellow) found them and stationed them at the upper end of the Senate House. After the preliminary arrangements of papers at the Vice-Chancellor's table, I, as Senior Wrangler, was led up first to receive the degree, and rarely has the Senate House rung with such applause as then filled it. For many minutes, after I was brought in front of the Vice-Chancellor, it was impossible to proceed with the ceremony on account of the uproar. I gave notice to the Smith's Prize Electors of my intention to 'sit' for that prize, and dined at Rothman's rooms with Drink-water, Buckle, and others. On Monday, Jan. 20th, I was examined by Professor Woodhouse, for Smith's Prize, from 10 to 1. I think that the only competitor was Jeffries. On Tuesday I was examined by Prof. Turton, 10 to 1, and on Wednesday by Prof. Lax, 10 to 1. On Thursday, Jan. 23rd, I went to Bury by coach, on one of the coldest evenings that I ever felt.

" Mr Peacock had once recommended me to sit for the Chancellor's medal (Classical Prize). But he now seemed to be cool in his advice, and I laid aside all thought of it."

It seems not out of place to insert here a copy of some "Cambridge Reminiscences" written by Airy, which will serve to explain the Acts and Opponencies referred to in the previous narrative, and other matters.

THE ACTS.

The examination for B.A. degrees was preceded, in my time, by keeping two Acts, in the Schools under the University Library: the second of them in the October term imme-

diately before the examination; the first (I think) in the
October term of the preceding year.

These Acts were reliques of the Disputations of the
Middle Ages, which probably held a very important place
in the discipline of the University. (There seems to be
something like them in some of the Continental Universities.)
The presiding authority was one of the Moderators. I apprehend that the word "Moderator" signified "President," in
which sense it is still used in the Kirk of Scotland; and that
it was peculiarly applied to the Presidency of the Disputations, the most important educational arrangement in the
University. The Moderator sent a summons to the "Respondent" to submit three subjects for argument, and to
prepare to defend them on a given day : he also named three
Opponents. This and all the following proceedings were
conducted in Latin. For my Act of 1822, Nov. 6, I submitted the following subjects :

"Recte statuit Newtonus in Principiis suis Mathematicis,
libro primo, sectione undecimâ."

"Recte statuit Woodius de Iride."

"Recte statuit Paleius de Obligationibus."

The Opponents named to attack these assertions were
Hamilton of St John's, Rusby of St Catharine's, Field of
Trinity. It was customary for the Opponents to meet at tea
at the rooms of the Senior Opponent, in order to discuss and
arrange their arguments; the Respondent was also invited,
but he was warned that he must depart as soon as tea would
be finished : then the three Opponents proceeded with their
occupation. As I have acted in both capacities, I am able to
say that the matter was transacted in an earnest and business-
like way. Indeed in the time preceding my own (I know not
whether in my own time) the assistance of a private tutor was
frequently engaged, and I remember hearing a senior M.A.
remark that my College Tutor (James D. Hustler) was the
best crammer for an Act in the University.

At the appointed time, the parties met in the Schools :

the Respondent first read a Latin Thesis on any subject (I think I took some metaphysical subject), but nobody paid any attention to it: then the Respondent read his first Dogma, and the first Opponent produced an argument against it, in Latin. After this there were repeated replies and rejoinders, all in vivâ voce Latin, the Moderator sometimes interposing a remark in Latin. When he considered that one argument was disposed of, he called for another by the words " Probes aliter." The arguments were sometimes shaped with considerable ingenuity, and required a clear head in the Respondent. When all was finished, the Moderator made a complimentary remark to the Respondent and one to the first Opponent (I forget whether to the second and third). In my Respondency of 1822, November 6, the compliment was, " Quæstiones tuas summo ingenio et acumine defendisti, et in rebus mathematicis scientiam planè mirabilem ostendisti.' In an Opponency (I forget when) the compliment was, " Magno ingenio argumenta tua et construxisti et defendisti."

The Acts of the high men excited much interest among the students. At my Acts the room was crowded with undergraduates.

I imagine that, at a time somewhat distant, the maintenance of the Acts was the only regulation by which the University acted on the studies of the place. When the Acts had been properly kept, license was given to the Father of the College to present the undergraduate to the Vice-Chancellor, who then solemnly admitted him " ad respondendum Quæstioni." There is no appearance of collective examination before this presentation: what the " Quæstio " might be, I do not know. Still the undergraduate was not B.A. The Quæstio however was finished and approved before the day of a certain Congregation, and then the undergraduate was declared to be " actualiter in artibus Baccalaureum."

Probably these regulations were found to be insufficient

for the control of education, and the January examination was instituted. I conjecture this to have been at or shortly before the date of the earliest Triposes recorded in the Cambridge Calendar, 1748.

The increasing importance of the January examination naturally diminished the value of the Acts in the eyes of the undergraduates; and, a few years after my M.A. degree, it was found that the Opponents met, not for the purpose of concealing their arguments from the Respondent, but for the purpose of revealing them to him. This led to the entire suppression of the system. The most active man in this suppression was Mr Whewell: its date must have been near to 1830.

The shape in which the arguments were delivered by an Opponent, reading from a written paper, was, " Si (quoting something from the Respondent's challenge), &c., &c. Cadit Quæstio; Sed (citing something else bearing on the subject of discussion), Valet Consequentia; Ergo (combining these to prove some inaccuracy in the Respondent's challenge), Valent Consequentia et Argumentum." Nobody pretended to understand these mystical terminations.

Apparently the original idea was that several Acts should be kept by each undergraduate; for, to keep up the number (as it seemed), each student had to gabble through a ridiculous form " Si quæstiones tuæ falsæ sint, Cadit Quæstio:—sed quæstiones tuæ falsæ sunt, Ergo valent Consequentia et Argumentum." I have forgotten time and place when this was uttered.

THE SENATE-HOUSE EXAMINATION.

The Questionists, as the undergraduates preparing for B.A. were called in the October term, were considered as a separate body; collected at a separate table in Hall, attending no lectures, but invited to attend a system of trial examinations conducted by one of the Tutors or Assistant-Tutors.

From the Acts, from the annual College examinations, and (I suppose) from enquiries in the separate Colleges, the Moderators acquired a general idea of the relative merits of the candidates for honours. Guided by this, the candidates were divided into six classes. The Moderators and Assistant Examiners were provided each with a set of questions in manuscript (no printed papers were used for Honours in the Senate House; in regard to the οἱ πολλοὶ I cannot say). On the Monday on which the examination began, the Father of the College received all the Questionists (I believe), at any rate all the candidates for honours, at breakfast in the Combination Room at 8 o'clock, and marched them to the Senate House. My place with other honour-men was in the East Gallery. There one Examiner took charge of the 1st and 2nd classes united, another Examiner took the 3rd and 4th classes united, and a third took the 5th and 6th united. On Tuesday, one Examiner took the 1st class alone, a second took the 2nd and 3rd classes united, a third took the 4th and 5th classes united, and a fourth took the 6th class alone. On Wednesday, Thursday, and Friday the changes were similar. And, in all, the questioning was thus conducted. The Examiner read from his manuscript the first question. Those who could answer it proceeded to write out their answers, and as soon as one had finished he gave the word "Done"; then the Examiner read out his second question, repeating it when necessary for the understanding by those who took it up more lately. And so on. I think that the same process was repeated in the afternoon; but I do not remember precisely. In this manner the Examination was conducted through five days (Monday to Friday) with no interruption except on Friday afternoon. It was principally, perhaps entirely, bookwork.

But on two *evenings* there were printed papers of problems: and the examination in these was conducted just as in the printed papers of the present day: but in the private College Rooms of the Moderators. And there, wine and other refreshments were offered to the Examinees. How this singular custom began, I know not.

The order of merit was worked out on Friday afternoon and evening, and was in some measure known through the University late in the evening. I remember Mr Peacock coming to a party of Examinees and giving information on several places. I do not remember his mentioning mine (though undoubtedly he did) but I distinctly remember his giving the Wooden Spoon. On the Saturday morning at 8 o'clock the manuscript list was nailed to the door of the Senate-House. The form of further proceedings in the presentation for degree (ad respondendum quæstioni) I imagine has not been much altered. The kneeling before the Vice-Chancellor and placing hands in the Vice-Chancellor's hands were those of the old form of doing homage.

The form of examination which I have described was complicated and perhaps troublesome, but I believe that it was very efficient, possibly more so than the modern form (established I suppose at the same time as the abolition of the Acts). The proportion of questions now answered to the whole number set is ridiculously small, and no accurate idea of relative merit can be formed from them.

THE COLLEGE HALL.

When I went up in 1819, and for several years later, the dinner was at ¼ past 3. There was no supplementary dinner for special demands. Boat-clubs I think were not invented, even in a plain social way, till about 1824 or 1825; and not in connection with the College till some years later. Some of the senior Fellows spoke of the time when dinner was at 2, and regretted the change.

There was supper in Hall at 9 o'clock: I have known it to be attended by a few undergraduates when tired by examinations or by evening walks; and there were always some seniors at the upper table: I have occasionally joined them, and have had some very interesting conversations. The supper was cold, but hot additions were made when required.

One little arrangement amused me, as shewing the ecclesiastical character of the College. The Fasts of the Church were to be strictly kept, and there was to be no dinner in Hall. It was thus arranged. The evening chapel service, which was usually at $5\frac{1}{2}$ (I think), was held at 3; and at 4 the ordinary full meal was served in Hall, but as it followed the chapel attendance it was held to be supper; and there was no subsequent meal.

There were no chairs whatever in Hall, except the single chair of the vice-master at the head of the table on the dais and that of the senior dean at the table next the East wall. All others sat on benches. And I have heard allusions to a ludicrous difficulty which occurred when some princesses (of the Royal Family) dined in the Hall, and it was a great puzzle how to get them to the right side of the benches.

The Sizars dined after all the rest; their dinner usually began soon after 4. For the non-foundationists a separate dinner was provided, as for pensioners. But for the foundationists, the remains of the Fellows' dinner were brought down; and I think that this provision was generally preferred to the other.

The dishes at all the tables of undergraduates were of pewter, till a certain day when they were changed for porcelain. I cannot remember whether this was at the time when they became Questionists (in the October Term), or at the time when they were declared "actualiter esse in artibus Baccalaureos" (in the Lent Term).

Up to the Questionist time the undergraduate Scholars had no mixture whatever; they were the only pure table in the Hall: and I looked on this as a matter very valuable for the ultimate state of the College society. But in the October term, those who were to proceed to B.A. were drafted into the mixed body of Questionists: and they greatly disliked the change. They continued so till the Lent Term, when they were formally invited by the Bachelor Scholars to join the upper table.

MATHEMATICAL SUBJECTS OF STUDY AND EXAMINATION.

In the October Term 1819, the only books on Pure Mathematics were:—Euclid generally, Algebra by Dr Wood (formerly Tutor, but in 1819 Master, of St John's College), Vince's Fluxions and Dealtry's Fluxions, Woodhouse's and other Trigonometries. Not a whisper passed through the University generally on the subject of Differential Calculus; although some papers (subsequently much valued) on that subject had been written by Mr Woodhouse, fellow of Caius College; but their style was repulsive, and they never took hold of the University. Whewell's Mechanics (1819) contains a few and easy applications of the Differential Calculus. The books on applied Mathematics were Wood's Mechanics, Whewell's Mechanics, Wood's Optics, Vince's Hydrostatics, Vince's Astronomy, Woodhouse's Plane Astronomy (perhaps rather later), The First Book of Newton's Principia: I do not remember any others. These works were undoubtedly able; and for the great proportion of University students going into active life, I do not conceal my opinion that books constructed on the principles of those which I have cited were more useful than those exclusively founded on the more modern system. For those students who aimed at the mastery of results more difficult and (in the intellectual sense) more important, the older books were quite insufficient. More aspiring students read, and generally with much care, several parts of Newton's Principia, Book I., and also Book III. (perhaps the noblest example of geometrical form of cosmical theory that the world has seen). I remember some questions from Book III. proposed in the Senate-House Examination 1823.

In the October term 1819, I went up to the University.
The works of Wood and Vince, which I have mentioned, still
occupied the lecture-rooms.   But a great change was in
preparation for the University Course of Mathematics.
During the great Continental war, the intercourse between
men of science in England and in France had been most
insignificant.   But in the autumn of 1819, three members of
the Senate (John Herschel, George Peacock, and Charles
Babbage) had entered into the mathematical society of Paris,
and brought away some of the works on Pure Mathematics
(especially those of Lacroix) and on Mechanics (principally
Poisson's).   In 1820 they made a translation of Lacroix's
Differential Calculus; and they prepared a volume of Examples of the Differential and Integral Calculus.   These
were extensively studied : but the form of the College Examinations or the University Examinations was not, I think,
influenced by them in the winter 1820—1821 or the two
following terms.   But in the winter 1821—1822 Peacock was
one of the Moderators; and in the Senate-House Examination, January 1822, he boldly proposed a Paper of important
questions entirely in the Differential Calculus.   This was
considered as establishing the new system in the University.
In January 1823, I think the two systems were mingled.
Though I was myself subject to that examination, I grieve to
say that I have forgotten much of the details, except that I
well remember that some of the questions referred to Newton,
Book III. on the Lunar Theory.   To these I have already
alluded.

No other work occurs to me as worthy of mention, except
Woodhouse's Lunar Theory, entirely founded on the Differential Calculus.   The style of this book was not attractive,
and it was very little read.

CHAPTER III.

At Trinity College, Cambridge, from his taking his B.A. Degree to his taking charge of the Cambridge Observatory as Plumian Professor.

From January 18th, 1823, to March 15th, 1828.

"On Jan. 30th, 1823, I returned to Cambridge. I had already heard that I had gained the 1st Smith's Prize, and one of the first notifications to me on my return was that the Walker's good-conduct prize of £10 was awarded to me.

" I remember that my return was not very pleasant, for our table in hall was half occupied by a set of irregular men who had lost terms and were obliged to reside somewhat longer in order to receive the B.A. degree. But at the time of my completing the B.A. degree (which is not till some weeks after the examination and admission) I with the other complete bachelors was duly invited to the table of the B.A. scholars, and that annoyance ended.

"The liberation from undergraduate study left me at liberty generally to pursue my own course (except so far as it was influenced by the preparation for fellowship examination), and also left me at liberty to earn more money, in the way usual with the graduates, by taking undergraduate pupils. Mr Peacock recommended me to take only four, which occupied me four hours every day, and for each of them I received 20 guineas each term. My first pupils, for the Lent and Easter terms, were Williamson (afterwards Head Master of Westminster School), James Parker (afterwards Q.C. and

Vice-Chancellor), Bissett, and Clinton of Caius.   To all these
I had been engaged before taking my B.A. degree.

"I kept up classical subjects.   I have a set of notes on the
Πλοῦτος and Νεφέλαι of Aristophanes, finished on Mar. 15th,
1823, and I began my daily writing of Latin as usual on
Feb. 8th.   In mathematics I worked very hard at Lunar and
Planetary Theories.   I have two MS. books of Lunar Theory
to the 5th order of small quantities, which however answered
no purpose except that of making me perfectly familiar with
that subject.   I worked well, upon my quires, the figure of
Saturn supposed homogeneous as affected by the attraction
of his ring, and the figure of the Earth as heterogeneous, and
the Calculus of Variations.   I think it was now that I wrote
a MS. on constrained motion.

"On Mar. 17th, 1823, I was elected Fellow of the Cambridge Philosophical Society.   On May 9th a cast of my head
was taken for Dr Elliotson, an active phrenologist, by Deville,
a tradesman in the Strand.

"I had long thought that I should like to visit Scotland,
and on my once saying so to my mother, she (who had a
most kindly recollection of Alnwick) said in a few words that
she thought I could not do better.   I had therefore for some
time past fully determined that as soon as I had sufficient
spare time and money enough I would go to Scotland.   The
interval between the end of Easter Term and the usual
beginning with pupils in the Long Vacation offered sufficient
time, and I had now earned a little money, and I therefore
determined to go, and invited my sister to accompany me.
I had no private introductions, except one from James Parker
to Mr Reach, a writer of Inverness : some which Drinkwater
sent being too late.   On May 20th we went by coach to
Stamford ; thence by Pontefract and Oulton to York, where
I saw the Cathedral, which *then* disappointed me, but I suppose
that we were tired with the night journey.   Then by Newcastle
to Alnwick, where we stopped for the day to see my birth-
place.   On May 24th to Edinburgh.   On this journey I

remember well the stone walls between the fields, the place (in Yorkshire) where for the first time in my life I saw rock, the Hambleton, Kyloe, Cheviot and Pentland Hills, Arthur's Seat, but still more strikingly the revolving Inch Keith Light. At Edinburgh I hired a horse and gig for our journey in Scotland, and we drove by Queensferry to Kinross (where for the first time in my life I saw clouds on the hills, viz. on the Lomond Hills), and so to Perth. Thence by Dunkeld and Killicrankie to Blair Athol (the dreariness of the Drumochter Pass made a strong impression on me), and by Aviemore (where I saw snow on the mountains) to Inverness. Here we received much kindness and attention from Mr Reach, and after visiting the Falls of Foyers and other sights we went to Fort Augustus and Fort William. We ascended Ben Nevis, on which there was a great deal of snow, and visited the vitrified fort in Glen Nevis. Then by Inverary to Tarbet, and ascended Ben Lomond, from whence we had a magnificent view. We then passed by Loch Achray to Glasgow, where we found James Parker's brother (his father, of the house of Macinroy and Parker, being a wealthy merchant of Glasgow). On June 15th to Mr Parker's house at Blochairn, near Glasgow (on this day I heard Dr Chalmers preach), and on the 17th went with the family by steamer (the first that I had seen) to Fairly, near Largs. I returned the gig to Edinburgh, visited Arran and Bute, and we then went by coach to Carlisle, and by Penrith to Keswick (by the old road : never shall I forget the beauty of the approach to Keswick). After visiting Ambleside and Kendal we returned to Cambridge by way of Leeds, and posted to Bury on the 28th June. The expense of this expedition was about £81. It opened a completely new world to me.

"I had little time to rest at Bury. In the preceding term Drinkwater, Buckle, and myself, had engaged to go somewhere into the country with pupils during the Long Vacation (as was customary with Cambridge men). Buckle however changed his mind. Drinkwater went to look for a place,

4—2

fixed on Swansea, and engaged a house (called the Cambrian
Hotel, kept by a Captain Jenkins). On the morning of
July 2nd I left Bury for London and by mail coach to
Bristol. On the morning of July 3rd by steamer to Swansea,
and arrived late at night. I had then five pupils: Parker,
Harman Lewis (afterwards Professor in King's College, Lon-
don), Pierce Morton, Gibson, and Guest of Caius (afterwards
Master of the College). Drinkwater had four, viz. two
Malkins (from Bury), Elphinstone (afterwards M.P.), and
Farish (son of Professor Farish). We lived a hard-working
strange life. My pupils began with me at six in the morning:
I was myself reading busily. We lived completely *en famille*,
with two men-servants besides the house establishment. One
of our first acts was to order a four-oared boat to be built,
fitted with a lug-sail: she was called the Granta of Swansea.
In the meantime we made sea excursions with boats borrowed
from ships in the port. On July 23rd, with a borrowed boat,
we went out when the sea was high, but soon found our boat
unmanageable, and at last got into a place where the sea was
breaking heavily over a shoal, and the two of the crew who
were nearest to me (A. Malkin and Lewis), one on each side,
were carried out: they were good swimmers and we recovered
them, though with some trouble: the breaker had passed
quite over my head: we gained the shore and the boat was
taken home by land. When our own boat was finished, we
had some most picturesque adventures at the Mumbles,
Aberavon, Caswell Bay, Ilfracombe, and Tenby. From all
this I learnt navigation pretty well. The mixture of hard
study and open-air exertion seemed to affect the health of
several of us (I was one): we were covered with painful boils.

"My Latin-writing began again on July 25th: I have notes
on Demosthenes, Lucretius, and Greek History. In mathematics I find Chances, Figure of the Earth with variable
density, Differential Equations, Partial Differentials, sketch
for an instrument for shewing refraction, and Optical instruments with effects of chromatic aberration. In August there

occurred an absurd quarrel between the Fellows of Trinity and the undergraduates, on the occasion of commencing the building of King's Court, when the undergraduates were not invited to wine, and absented themselves from the hall.

"There were vacant this year (1823) five fellowships in Trinity College. In general, the B.A.'s of the first year are not allowed to sit for fellowships: but this year it was thought so probable that permission would be given, that on Sept. 2nd Mr Higman, then appointed as Tutor to a third 'side' of the College, wrote to me to engage me as Assistant Mathematical Tutor in the event of my being elected a Fellow on Oct. 1st, and I provisionally engaged myself. About the same time I had written to Mr Peacock, who recommended me to sit, and to Mr Whewell, who after consultation with the Master (Dr Wordsworth), discouraged it. As there was no absolute prohibition, I left Swansea on Sept. 11th (before my engagement to my pupils was quite finished) and returned to Cambridge by Gloucester, Oxford, and London. I gave in my name at the butteries as candidate for fellowship, but was informed in a day or two that I should not be allowed to sit. On Sept. 19th I walked to Bury.

"I walked back to Cambridge on Oct. 17th, 1823. During this October term I had four pupils: Neate, Cankrein, Turner (afterwards 2nd wrangler and Treasurer of Guy's Hospital), and William Hervey (son of the Marquis of Bristol). In the Lent term I had four (Neate, Cankrein, Turner, Clinton). In the Easter term I had three (Neate, Cankrein, Turner).

"My daily writing of Latin commenced on Oct. 27th. In November I began re-reading Sophocles with my usual care. In mathematics I find investigations of Motion in a resisting medium, Form of Saturn, Draft of a Paper about an instrument for exhibiting the fundamental law of refraction (read at the Philosophical Society by Mr Peacock on Nov. 10th, 1823), Optics, Solid Geometry, Figure of the Earth with variable density, and much about attractions. I also in this term wrote a MS. on the Calculus of Variations, and one on

Wood's Algebra, 2nd and 4th parts. I have also notes of the temperature of mines in Cornwall, something on the light of oil-gas, and reminiscences of Swansea in a view of Oswick Bay. In November I attended Professor Sedgwick's geological lectures.

"At some time in this term I had a letter from Mr South (to whom I suppose I had written) regarding the difficulty of my telescope: he was intimately acquainted with Tulley, and I suppose that thus the matter had become more fully known to him. He then enquired if I could visit him in the winter vacation. I accordingly went from Bury, and was received by him at his house in Blackman Street for a week or more with great kindness. He introduced me to Sir Humphrey Davy and many other London savans, and shewed me many London sights and the Greenwich Observatory. I also had a little practice with his own instruments. He was then on intimate terms with Mr Herschel (afterwards Sir John Herschel), then living in London, who came occasionally to observe double stars. This was the first time that I saw practical astronomy. It seems that I borrowed his mountain barometer. In the Lent term I wrote to him regarding the deduction of the parallax of Mars, from a comparison of the relative positions of Mars and 46 Leonis, as observed by him and by Rumker at Paramatta. My working is on loose papers. I see that I have worked out perfectly the interpolations, the effects of uncertainty of longitude, &c., but I do not see whether I have a final result.

"In Jan. 1824, at Playford, I was working on the effects of separating the two lenses of an object-glass, and on the kind of eye-piece which would be necessary: also on spherical aberrations and Saturn's figure. On my quires at Cambridge I was working on the effects of separating the object-glass lenses, with the view of correcting the secondary spectrum: and on Jan. 31st I received some numbers (indices of refraction) from Mr Herschel, and reference to Fraunhofer's numbers.

"About this time it was contemplated to add to the Royal

Observatory of Greenwich two assistants of superior education. Whether this scheme was entertained by the Admiralty,
the Board of Longitude, or the Royal Society, I do not know.
Somehow (I think through Mr Peacock) a message from Mr
Herschel was conveyed to me, acquainting me of this, and
suggesting that I should be an excellent person for the principal place. To procure information, I went to London on
Saturday, Feb. 7th, sleeping at Mr South's, to be present at
one of Sir Humphrey Davy's Saturday evening soirées (they
were then held every Saturday), and to enquire of Sir H.
Davy and Dr Young. When I found that succession to the
post of Astronomer Royal was not considered as distinctly a
consequence of it, I took it coolly, and returned the next
night. The whole proposal came to nothing.

"At this time I was engaged upon differential equations,
mountain barometer problem and determination of the height
of the Gogmagogs and several other points, investigations
connected with Laplace's calculus, spherical aberration in
different planes, geology (especially regarding Derbyshire,
which I proposed to visit), and much of optics. I wrote a
draft of my Paper on the figure of Saturn, and on Mar. 15th,
1824, it was read at the Philosophical Society under the title
of 'On the figure assumed by a fluid homogeneous mass,
whose particles are acted on by their mutual attraction, and
by small extraneous forces,' and is printed in their Memoirs.
I also wrote a draft of my Paper on Achromatic Eye-pieces,
and on May 17th, 1824, it was read at the Philosophical
Society under the title of 'On the Principles and Construction of the Achromatic Eye-pieces of Telescopes, and on the
Achromatism of Microscopes,' including also the effects of
separating the lenses of the object-glass. It is printed in
their Memoirs.

"Amongst miscellaneous matters I find that on Mar. 22nd
of this year I began regularly making extracts from the
books of the Book Society, a practice which I continued to
March 1826. On Mar. 27th, a very rainy day, I walked to

Bury to attend the funeral of my uncle William Biddell, near Diss, and on Mar. 30th I walked back in rain and snow. On Feb. 24th I dined with Cubitt in Cambridge. On May 21st I gave a certificate to Rogers (the assistant in Crosse's school, and my instructor in mathematics), which my mother amplified much, and which I believe procured his election as master of Walsall School. On June 23rd I went to Bury. The speeches at Bury School, which I wished to attend, took place next day."

At this point of his Autobiography the writer continues, "Now came one of the most important occurrences in my life." The important event in question was his acquaintance with Richarda Smith, the lady who afterwards became his wife. The courtship was a long one, and in the Autobiography there are various passages relating to it, all written in the most natural and unaffected manner, but of somewhat too private a nature for publication. It will therefore be convenient to digress from the straight path of the narrative in order to insert a short memoir of the lady who was destined to influence his life and happiness in a most important degree.

Richarda Smith was the eldest daughter of the Rev. Richard Smith, who had been a Fellow of Trinity College, Cambridge, but was at this time Private Chaplain to the Duke of Devonshire, and held the small living of Edensor, near Chatsworth, in Derbyshire. He had a family of two sons and seven daughters, whom he had brought up and educated very carefully. Several of his daughters were remarkable both for their beauty and accomplishments. Richarda Smith was now in her 20th year, and the writer of the Autobiography records that "at Matlock we received great attention from Mr Chenery: in speaking of Mr Smith I remember his saying that Mr Smith had a daughter whom the Duke of Devonshire declared to be the most beautiful girl he ever saw." This was before he had made the acquaintance of the family. Airy was at this time on a walking tour in Derbyshire with his brother William, and they were received at Edensor by Mr Smith, to whom he had letters of introduction. He seems to have fallen in love with Miss Smith " at first sight," and within two days of first seeing her he made her an offer of marriage. Neither his means nor his prospects at that time permitted the least idea of an immediate marriage, and Mr Smith would not hear of any engagement. But he never had the least doubt as to the wisdom of the choice that he had made : he worked steadily on, winning fame and position, and recommending his suit from time to time to Miss Smith as opportunity offered, and finally married her, nearly six years after his first proposal. His constancy had its reward, for he gained a most charming and affectionate wife. As he records at the time of his marriage, " My wife was aged between 25 and 26, but she scarcely appeared more than 18 or 20. Her beauty and accomplishments, her skill and fidelity in sketching, and above all her exquisite singing of ballads, made a great sensation in Cambridge."

Their married life lasted 45 years, but the last six years were saddened by the partial paralysis and serious illness of Lady Airy. The entire correspondence between them was most carefully preserved, and is a record of a most happy union. The letters were written during his numerous journeys and excursions on business or pleasure, and it is evident that his thoughts were with her from the moment of their parting. Every opportunity of writing was seized with an energy and avidity that shewed how much his heart was in the correspondence. Nothing was too trivial or too important to communicate to his wife, whether relating to family or business matters. The letters on both sides are always full of affection and sympathy, and are written in that spirit of confidence which arises from a deep sense of the value and necessity of mutual support in the troubles of life. And with his active and varied employments and his numerous family there was no lack of troubles. They were both of them simple-minded, sensible, and practical people, and were very grateful for such comforts and advantages as they were able to command, but for nothing in comparison with their deep respect and affection for one another.

Both by natural ability and education she was well qualified to enter into the pursuits of her husband, and in many cases to assist him. She always welcomed her husband's friends, and by her skill and attractive courtesy kept them well together. She was an admirable letter-writer, and in the midst of her numerous domestic distractions always found time for the duties of correspondence. In conversation she was very attractive, not so much from the wit or brilliancy of her remarks as from the brightness and interest with which she entered into the topics under discussion, and from the unfailing grace and courtesy with which she attended to the views of others. This was especially recognized by the foreign astronomers and men of science who from time to time stayed as guests at the Observatory and to whom she acted as hostess. Although she was not an accomplished linguist yet she was well able to express herself in French and German, and her natural good sense and kindliness placed her guests at their ease, and made them feel themselves (as indeed they were) welcomed and at home.

Her father, the Rev. Richard Smith, was a man of most cultivated mind, and of the highest principles, with a keen enjoyment of good society, which the confidence and friendship of his patron the Duke of Devonshire amply secured to him, both at Chatsworth and in London. He had a deep attachment to his Alma Mater of Cambridge, and though not himself a mathematician he had a great respect for the science of mathematics and for eminent mathematicians. During the long courtship already related Mr Smith conceived the highest respect for Airy's character, as well as for his great repute and attainments, and expressed his lively satisfaction at his daughter's marriage. Thus on January 20th, 1830, he wrote to his intended son-in-law as follows: " I have

little else to say to you than that I continue with heartfelt
satisfaction to reflect on the important change about to take
place in my dear daughter's situation. A father must not
allow himself to dilate on such a subject: of course I feel
confident that you will have no reason to repent the irre-
vocable step you have taken, but from the manner in which
Richarda has been brought up, you will find such a helpmate
in her as a man of sense and affection would wish to have,
and that she is well prepared to meet the duties and trials
(for such must be met with) of domestic life with a firm and
cultivated mind, and the warm feelings of a kind heart. Her
habits are such as by no means to lead her to expensive
wishes, nor will you I trust ever find it necessary to neglect
those studies and pursuits upon which your reputation and
subsistence are chiefly founded, to seek for idle amusements
for your companion. I must indulge no further in speaking
of her, and have only at present to add that I commit in full
confidence into your hands the guardianship of my daughter's
happiness." And on April 5th, 1830, shortly after their
marriage, he wrote to his daughter thus: "If thinking of you
could supply your place amongst us you would have been
with us unceasingly, for we have all of us made you the
principal object of our thoughts and our talk since you left
us, and I travelled with you all your journey to your present
delightful home. We had all but one feeling of the purest
pleasure in the prospect of the true domestic comfort to
which we fully believe you to be now gone, and we rejoice
that all your endearing qualities will now be employed to
promote the happiness of one whom we think so worthy of
them as your dear husband, who has left us in the best
opinion of his good heart, as well as his enlightened and
sound understanding. His late stay with us has endeared
him to us all. Never did man enter into the married state
from more honourable motives, or from a heart more truly
seeking the genuine happiness of that state than Mr Airy,
and he will, I trust, find his reward in you from all that a

good wife can render to the best of husbands, and his happiness be reflected on yourself." It would be difficult to find
letters of more genuine feeling and satisfaction, or more
eloquently expressed, than these.

   The narrative of the Autobiography will now be resumed.
   " I had been disappointed two years before of an expedition to Derbyshire. I had wished still to make it, and my
brother wished to go: and we determined to make it this
year (1824). We were prepared with walking dresses and
knapsacks. I had well considered every detail of our route,
and was well provided with letters of introduction, including
one to the Rev. R. Smith of Edensor. On June 29th we
started by coach to Newmarket and walked through the Fens
by Ramsay to Peterborough. Then by Stamford and Ketton
quarries to Leicester and Derby. Here we were recognized
by a Mr Calvert, who had seen me take my degree, and he
invited us to breakfast, and employed himself in shewing us
several manufactories, &c. to which we had been denied
access when presenting ourselves unsupported. We then
went to Belper with an introduction from Mr Calvert to
Jedediah Strutt: saw the great cotton mills, and in the
evening walked to Matlock. Up to this time the country of
greatest interest was the region of the fens about Ramsay (a
most remarkable district), but now began beauty of scenery.
On July 9th we walked by Rowsley and Haddon Hall over
the hills to Edensor, where we stayed till the 12th with
Mr Smith. We next visited Hathersage, Castleton, and
Marple (where I wished to see the canal aqueduct), and
went by coach to Manchester, and afterwards to Liverpool.
Here Dr Traill recommended us to see the Pontycyssylte
Aqueduct, and we went by Chester and Wrexham to
Rhuabon, saw the magnificent work, and proceeded to
Llangollen. Thence by Chester and Northwich (where we
descended a salt-mine) to Macclesfield. Then to the Ecton
mine (of which we saw but little) through Dovedale to

Ashbourn, and by coach to Derby. On July 24th to Birmingham, where we found Mr Guest, lodged in his house, and were joined by my pupil Guest. Here we were fully employed in visiting the manufactures, and then went into the iron country, where I descended a pit in the Staffordshire Main. Thence by coach to Cambridge, where I stopped to prepare for the Fellowship Examination.

"I had two pupils in this portion of the Long Vacation, Turner and Dobbs. On August 2nd my writing of Latin began regularly as before. My principal mathematics on the quires are Optics. On August 25th I made experiments on my left eye, with good measures, and on Aug. 26th ordered a cylindrical lens of Peters, a silversmith in the town, which I believe was never made. Subsequently, while at Playford, I ordered cylindrical lenses of an artist named Fuller, living at Ipswich, and these were completed in November, 1824.

"My letter to the Examiners, announcing my intention of sitting for Fellowship (which like all other such documents is preserved on my quires) was delivered on Sept. 21st. The Examination took place on Sept. 22nd and the two following days. On Oct. 1st, 1824, at the usual hour of the morning, I was elected Fellow. There were elected at the same time T. B. Macaulay (afterwards Lord Macaulay), who was a year senior to me in College, and I think Field of my own year. I drew up my letter of acknowledgment to the Electors. On Oct. 2nd at 9 in the morning I was admitted Fellow with the usual ceremonies, and at 10 I called on the Electors with my letter of acknowledgment. I immediately journeyed to Derbyshire, paid a visit at Edensor, and returned by Sheffield.

"On Oct. 11th (it having been understood with Mr Higman that my engagement as Assistant Mathematical Tutor stood) the Master sent for me to appoint me and to say what was expected as duty of the office. He held out to me the prospect of ultimately succeeding to the Tutorship, and I told him that I hoped to be out of College before that time.

"About this time the 'Athenæum,' a club of a scientific character, was established in London, and I was nominated on it, but I declined. (Oct. 14th). In this year (1824) I commenced account with a banker by placing £110 in the hands of Messrs Mortlock and Co. On Oct. 16th I walked to Bury, and after a single day's stay there returned to Cambridge.

"On Oct. 23rd, 1824, I began my lectures as Mathematical Assistant Tutor. I lectured the Senior Sophs and Junior Sophs on Higman's side. The number of Senior Sophs was 21. Besides this I took part in the 'Examinations of the Questionists,' a series of exercises for those who were to take the Bachelor's degree in the next January. I examined in Mechanics, Newton, and Optics. I had also as private pupils Turner, Dobbs, and Cooper. I now ceased from the exercise which I had followed with such regularity for five years, namely that of daily writing Latin. In its stead I engaged a French Master (Goussel) with whom I studied French with reasonable assiduity for the three terms to June, 1825.

"Among mathematical investigations I find: Theory of the Moon's brightness, Motion of a body in an ellipse round two centres of force, Various differential equations, Numerical computation of $\sin \pi$ from series, Numerical computation of sines of various arcs to 18 decimals, Curvature of surfaces in various directions, Generating functions, Problem of sound. I began in the winter a Latin Essay as competing for the Middle Bachelors' Prize, but did not proceed with it. I afterwards wished that I had followed it up: but my time was fully occupied.

"On Jan. 28th, 1825, I started for Edensor, where I paid a visit, and returned on Feb. 2nd. On Feb. 4th I wrote to Mr Clarkson, asking his advice about a profession or mode of life (the cares of life were now beginning to press me heavily, and continued to do so for several years). He replied very kindly, but his answer amounted to nothing.

About the same time I had some conversation of the same kind with Mr Peacock, which was equally fruitless.

"On Feb. 4th I have investigations of the density of light near a caustic (on the theory of emissions). On Feb. 5th I finished a Paper about the defect in my eye, which was communicated to the Cambridge Philosophical Society on Feb. 21st. Mr Peacock or Mr Whewell had some time previously applied to me to write a Paper on Trigonometry for the Encyclopædia Metropolitana, and I had been collecting some materials (especially in regard to its history) at every visit to London, where I read sometimes at the British Museum: also in the Cambridge libraries. I began this Paper (roughly) on Feb. 8th, and finished it on Mar. 3rd. The history of which I speak, by some odd management of the Editors of the Encyclopædia, was never published. The MS. is now amongst the MSS. of the Royal Observatory, Greenwich. Other subjects on my quires are: Theory of musical concords, many things relating to trigonometry and trigonometrical tables, achromatic eye-pieces, equation to the surface bounding the rays that enter my left eye, experiments on percussion. Also notes on Cumberland and Wales (I had already proposed to myself to take a party of pupils in the Long Vacation to Keswick), and notes on history and geology.

"I had been in correspondence with Dr Malkin (master of Bury School), who on Feb. 8th sent a certificate for my brother William, whom I entered at Trinity on Peacock's side. On Mar. 25th I changed my rooms, quitting those on the ground-floor east side of Queen Mary's Gate for first-floor rooms in Neville's Court, south side, the easternmost rooms. In this term my lectures lasted from Apr. 18th to May 14th. Apparently I had only the Senior Sophs, 19 in number, and the same four pupils (Turner, Dobbs, Cooper, Hovenden) as in the preceding term. The only scientific subjects on which I find notes are, a Paper on the forms of the Teeth of Wheels, communicated to the Philosophical

Society on May 2nd; some notes about Musical Concords, and some examination of a strange piece of Iceland Spar. On Apr. 29th I was elected to the Northern Institution (of Inverness); the first compliment that I received from an extraneous body.

"On May 14th I have a most careful examination of my money accounts, to see whether I can make an expedition with my sister into Wales. My sister came to Cambridge, and on Monday, May 23rd, 1825, we started for Wales, equipped in the lightest way for a walking expedition. We went by Birmingham to Shrewsbury: then to the Pontycyssylte Aqueduct and by various places to Bala, and thence by Llanrwst to Conway. Here the suspension bridge was under construction: the mole was made and the piers, but nothing else. Then on to Bangor, where nine chains of the suspension bridge were in place, and so to Holyhead. Then by Carnarvon to Bethgelert, ascending Snowdon by the way, and in succession by Festiniog, Dolgelly, and Aberystwyth to Hereford (the first time that I had visited it since my father left it). From thence we went by coach to London, and I went on to Cambridge on the 23rd of June.

"I had arranged to take a party of pupils to Keswick, and to take my brother there. Mr Clarkson had provided me with introductions to Mr Southey and Mr Wordsworth. On Wednesday, June 29th, 1825, we started, and went by Leicester, Sheffield, Leeds, and Kendal, to Keswick, calling at Edensor on the way. My pupils were Cleasby, Marshman, Clinton, Wigram, Tottenham, and M. Smith. At Keswick I passed three months very happily. I saw Mr Southey's family frequently, and Mr Wordsworth's occasionally. By continual excursions in the neighbourhood, and by a few excursions to places as distant as Bowness, Calder Bridge, &c. (always climbing the intermediate mountains), I became well acquainted with almost the whole of that beautiful country, excepting some of the S. W. dales. A geological hammer and a mountain barometer were very interesting

companions. I had plenty of work with my pupils: I worked
a little Lunar Theory, a little of Laplace's Equations, something of the Figure of the Earth, and I wrote out very carefully my Trigonometry for the Encyclopædia Metropolitana.
I read a little of Machiavelli, and various books which I borrowed of Mr Southey. On Friday, Sept. 30th, my brother and
I left for Kendal, and after a stay of a few days at Edensor,
arrived at Cambridge on Oct. 11th.

"On Oct. 21st my Lectures to the Junior Sophs began,
39 names, lasting to Dec. 13th. Those to the Senior Sophs,
16 names, Oct 29th to Dec. 10th. I also examined Questionists as last year. I have notes about a Paper on the connection of impact and pressure, read at the Philosophical
Society on Nov. 14th, but not printed, dipping-needle problems, curve described round three centres of force, barometer
observations, theory of the Figure of the Earth with variable
density, and effect on the Moon, correction to the Madras
pendulum, wedge with friction, spots seen in my eyes, density
of rays near a caustic. In this term I accomplished the preparation of a volume of Mathematical Tracts on subjects
which, either from their absolute deficiency in the University
or from the unreadable form in which they had been presented, appeared to be wanted. The subjects of my Tracts
were, Lunar Theory (begun Oct. 26th, finished Nov. 1st),
Figure of the Earth (1st part finished Nov. 18th), Precession
and Nutation (my old MS. put in order), and the Calculus of
Variations. I applied, as is frequently done, to the Syndicate
of the University Press for assistance in publishing the work;
and they agreed to give me paper and printing for 500 copies.
This notice was received from Professor Turton on Nov. 29th,
1825. It was probably also in this year that I drew up an
imperfect 'Review' of Coddington's Optics, a work which
deserved severe censure: my review was never finished.

"In the Long Vacation at Keswick I had six pupils at
£42 each. In the October term I had Marshman and
Ogilby at £105 for three terms, and Dobbs at £75 for three

A. B.   5

terms. I had, at Mr Peacock's suggestion, raised my rate
from 60 to 100 guineas for three terms: this prevented some
from applying to me, and induced some to withdraw who had
been connected with me: but it did me no real hurt, for
engrossment by pupils is the worst of all things that can
happen to a man who hopes to distinguish himself. On Dec.
17th I went to Bury, and returned to Cambridge on Jan.
30th, 1826.

"I have the attendance-bills of my Lectures to Senior
Sophs (16) from Feb. 3rd to Feb. 23rd, and to Freshmen (40)
from Feb. 27th to Mar. 15. It would appear that I gave but
one college-lecture per day (my belief was that I always had
two). The tutor's stipend per term was £50. On my quires
I find, Investigations for the ellipticity of a heterogeneous
spheroid when the density is expressed by $\frac{\sin qc}{qc}$ (the remarkable properties of which I believe I discovered entirely myself,
although they had been discovered by other persons), Theoretical Numbers for precession, nutation, &c., some investigations using Laplace's Y, hard work on the Figure of the
Earth to the 2nd order, 'Woodhouse's remaining apparatus,'
Notes about Lambton's and Kater's errors, Depolarization,
Notes of Papers on depolarization in the Phil. Trans., Magnetic Investigations for Lieut. Foster, Isochronous Oscilla-
tions in a resisting medium, Observations on a strange piece
of Iceland Spar. On Mar. 7th forwarded Preface and Title
Page for my Mathematical Tracts.

"Some time in this term I began to think of the possibility of observing the diminution of gravity in a deep mine,
and communicated with Whewell, who was disposed to join
in experiments. My first notion was simply to try the rate
of a clock, and the Ecton mine was first thought of. I made
enquiries about the Ecton mine through Mr Smith (of Edensor), and visited the mine, but in the meantime Whewell had
made enquiries in London and found (principally from Dr
Paris) that the mine of Dolcoath near Camborne in Cornwall

would be a better place for the experiment. Dr Paris wrote
to me repeatedly, and ultimately we resolved on trying it
there. In my papers on Mar. 21st are various investigations
about attractions in both mines. On Apr. 3rd I went to
London, principally to arrange about Dolcoath, and during
April and May I was engaged in correspondence with Sir H.
Davy (President of the Royal Society), Mr Herschel, and Dr
Young (Secretary of the Board of Longitude) about the loan
of instruments and pendulums. On Apr. 23rd I was prac-
tising pendulum-observations (by coincidence); and about
this time repeatedly practised transits with a small instrument lent by Mr Sheepshanks (with whom my acquaintance
must have begun no long time before) which was erected
under a tent in the Fellows' Walks. On my quires I find
various schemes for graduating thermometers for pendulum
experiments.

"I find also Notes of examination of my brother William,
who had come to College last October ; and a great deal of
correspondence with my mother and sister and Mr Case, a
lawyer, about a troublesome business with Mr Cropley, an
old friend of G. Biddell, to whom my father had lent £500
and whose affairs were in Chancery.

" My lectures in this term were to the Junior Sophs from
Apr. 10th to May 13th : they were six in number and not
very regular. On Apr. 28th I sent to Mawman the copy of
my Trigonometry for the Encyclopædia Metropolitana, for
which I received £42. I received notice from the Press Syndicate that the price of my Mathematical Tracts was fixed at
6s. 6d.: I sold the edition to Deighton for £70, and it was
immediately published. About this time I have letters from
Mr Herschel and Sir H. Davy about a Paper to be presented
to the Royal Society—I suppose about the Figure of the
Earth to the 2nd order of ellipticity, which was read to the
Royal Society on June 15th.

" On Saturday, May 13th, 1826, I went to London on the
way to Dolcoath, and received four chronometers from the

Royal Observatory, Greenwich. I travelled by Devonport and Falmouth to Camborne, where I arrived on May 20th and dined at the count-house dinner at the mine. I was accompanied by Ibbotson, who was engaged as a pupil, and intended for an engineer. On May 24th Whewell arrived, and we took a pendulum and clock down, and on the 30th commenced the observation of coincidences in earnest. This work, with the changing of the pendulums, and sundry short expeditions, occupied nearly three weeks. We had continued the computation of our observations at every possible interval. It is to be understood that we had one detached pendulum swinging in front of a clock pendulum above, and another similarly mounted below; and that the clocks were compared by chronometers compared above, carried down and compared, compared before leaving, and brought up and compared. The upper and lower pendulums had been interchanged. It was found now that the reliance on the steadiness of the chronometers was too great; and a new method was devised, in which for each series the chronometers should make four journeys and have four comparisons above and two below. This arrangement commenced on the 19th June and continued till the 26th. On the 26th we packed the lower instruments, intending to compare the pendulum directly with the upper one, and sent them up the shaft: when an inexplicable occurrence stopped all proceedings. The basket containing all the important instruments was brought up to the surface (in my presence) on fire; some of the instruments had fallen out with their cases burning. Whether a superstitious miner had intentionally fired it, or whether the snuff of a candle had been thrown into it, is not known. Our labour was now rendered useless. On the 28th I packed up what remained of instruments, left for Truro, and arrived at Bury on July 1st. During our stay in Cornwall I had attended a 'ticketing' or sale of ore at Camborne, and we had made expeditions to the N. W. Coast, to Portreath and Illogan, to Marazion and St Michael's Mount, and to Penzance and the Land's End. On July 3rd I saw Mr Cropley in Bury gaol, and went to Cambridge. On the 4th I was admitted A.M., and on the 5th was admitted Major Fellow.

"I had engaged with four pupils to go to Orléans in this Long Vacation: my brother William was also to go. One of my pupils, Dobbs, did not join: the other three were Tinkler, Ogilby, and Ibbotson. We left London on July 9th, and travelled by Brighton, Dieppe, Rouen, and Paris to Orléans. At Paris I saw Bouvard, Pouillet, Laplace and Arago. I had introductions from Mr Peacock, Mr South, Mr Herschel, Dr Young; and from Professor Sedgwick to an English resident, Mr Underwood. On the 19th I was established in the house of M. Lagarde, Protestant Minister. Here I received my pupils. On the 28th I commenced Italian with an Italian master: perhaps I might have done more prudently in adhering to French, for I made no great progress. On Aug. 2nd I saw a murderer guillotined in the Place Martroi. The principal investigations on my quires are—Investigations about pendulums, Calculus of Variations, Notes for the Figure of the Earth (Encyc. Metrop.) and commencement of the article, steam-engine machinery, &c. I picked up various French ballads, read various books, got copies of the Marseillaise (this I was obliged to obtain rather secretly, as the legitimist power under Charles X. was then at its height) and other music, and particulars of farm wages for Whewell and R. Jones. The summer was intensely hot, and I believe that the heat and the work in Dolcoath had weakened me a good deal. The family was the old clergyman, his wife, his daughter, and finally his son. We lived together very amicably. My brother lodged in a Café in the Place Martroi; the others in different families. I left Orléans on Sept. 30th for Paris. Here I attended the Institut, and was present at one of Ampère's Lectures. I arrived at Cambridge on Oct. 14th.

"On Oct. 16th Whewell mentioned to me that the Lucasian Professorship would be immediately vacated by

Turton, and encouraged me to compete for it. Shortly afterwards Mr Higman mentioned the Professorship, and Joshua
King (of Queens') spoke on the restriction which prevented
College tutors or Assistant tutors from holding the office.
About this time Mr Peacock rendered me a very important
service. As the emolument of the Lucasian Professorship
was only £99, and that of the Assistant Tutorship £150, I
had determined to withdraw from the candidature. But Mr
Peacock represented to me the advantage of position which
would be gained by obtaining the Professorship (which I then
instantly saw), and I continued to be a candidate. I wrote
letters to the Heads of Colleges (the electors) and canvassed
them personally. Only Dr Davy, the Master of Caius
College, at once promised me his vote. Dr French, Master
of Jesus College, was a candidate; and several of the Heads
had promised him their votes. Mr Babbage, the third candidate, threatened legal proceedings, and Dr French withdrew.
The course was now open for Mr Babbage and me.

"In the meetings of the Philosophical Society a new
mode of proceeding was introduced this term. To enliven
the meetings, private members were requested to give oral
lectures. Mine was the second, I think, and I took for
subject The Machinery of the Steam Engines in the Cornish
mines, and especially of the Pumping Engines and Pumps.
It made an excellent lecture: the subjects were at that time
undescribed in books, and unknown to engineers in general
out of Cornwall.

"My College lectures seem to have been, Oct. 21st to
Dec. 14th to 31 Junior Sophs, Dec. 4th to 12th to 12 Senior
Sophs. I assisted at the examinations of the Questionists.
I had no private pupils. On Nov. 26th I communicated to
the Cambridge Philosophical Society a Paper on the Theory
of Pendulums, Balances, and Escapements: and I find applications of Babbage's symbolism to an escapement which I
proposed. I have various investigations about the Earth,
supposed to project at middle latitudes above the elliptical

form. In November an account of the Dolcoath failure (by Whewell) was given to the Royal Society.

"At length on Dec. 7th, 1826, the election to the Lucasian Professorship took place: I was elected (I think unanimously) and admitted. I believe that this gave great satisfaction to the University in general. My uncle, Arthur Biddell, was in Cambridge on that evening, and was the first of my friends who heard of it. On the same page of my quires on which this is mentioned, there is a great list of apparatus to be constructed for Lucasian Lectures, notes of experiments with Atwood's Machine, &c. In December, correspondence with Dollond about prisms. I immediately issued a printed notice that I would give professorial lectures in the next Term.

" On Dec. 13th I have a letter from Mr Smith informing me of the dangerous illness (fever) which had attacked nearly every member of his family, Richarda worst of all. On Dec. 23rd I went to Bury. The affairs with Cropley had been settled by the sale of his property under execution, and my father did not lose much of his debt. But he had declined much in body and mind, and now had strange hallucinations.

" The commencement of 1827 found me in a better position (not in money but in prospects) than I had before stood in: yet it was far from satisfactory. I had resigned my Assistant Tutorship of £150 per annum together with the prospect of succeeding to a Tutorship, and gained only the Lucasian Professorship of £99 per annum. I had a great aversion to entering the Church: and my lay fellowship would expire in 7 years. My prospects in the law or other professions might have been good if I could have waited: but then I must have been in a state of starvation probably for many years, and marriage would have been out of the question: I much preferred a moderate income in no long time, and I am sure that in this I judged rightly for my happiness. I had now in some measure taken science as my line (though not irrevocably), and I thought it best to work it well, for a time at least, and wait for accidents.

"The acceptance of the Lucasian Professorship prevented me from being pressed by Sedgwick (who was Proctor this year) to take the office of moderator: which was a great relief to me. As Lucasian Professor I was ipso facto Member of the Board of Longitude. A stipend of £100 a year was attached to this, on condition of attending four meetings: but I had good reason (from intimations by South and other persons in London) for believing that this would not last long. The fortnightly notices of the meetings of the Board were given on Jan. 18th, Mar. 22nd, May 24th and Oct. 18th.

"On Jan. 2nd, 1827, I came from London to Bury. I found my father in a very declining state (the painful rheumatism of some years had changed to ulcerations of the legs, and he was otherwise helpless and had distressing hallucinations). On Jan. 8th I walked to Cambridge. At both places I was occupied in preparations for the Smith's Prize Examination and for lectures (for the latter I obtained at Bury gaol some numerical results about tread-mills).

"Of the Smith's Prize I was officially an Examiner: and I determined to begin with—what had never been done before—making the examination public, by printing the papers of questions. The Prize is the highest Mathematical honour in the University: the competitors are incepting Bachelors of Arts after the examination for that Degree. My day of examination (apparently) was Jan. 21st. The candidates were Turner, Cankrein, Cleasby, and Mr Gordon. The first three had been my private pupils: Mr Gordon was a Fellow-commoner of St Peter's College, and had just passed the B.A. examination as Senior Wrangler, Turner being second. My situation as Examiner was rather a delicate one, and the more so as, when I came to examine the papers of answers, Turner appeared distinctly the first. Late at night I carried the papers to Whewell's rooms, and he on inspection agreed with me. The other examiners (Professors Lax and Woodhouse, Lowndean and Plumian Professors) generally supported me: and Turner had the honour of First Smith's Prize.

"On Jan. 30th my mother wrote, asking if I could see Cropley in London, where he was imprisoned for contempt of Chancery. I attended the meeting of the Board of Longitude on Feb. 1st, and afterwards visited Cropley in the Fleet Prison. He died there, some time later. It was by the sale of his effects under execution that my father's debt was paid.

"On Feb. 15th I communicated to the Royal Society a Paper on the correction of the Solar Tables from South's observations. I believe that I had alluded to this at the February meeting of the Board of Longitude, and that in consequence Mr Pond, the Astronomer Royal, had been requested to prepare the errors of the Sun's place from the Greenwich observations: which were supplied some months later. With the exception of South's Solar Errors, and some investigations about dipping-needles, I do not find anything going on but matters connected with my approaching lectures. There are bridges, trusses, and other mechanical matters, theoretical and practical, without end. Several tradesmen in Cambridge and London were well employed. On Feb. 13th I have a letter from Cubitt about groins: I remember studying those of the Custom-house and other places. On Feb. 20th my Syllabus of Lectures was finished: this in subsequent years was greatly improved. I applied to the Royal Society for the loan of Huyghens's object-glass, but they declined to lend it. About this time I find observations of the spectrum of Sirius.

"There had been no lectures on Experimental Philosophy (Mechanics, Hydrostatics, Optics) for many years. The University in general, I believe, looked with great satisfaction to my vigorous beginning: still there was considerable difficulty about it. There was no understood term for the Lectures: no understood hour of the day: no understood lecture room. I began this year in the Lent Term, but in

all subsequent years I took the Easter Term, mainly for the
chance of sunlight for the optical experiments, which I soon
made important. I could get no room but a private or
retiring room (not a regular lecture room) in the buildings at
the old Botanic Garden: in following years I had the room
under the University Library. The Lectures commenced on
some day in February 1827: I think that the number who
attended them was about 64. I remember very well that the
matter which I had prepared as an Introductory Lecture did
not last above half the time that I had expected, but I
managed very well to fill up the hour. On another occasion
I was so ill-prepared that I had contemplated giving notice
that I was unable to complete the hour's lecture, but I saw in
the front row some strangers, introduced by some of my
regular attendants, very busy in taking notes, and as it was
evident that a break-down now would not do, I silently
exerted myself to think of something, and made a very good
lecture.

"On Mar. 1st, as official examiner, I received notices
from 14 candidates for Bell's Scholarships, and prepared my
Paper of questions. I do not remember my day of examination; but I had all the answers to all the examiners' questions in my hands, when on Mar. 27th I received notice that
my father had died the preceding evening. This stopped my
Lectures: they were concluded in the next term. I think
that I had only Mechanics and imperfect Optics this term,
no Hydrostatics; and that the resumed Lectures were principally Optical. They terminated about May 14th.

"With my brother I at once went to Bury to attend my
father's funeral. He was buried on Mar. 31st, 1827, in the
churchyard of Little Whelnetham, on the north side of the
church. Shortly afterwards I went to London, and on Apr.
5th I attended a meeting of the Board of Longitude, at which
Herschel produced a Paper regarding improvements of the
Nautical Almanac. Herschel and I were in fact the leaders
of the reforming party in the Board of Longitude: Dr Young

the Secretary resisted change as much as possible. After the meeting I went to Cambridge.   I find then calculations of achromatic eye-pieces for a very nice model with silk threads of various colours which I made with my own hands for my optical lectures.

"On Apr. 7th Herschel wrote to me that the Professorship held by Dr Brinkley (then appointed Bishop of Cloyne) at Dublin would be vacant, and recommended it to my notice, and sent me some introductions. I reached Dublin on Apr. 15th, where I was received with great kindness by Dr Brinkley and Dr MacDonnell (afterwards Provost). I there met the then Provost Dr Bartholomew Lloyd, Dr Lardner, Mr Hamilton (afterwards Sir W. R. Hamilton) and others. In a few days I found that they greatly desired to appoint Hamilton if possible (they did in fact overcome some difficulties and appoint him in a few months), and that they would not make such an augmentation as would induce me to offer myself as a candidate, and I withdrew. I have always remembered with gratitude Dr MacDonnell's conduct, in carefully putting me on a fair footing in this matter. I returned by Holyhead, and arrived at Birmingham on Apr. 23rd. While waiting there and looking over some papers relating to the spherical aberration of eye-pieces, in which I had been stopped some time by a geometrical difficulty, I did in the coffee-room of a hotel overcome the difficulty; and this was the foundation of a capital paper on the Spherical Aberration of Eye-pieces. This paper was afterwards presented to the Cambridge Philosophical Society.

"About this time a circumstance occurred of a disagreeable nature, which however did not much disconcert me. Mr Ivory, who had a good many years before made himself favourably known as a mathematician, especially by his acquaintance with Laplace's peculiar analysis, had adopted (as not unfrequently happens) some singular hydrostatical theories. In my last Paper on the Figure of the Earth, I had said that I could not receive one of his equations. In

the Philosophical Magazine of May he attacked me for this
with great heat. On May 8th I wrote an answer, and I
think it soon became known that I was not to be attacked
with impunity.

   "Long before this time there had been some proposal
about an excursion to the Lake District with my sister, and
I now arranged to carry it out. On May 23rd I went to
Bury and on to Playford: while there I sketched the Cumberland excursion. On June 5th I went to London, I believe
to the Visitation of the Greenwich Observatory to which I
was invited. I also attended the meeting of the Board of
Longitude. I think it was here that Pond's Errors of the
Sun's place in the Nautical Almanac from Greenwich Observations were produced. On June 7th I went by coach to
Rugby, where I met my sister, and we travelled to Edensor.
We made a number of excursions in Derbyshire, and then
passed on by Penrith to Keswick, where we arrived on June
22nd. From Keswick we made many excursions in the
Lake District, visited Mr Southey and Mr Wordsworth,
descended a coal mine at Whitehaven, and returned to
Edensor by the way of Ambleside, Kendal, and Manchester.
With sundry excursions in Derbyshire our trip ended, and
we returned to Cambridge on the 21st July.

   "During this Long Vacation I had one private pupil,
Crawford, the only pupil this year, and the last that I ever
had. At this time there is on my papers an infinity of optical investigations: also a plan of an eye-piece with a concave
lens to destroy certain aberrations. On Aug. 20th I went to
Woodford to see Messrs Gilbert's optical works. From Aug.
13th I had been preparing for the discussion of the Greenwich Solar Errors, and I had a man at work in my rooms,
engaged on the calculation of the Errors. I wrote to Bouvard
at Paris for observations of the sun, but he recommended me
to wait for the Tables which Bessel was preparing. I was
busy too about my Lectures: on Sept. 29th I have a set of
plans of printing presses from Hansard the printer (who in a

visit to Cambridge had found me making enquiries about them), and I corresponded with Messrs Gilbert about optical constructions, and with W. and S. Jones, Eastons, and others about pumps, hydraulic rams, &c. On Sept. 25th occurred a very magnificent Aurora Borealis.

"I do not find when the investigation of Corrections of Solar Elements was finished, or when my Extracts from Burckhardt, Connaissance des Temps 1816, were made. But these led me to suspect an unknown inequality in the Sun's motion. On Sept. 27th and 28th I find the first suspicions of an inequality depending on 8 × mean longitude of Venus – 13 × mean longitude of Earth. The thing appeared so promising that I commenced the investigation of the pertur- bation related to this term, and continued it (a very laborious work) as fast as I was able, though with various interruptions, which in fact were necessary to keep up my spirits. On Oct. 30th I went to London for the Board of Longitude meeting. Here I exhibited the results of my Sun investiga- tions, and urged the correction of the elements used in the Nautical Almanac. Dr Young objected, and proposed that Bouvard should be consulted. Professor Woodhouse, the Plumian Professor, was present, and behaved so captiously that some members met afterwards to consider how order could be maintained. I believe it was during this visit to London that I took measures of Hammersmith Suspension Bridge for an intended Lecture-model. Frequently, but not always, when in London, I resided at the house of Mr Sheep- shanks and his sister Miss Sheepshanks, 30 Woburn Place. My quires, at this time, abound with suggestions for lectures and examinations.

"On some day about the end of November or beginning of December 1827, when I was walking with Mr Peacock near the outside gate of the Trinity Walks, on some mention of Woodhouse, the Plumian Professor, Mr Peacock said that he was never likely to rise into activity again (or using some expression importing mortal illness). Instantly there had passed through my mind the certainty of my succeeding him, the good position in which I stood towards the University, the probability of that position being improved by improved lectures, &c., &c., and by increased reputation from the matters in which I was now engaged, the power of thus commanding an increase of income. I should then have, independent of my Fellowship, some competent income, and a house over my head. I was quite aware that some time might elapse, but now for the first time I saw my way clearly. The care of the Observatory had been for two or three years attached to the Plumian Professorship. A Grace was immediately prepared, entrusting the temporary care of the Observatory to Dr French, to me, Mr Catton, Mr Sheepshanks, and Mr King (afterwards Master of Queens' College). On Dec. 6th I have a note from Mr King about going to the Observatory.

"On Dec. 6th my Paper on corrections of the elements of the Solar Tables was presented to the Royal Society. On Dec. 9th, at 1 h. 4 m. a.m. (Sunday morning), I arrived at the result of my calculations of the new inequality. I had gone through some fluctuations of feeling. Usually the important part of an inequality of this kind depends entirely on the eccentricities of the orbits, but it so happened that from the positions of the axes of the orbits, &c., these terms very nearly destroyed each other. After this came the consideration of inclinations of orbits; and here were sensible terms which were not destroyed. Finally I arrived at the result that the inequality would be about 3″; just such a magnitude as was required. I slipped this into Whewell's door. This is, to the time of writing (1853), the last improvement of any importance in the Solar Theory. Some little remaining work went on to Dec. 14th, and then, being thoroughly tired, I laid by the work for revision at some future time. I however added a Postscript to my Royal Society Paper on Solar Errors, notifying this result.

"On Dec. 19th I went to Bury. While there I heard from

Whewell that Woodhouse was dead.   I returned to Cambridge and immediately made known that I was a candidate for the now vacant Plumian Professorship.   Of miscellaneous scientific business, I find that on Oct. 13th Professor Barlow of Woolwich prepared a memorial to the Board of Longitude concerning his fluid telescope (which I had seen at Woodford), which was considered on Nov. 1st, and I had some correspondence with him in December.   In June and August my Trigonometry was printing.

"On Jan. 5th, 1828, I came from London.   It seems that I had been speculating truly 'without book' on perturbations of planetary elements, for on Jan. 17th and 18th I wrote a Paper on a supposed error of Laplace, and just at the end I discovered that he was quite right : I folded up the Paper and marked it 'A Lesson.'   I set two papers of questions for Smith's Prizes (there being a deficiency of one Examiner, viz. the Plumian Professor).

"Before the beginning of 1828 Whewell and I had determined on repeating the Dolcoath experiments.   On Jan. 8th I have a letter from Davies Gilbert (then President of the Royal Society) congratulating me upon the Solar Theory, and alluding to our intended summer's visit to Cornwall. We had somehow applied to the Board of Longitude for pendulums, but Dr Young wished to delay them, having with Capt. Basil Hall concocted a scheme for making Lieut. Foster do all the work : Whewell and I were indignant at this, and no more was said about it.   On Jan. 24th Dr Young, in giving notice of the Board of Longitude meeting, informs me that the clocks and pendulums are ready.

"I had made known that I was a candidate for the Plumian Professorship, and nobody thought it worth while to oppose me.   One person at least (Earnshaw) had intended to compete, but he called on me to make certain that I was a candidate, and immediately withdrew.   I went on in quality of Syndic for the care of the Observatory, ingrafting myself into it.   But meantime I told everybody that the salary (about £300) was not sufficient for me; and on Jan. 20th I drafted a manifesto or application to the University for an increase of salary. The day of election to the Professorship was Feb. 6th. As I was officially (as Lucasian Professor) an elector, I was present, and I explained to the electors that I could not undertake the responsibility of the Observatory without augmentation of income, and that I requested their express sanction to my application to the University for that purpose. They agreed to this generally, and I was elected. I went to London immediately to attend a meeting of the Board of Longitude and returned on Feb. 8th. On Feb. 15th I began my Lectures (which, this year, included Mechanics, Optics, Pneumatics, and Hydrostatics) in the room below the University Library. The number of names was 26. The Lectures terminated on Mar. 22nd.

"On Feb. 25th I received from Mr Pond information on the emoluments at Greenwich Observatory. I drew up a second manifesto, and on Feb. 26th I wrote and signed a formal copy for the Plumian electors. On Feb. 27th I met them at Caius Lodge (the Master, Dr Davy, being Vice-Chancellor). I read my Paper, which was approved, and their sanction was given in the form of a request to the Vice-Chancellor to permit the paper to be printed and circulated. My paper, with this request at the head, was immediately printed, and a copy was sent to every resident M.A. (more than 200 went out in one day). The statement and composition of the paper were generally approved, but the University had never before been taken by storm in such a manner, and there was some commotion about it. I believe that very few persons would have taken the same step. Mr Sheepshanks wrote to me on Mar. 7th, intimating that it was desperate. I had no doubt of success. Whewell told me that some people accused me of bad faith, in omitting allusion to the £100 a year received as Member of the Board of Longitude, and to the profits of Lectures. I wrote him a note, telling him that I had most certain information of the intention to dissolve the Board of Longitude (which was done in less than six months), and that by two years' Lectures I had gained £45 (the expenses being £200, receipts £245). This letter was sent to the complaining people, and no more was said. By the activity of Sheepshanks and the kindness of Dr Davy the business gradually grew into shape, and on Mar. 21st a Grace passed the Senate for appointing a Syndicate to consider of augmentation. Sheepshanks was one of the Syndicate, and was understood to represent, in some measure, my interests. The progress of the Syndicate however was by no means a straightforward one. Members of the Senate soon began to remark that before giving anything they ought to know the amount of the University revenue, and another Syndicate was then appointed to enquire and report upon it. It was more than a year before my Syndicate could make their recommendation : however, in fact, I lost nothing by that delay, as I was rising in the estimation of the University.

"The Observatory house was furnished, partly from Woodhouse's sale, and partly from new furniture. My mother and sister came to live with me there. On Mar. 15th 1828 I began the Observatory Journal; on Mar. 27th I slept at the Observatory for the first time, and on Apr. 15th I came to reside there permanently, and gave up my college rooms."

CHAPTER IV.

At Cambridge Observatory. From his taking charge of the Cambridge Observatory to his residence at Greenwich Observatory as Astronomer Royal.

From March 15th 1828 to Jan. 1st 1836.

1828

" I attended a meeting of the Board of Longitude on Apr. 3rd. And again on June 4th; this was the last meeting: Sheepshanks had previously given me private information of the certainty of its dissolution.—On Apr. 4th I visited Mr Herschel at Slough, where one evening I saw Saturn with his 20-foot telescope, the best view of it that I have ever had.—In June I attended the Greenwich Observatory Visitation.—Before my election (as Plumian Professor) there are various schemes on my quires for computation of transit corrections, &c. After Apr. 15th there are corrections for deficient wires, inequality of pivots, &c. And I began a book of proposed regulations for observations. In this are plans for groups of stars for R.A. (the Transit Instrument being the only one finished): order of preference of classes of observations: no reductions to be made after dinner, or on Sunday: no loose papers: observations to be stopped if reductions are two months in arrear: stars selected for parallax.—The reduction of transits begins on Apr. 15th. On May 15th Mr Pond sent me some moon-transits to aid in determining my longitude.—Dr Young, in a letter to me of

May 7th, enquires whether I will accept a free admission to the Royal Society, which I declined. On May 9th I was elected to the Astronomical Society.—Towards the end of the year I observed Encke's Comet: and determined the latitude of the Observatory with Sheepshanks's repeating circle.—On my papers I find a sketch of an Article on the Figure of the Earth for the Encyclopædia Metropolitana.

"As early as Feb. 23rd I had been in correspondence with T. Jones, the instrument-maker, about pendulums for a repetition of the Dolcoath Experiments. Invitations had been received, and everything was arranged with Whewell. Sheepshanks, my brother, and Mr Jackson of Ipswich (Caius Coll.) were to go, and we were subsequently joined by Sedgwick, and Lodge (Magdalene Coll.). On July 3rd Sheepshanks and I started by Salisbury, taking Sherborne on our way to look at the church, which had alarmed the people by signs of a crack, and arrived at Camborne on July 8th. On the 14th we set up the pendulums, and at once commenced observations, our plan being, to have no intermission in the pendulum observations, so that as soon as the arc became too small a fresh series was started. On July 29th we raised the instruments, and Sheepshanks, who managed much of the upper operations, both astronomical and of pendulums, mounted the pendulums together in his observatory. We went on with our calculations, and on August 8th, on returning from a visit to John Williams at Barncoose, we heard that there was a 'run' in Dolcoath, that is a sinking of the whole mass of rock where it had been set free by the mine excavations: probably only a few inches, but enough to break the rock much and to stop the pumps. On Aug. 10th the calculations of our observations shewed that there was something wrong, and on the 13th I perceived an anomaly in the form of the knife edge of one pendulum, and of its agate planes, and suggested cautions for repeating the observations. We determined at once to repeat them : and as the water was rising in the mine there was no time to be lost. We again sent the instruments down, and made observations on the 16th, 17th and 18th. On the 19th I sent the instruments up, for the water was near our station, and Sedgwick, Whewell, and I went on a geological expedition to the Lizard. On our return we met Sheepshanks and the others, and found the results of the last observations unsatisfactory. The results of comparing the pendulums were discordant, and the knife edge of the faulty pendulum had very sensibly altered. We now gave up observations, with the feeling that our time had been totally lost, mainly through the fault of the maker of the pendulum (T. Jones). On the 28th we made an expedition to Penzance and other places, and arrived at Cambridge on the 17th of September.

"In the course of the work at Dolcoath we made various expeditions as opportunity offered. Thus we walked to Carn Brea and witnessed the wrestling, the common game of the country. On another occasion Sedgwick, Whewell, and I had a capital geological expedition to Trewavas Head to examine granite veins. We visited at Pendarves and Trevince, and made the expedition to the Lizard already referred to, and saw many of the sights in the neighbourhood. After visiting Penzance on the conclusion of our work we saw Cape Cornwall (where Whewell overturned me in a gig), and returned homewards by way of Truro, Plymouth (where we saw the watering-place and breakwater: also the Dock-yard, and descended in one of the working diving-bells), Exeter, Salisbury, and Portsmouth. In returning from Camborne in 1826 I lost the principal of our papers. It was an odd thing that, in going through Exeter on our way to Camborne in 1828, I found them complete at Exeter, identified to the custodian by the dropping out of a letter with my address.

"On my return to Cambridge I was immediately im-mersed in the work of the Observatory. The only instru-ment then mounted at the Observatory was the Transit. I had no Assistant whatever.—A Mr Galbraith of Edinburgh had questioned something in one of my Papers about the Figure of the Earth. I drew up a rather formal answer to it: Whewell saw my draft and drew up a much more pithy one, which I adopted and sent to the Philosophical Magazine.— For comparing our clocks at the upper and lower stations of Dolcoath we had borrowed from the Royal Observatory, Greenwich, six good pocket chronometers: they were still in the care of Mr Sheepshanks. I arranged with him that they should be sent backwards and forwards a few times for determining the longitude of Cambridge Observatory. This was done on Oct. 21st, 22nd, 23rd: the result was 23^s·54, and this has been used to the present time (1853). It evinced an error in the Trigonometrical Survey, the origin of which was found, I think, afterwards (Dr Pearson in a letter of Dec. 17th spoke of the mistake of a may-pole for a signal-staff). I drew up a Paper on this, and gave it to the Cambridge Philosophical Society on Nov. 24th. (My only academical Paper this year.)—I had several letters from Dr Young, partly supplying me with calculations that I wanted, partly on reform or extension of the Nautical Almanac (which Dr Young resisted as much as possible). He considered me very unfairly treated in the dissolution of the Board of Longitude: Professor Lax wished me to join in some effort for its restoration, but I declined.

"As my reduction of observations was kept quite close, I now began to think of printing. In regard to the form I determined to adopt a plan totally different from that of any other observations which I had seen. The results were to be the important things: I was desirous of suppressing the separate wires of transits. But upon consulting Herschel and other persons they would not agree to it, and I assented to keeping them. I applied to the Press Syndicate to print the work, and on Nov. 10th at the request of T. Musgrave (afterwards Archbishop of York) I sent a specimen of my MS.: on Nov. 11th they granted 250 copies, and the printing soon commenced."

1829

"During a winter holiday at Playford I wrote out some investigations about the orbits of comets, and on Jan. 23rd 1829 I returned to Cambridge. The Smith's Prize Examination soon followed, in which I set a Paper of questions as usual. On Feb. 18th I made notes on Liesganig's geodetic work at the British Museum.

"I was naturally anxious now about the settlement of my salary and of the Observatory establishment. I do not know when the Syndicate made their Report, but it must have been in the last term of 1828. It recommended that the salary should be annually made up (by Grace) to £500: that an Assistant should be appointed with the assent of the Vice-Chancellor and dismissable by the Plumian Professor: and that a Visiting Syndicate should be appointed, partly official and partly of persons to be named every year by Grace. The Grace for adopting this Report was to be offered to the Senate on Feb. 27th. The passing of the Grace was exposed to two considerable perils. First, I found out (just in time) that a Senior Fellow of Trinity (G. A. Browne) was determined to oppose the whole, on account of the insignificant clause regarding dismissal of Assistants, which he regarded as tyrannical. I at once undertook that that clause should be rejected. Secondly, by the absurd constitution of the 'Caput' at Cambridge, a single M.A. had the power of stopping any business whatever, and an M.A. actually came to the Senate House with the intention of throwing out all the Graces on various business that day presented to the Senate. Luckily he mistook the hour, and came at 11 instead of 10, and found that all were dispatched. The important parts of the Grace passed without any opposition: but I mustered some friends who negatived that part which had alarmed G. A. Browne, and it was corrected to his satisfaction by a new Grace on Mar. 18th. I was now almost set at rest on one of the great objects of my life: but not quite. I did not regard, and I determined not to regard, the addition to my salary as absolutely certain until a payment had been actually made to me: and I carefully abstained, for the present, from taking any steps based upon it. I found for Assistant at the Observatory an old Lieutenant of the Royal Navy, Mr Baldrey, who came on Mar. 16.

"On May 4th I began lectures: there were 32 names. The Lectures were improving, especially in the optical part. I do not find note of the day of termination.—I do not know the actual day of publication of my first small volume of Cambridge Observations, 1828, and of circulation. The date of the preface is Apr. 27th 1829. I have letters of approval of it from Davies Gilbert, Rigaud, and Lax. The system which I endeavoured to introduce into printed astronomical observations was partially introduced into this volume, and was steadily improved in subsequent volumes. I think that I am justified, by letters and other remarks, in believing that this introduction of an orderly system of exhibition, not merely of observations but of the steps for bringing them to a practical result—quite a novelty in astronomical publications—had a markedly good effect on European astronomy in general.—In Feb. and March I have letters from Young about the Nautical Almanac: he was unwilling to make any great change, but glad to receive any small assistance. South, who had been keeping up a series of attacks on Young, wrote to me to enquire how I stood in engagements of assistance to Young: I replied that I should assist Young whenever he asked me, and that I disapproved of South's course.—The date of the first visitation of the (Cambridge) Observatory must have been near May 11th: I invited South and Baily to my house; South and I were very near quarrelling about the treatment of Young.—In a few days after Dr Young died: I applied to Lord Melville for the superintendence of the Nautical Almanac: Mr Croker replied that it devolved legally upon the Astronomer Royal, and on May 30th Pond

wrote to ask my assistance when I could give any. On June
6th I was invited to the Greenwich Visitation, to which I
believe I went on the 10th.

" I had long desired to see Switzerland, and I wished now
to see some of the Continental Observatories. I was therefore glad to arrange with Mr Lodge, of Magdalene College
(perhaps 10 years senior to myself), to make a little tour.
Capt. W. H. Smyth and others gave me introductions. I met
Lodge in London, and we started for Calais on July 27th
1829. We visited a number of towns in Belgium (at Brussels
I saw the beginning of the Observatory with Quetelet), and
passed by Cologne, Frankfort, Fribourg, and Basle to Zurich.
Thus far we had travelled by diligence or posting : we now
procured a guide, and travelled generally on foot. From the
13th to the 31st August we travelled diligently through the
well-known mountainous parts of Switzerland and arrived at
Geneva on the 31st August. Here I saw M. Gautier, M.
Gambard, and the beginning of the Observatory. Mr Lodge
was now compelled to return to Cambridge, and I proceeded
alone by Chambéry to Turin, where I made the acquaintance
of M. Plana and saw the Observatory. I then made a tour
through north Italy, looking over the Observatories at Milan,
Padua, Bologna, and Florence. At Leghorn I took a passage
for Marseille in a xebeque, but after sailing for three days
the weather proved very unfavourable, and I landed at Spezia
and proceeded by Genoa and the Cornici Road to Marseille.
At Marseille I saw M. Gambart and the Observatory, and
passed by Avignon, Lyons, and Nevers to Orléans, where I
visited my old host M. Legarde. Thence by Paris, Beauvais,
and Calais to London and Cambridge, where I arrived on
the 30th October. I had started with more than £140 and
returned with 2s. 6d. The expedition was in many ways
invaluable to me.

"On my return I found various letters from scientific
men : some approving of my method for the mass of the
Moon : some approving highly of my printed observations,

especially D. Gilbert, who informed me that they had produced good effect (I believe at Greenwich), and Herschel.—
On Nov. 13th I gave the Royal Astronomical Society a
Paper about deducing the mass of the Moon from observations of Venus: on Nov. 16th a Paper to the Cambridge
Philosophical Society on a correction to the length of a ball-
pendulum: and on Dec. 14th a Paper on certain conditions
under which perpetual motion is possible.—The engravings
for my Figure of the Earth in the Encyclopædia Metropoli-
tana were dispatched at the end of the year. Some of the
Paper (perhaps much) was written after my return from the
Continent.—I began, but never finished, a Paper on the form
of the Earth supposed to be projecting at middle latitudes.
In this I refer to the printed Paper which Nicollet gave me at
Paris. I believe that the investigations for my Paper in the
Encyclopædia Metropolitana led me to think the supposition
unnecessary.—On Nov. 6th I was elected member of the
Geological Society.

"On Nov. 16th 1829 notice was given of a Grace to
authorize payment to me of £157. 9s. 1d., in conformity with
the regulations adopted on Feb. 27th, and on Nov. 18th the
Grace passed the Senate. On Nov. 19th the Vice-Chancellor
wrote me a note enclosing the cheque. On Nov. 23rd (practically the first day on which I could go) I went to London
and travelled to Edensor, where I arrived on the 26th. Here
I found Richarda Smith, proposed to her, and was accepted.
I stayed there a few days, and returned to Cambridge."

1830

"On Jan. 25th 1830 the Smith's Prize Paper was prepared.
I was (with my Assistant, Mr Baldrey) vigorously working
the Transit Instrument and its reductions, and gradually
forming a course of proceeding which has had a good effect
on European Astronomy. And I was preparing for my
marriage.

"On Mar. 11th I started with my sister to London, and arrived at Edensor on the afternoon of the 14th. On the 17th I started alone for Manchester and Liverpool. Through Mr Mason, a cotton-spinner at Calver, near Edensor, I had become acquainted with Mr John Kennedy of Manchester, and I had since 1824 been acquainted with Dr Traill of Liverpool. Amongst other things, I saw the works of the Manchester and Liverpool Railway, then advancing and exciting great interest, and saw George Stephenson and his son. On Mar. 24th I was married to Richarda Smith by her father in Edensor. We stopped at Edensor till Apr. 1st, and then started in chaises by way of Newark and Kettering (where we were in danger of being stopped by the snow), and arrived at Cambridge on Apr. 3rd.

"I was now busy in preparing for lectures, especially the part of the optical lectures which related to the theory of interferences and polarization. I think it was now that my wife drew some of my lecture pictures, exhibiting interference phenomena. My lectures began on Apr. 26th and finished on May 24th. The number of names was 50. They were considered an excellent course of lectures.

"May 9th is the date of my Preface to the 1829 Observations : all was then printed. Apparently I did not go to the Visitation of the Greenwich Observatory this year.—I was at this time pressing Tulley, the optician, about an object-glass for the Mural Circle.—A new edition of my 'Tracts' was wanted, and I prepared to add a Tract on the Undulatory Theory of Light in its utmost extent. The Syndicate of the University Press intimated through Dr Turton that they could not assist me (regarding the book as a second edition). On July 10th I have some negociation about it with Deighton the bookseller.—On May 18th I have a note from Whewell about a number of crystals of plagiedral quartz, in which he was to observe the crystalline indication, and I the optical phenomena.—The Report of the Syndicate for visiting the Observatory is dated June 18th : it is highly laudatory.—The

Proctor (Barnard of King's College) requested me to name
the Moderator for the next B.A. Examination: I named Mr
Challis.

"On June 14th my wife and I went, in company with Professor and Mrs Henslow, to London and Oxford; at Oxford
we were received in Christchurch College by Dr and Mrs
Buckland. My wife and I then went to Bedford to visit
Capt. and Mrs Smyth, and returned to Cambridge on the
23rd. On July 5th we went on a visit to my mother and
uncle at Playford. While there I took a drive with my uncle
into some parts near the valley of the Gipping, in which I
thought that the extent of the chalk was inadequately
exhibited on Greenough's map, and communicated my remarks to Buckland.

"I find letters from Dr Robinson and Col. Colby about
determining longitudes of certain observatories by fire signals:
I proposed chronometers as preferable. Also from Herschel,
approving of my second volume of observations: and from
F. Baily, disclaiming the origination of the attack on the old
Nautical Almanac (with which I suppose I had reproached
him). On July 30th I received a summons from South to a
committee for improving the Nautical Almanac; and subsequently a letter from Baily about Schumacher's taking
offence at a passage of mine in the Cambridge Observations,
on the comparative merits of Ephemerides, which I afterwards explained to his satisfaction.

"On Aug. 24th my wife and I started for Edensor, and
after a short stay there proceeded by Manchester to Cumberland, where we made many excursions. We returned by
Edensor, and reached Cambridge on Oct. 6th, bringing my
wife's sister Susanna on a visit. My mother had determined,
as soon as my intention of marriage was known to her, to
quit the house, although always (even to her death) enter-
taining the most friendly feelings and fondness for my wife.
It was also judged best by us all that my sister should not
reside with us as a settled inhabitant of the house. They

fixed themselves therefore at Playford in the farm-house of the Luck's Farm, then in the occupation of my uncle Arthur Biddell. On Oct. 21st I have a letter from my sister saying that they were comfortably settled there.

"In this month of October (principally, I believe) I made some capital Experiments on Quartz, which were treated mathematically in a Paper communicated in the next year to the Cambridge Philosophical Society. In some of these my wife assisted me, and also drew pictures.—On Nov. 15th the Grace for paying me £198. 13s. 8d. to make my income up to £500 passed the Senate.—I made three journeys to London to attend committees, one a committee on the Nautical Almanac, and one a Royal Society Committee about two southern observatories.—On Dec. 31st I have a letter from Maclear (medical practitioner and astronomer at Biggleswade) about occultations.—In this December I had a quartz object-glass by Cauchaix mounted by Dollond, and presented it to the Observatory.—In this December occurred the alarm from agrarian fires. There was a very large fire at Coton, about a mile from the Observatory. This created the most extraordinary panic that I ever saw. I do not think it is possible, without having witnessed it, to conceive the state of men's minds. The gownsmen were all armed with bludgeons, and put under a rude discipline for a few days."

1831

"On Jan. 4th I went with my wife, first to Miss Sheepshanks in London, at 30, Woburn Place, and next to the house of my wife's old friend, the Rev. John Courtney, at Sanderstead, near Croydon. I came to London on one day to attend a meeting of the new Board of Visitors of the Greenwich Observatory. Formerly the Board of Visitors consisted of the Council of the Royal Society with persons invited by them (in which capacity I had often attended). But a reforming party, of which South, Babbage, Baily and

Beaufort were prominent members, had induced the Admiralty to constitute a new Board, of which the Plumian Professor was a member. Mr Pond, the Astronomer Royal, was in a rather feeble state, and South seemed determined to bear him down : Sheepshanks and I did our best to support him. (I have various letters from Sheepshanks to this purpose.)—On Jan. 22nd we returned to Cambridge, and I set an Examination Paper for Smith's Prizes as usual.—On Jan. 30th I have a letter from Herschel about improving the arrangement of Pond's Observations. I believe that much of this zeal arose from the example of the Cambridge Observations.

"On Feb. 21st my Paper 'On the nature of the light in the two rays of Quartz' was communicated to the Philosophical Society: a capital piece of deductive optics. On Mar. 2nd I went to London, I suppose to attend the Board of Visitors (which met frequently, for the proposed reform of Pond's Observations, &c.). As I returned on the outside of the coach there occurred to me a very remarkable deduction from my ideas about the rays of Quartz, which I soon tried with success, and it is printed as an Appendix to the Paper above mentioned. On Mar. 6th my son George Richard was born."

Miscellaneous matters in the first half of this year are as follows :

"Faraday sends me a piece of glass for Amici (he had sent me a piece before).—On Apr. 9th I dispatched the Preface of my 1830 Observations: this implies that all was printed.—On Apr. 18th I began my Lectures and finished on May 24th. There were 49 names. A very good series of lectures.—I think it was immediately after this, at the Visitation of the Cambridge Observatory, that F. Baily and Lieut. Stratford were present, and that Sheepshanks went to Tharfield on the Royston Downs to fire powder signals to be seen at Biggleswade (by Maclear) and at Bedford (by Capt. Smyth) as well as by us at Cambridge.—On May 14th I received

£100 for my article on the Figure of the Earth from Baldwin the publisher of the Encyclopædia Metropolitana.—I attended the Greenwich Visitation on June 3rd.—On June 30th the Observatory Syndicate made their report: satisfactory.

"On July 6th 1831 I started with my wife and infant son for Edensor, and went on alone to Liverpool. I left for Dublin on the day on which the loss of the 'Rothsay Castle' was telegraphed, and had a bad voyage, which made me ill during my whole absence. After a little stay in Dublin I went to Armagh to visit Dr Robinson, and thence to Coleraine and the Giant's Causeway, returning by Belfast and Dublin to Edensor. We returned to Cambridge on Sept. 9th.

"Up to this time the Observatory was furnished with only one large instrument, namely the 10-foot Transit. On Feb. 24th of this year I had received from Thomas Jones (62, Charing Cross) a sketch of the stone pier for mounting the Equatoreal which he was commissioned to make: and the pier was prepared in the spring or summer. On Sept. 20th part of the instrument was sent to the Observatory; other parts followed, and Jones himself came to mount it. On Sept. 16th I received Simms's assurance that he was hastening the Mural Circle.—In this autumn I seriously took up the recalculation of my Long Inequality of Venus and the Earth, and worked through it independently; thus correcting two errors. On Nov. 10th I went to Slough, to put my Paper in the hands of Mr Herschel for communication to the Royal Society. The Paper was read on Nov. 24th.—This was the year of the first Meeting of the British Association at York. The next year's meeting was to be at Oxford, and on Oct. 17th I received from the Rev. W. Vernon Harcourt an invitation to supply a Report on Astronomy, which I undertook: it employed me much of the winter, and the succeeding spring and summer.—The second edition of my Tracts was ready in October. It contained, besides what was in the first edition, the Planetary Theory, and the Undulatory Theory of Light. The Profit was £80.—On Nov. 14th I presented to the Cambridge Philosophical Society a Paper 'On a remarkable modification of Newton's Rings': a pretty good Paper. —In November the Copley Medal was awarded to me by the Royal Society for my advances in Optics.—Amongst miscellaneous matters I was engaged in correspondence with Col. Colby and Capt. Portlock about the Irish Triangulation and its calculation. Also with the Admiralty on the form of publication of the Greenwich and Cape Observations."

1832

" In January my Examination Paper for Smith's Prizes was prepared as usual.—Two matters (in addition to the daily routine of Observatory work) occupied me at the beginning of this year. One was the translation of Encke's Paper in successive numbers of the Astronomische Nachrichten concerning Encke's Comet; the University Press printed this gratuitously, and I distributed copies, partly by the aid of Capt. Beaufort.—The other was the Report on Astronomy for the British Association, which required much labour. My reading for it was principally in the University Library (possibly some in London), but I borrowed some books from F. Baily, and I wrote to Capt. Beaufort about the possible repetition of Lacaille's Meridian Arc at the Cape of Good Hope. The Report appears to have been finished on May 2nd.—At this time the Reform Bill was under discussion, and one letter written by me (probably at Sheepshanks's request) addressed I think to Mr Drummond, Lord Althorp's secretary, was read in the House of Commons.

" Optics were not neglected. I have some correspondence with Brewster and Faraday. On Mar. 5th I gave the Cambridge Philosophical Society a Paper 'On a new Analyzer,' and on Mar. 19th one 'On Newton's Rings between two substances of different refractive powers,' both Papers satisfactory to myself.—On the death of Mr F. Fallows, astronomer at the Cape of Good Hope Observatory, the Admiralty

appointed Mr Henderson, an Edinburgh lawyer, who had
done some little things in astronomical calculation. On Jan.
10th I discussed with him observations to be made, and drew
up his Official Instructions which were sent on Jan. 16th.—
On Feb. 16th Sir James South writes that Encke's Comet is
seen : also that with his 12-inch achromatic, purchased at
Paris, and which he was preparing to mount equatoreally, he
had seen the disk of Aldebaran apparently bisected by the
Moon's limb.—Capt. Beaufort and D. Gilbert write in March
about instructions to Dunlop, the astronomer at Paramatta.
I sent a draft to Capt. Beaufort on Apr. 27th.

"The Preface to my 1831 Observations is dated Mar. 20th.
The distribution of the book would be a few weeks later.—
On May 7th I began my Lectures : 51 names : I finished on
May 29th.—The mounting of the Equatoreal was finished
some time before the Syndicate Visitation at the end of May,
but Jones's charge appeared to be exorbitant : I believe it
was paid at last, but it was considered unfair.—On June 2nd
I went to London : I presume to the Greenwich Visitation.—
I went to Oxford to the meeting of the British Association
(lodging I think with Prof. Rigaud at the Observatory) on
June 16th, and read part of my Report on Astronomy in the
Theatre.

"On June 26th I started with my wife for the Highlands of
Scotland. After a short stay at Edensor, we went by Carlisle
to Glasgow, and through the Lake District to Inverness.
Thence by Auchnanault to Balmacarra, where we were
received by Mr Lillingstone. After an expedition in Skye,
we returned to Balmacarra, and passed on to Invermoriston,
where we were received by Grant of Glenmoriston. We then
went to Fort William and Oban, and crossed over to Mull,
where we were received by Maclean of Loch Buy. We
returned to Oban and on to Edinburgh, where we made a
short stay. Then to Melrose, where we were received by
Sir D. Brewster, and by Edensor to Cambridge, where we
arrived on Sept. 17th.

"I received (at Edinburgh I believe) a letter from Arago, writing for the plans of our observing-room shutters.—Mr Vernon Harcourt wrote deprecating the tone of my Report on Astronomy as related to English Astronomers, but I refused to alter a word.—Sheepshanks wrote in September in great anxiety about the Cambridge Circle, for which he thought the pier ought to be raised: I would have no such thing, and arranged it much more conveniently by means of a pit. On Oct. 9th Simms says that he will come with the circle immediately, and Jones on Sept. 29th says that he will make some alteration in the equatoreal: thus there was at last a prospect of furnishing the Observatory properly.—On Oct. 9th, I have Encke's thanks for the translation of the Comet Paper.—One of the desiderata which I had pointed out in my Report on Astronomy was the determination of the mass of Jupiter by elongations of the 4th satellite: and as the Equatoreal of the Cambridge Observatory was on the point of coming into use, I determined to employ it for this purpose. It was necessary for the reduction of the observations that I should prepare Tables of the motion of Jupiter's 4th Satellite in a form applicable to computations of differences of right-ascension. The date of my Tables is Oct. 3rd, 1832.—In October the Observatory Syndicate made their Report: quite satisfactory.

"On Oct. 20th Sheepshanks wrote asking my assistance in the Penny Cyclopædia: I did afterwards write 'Gravitation' and 'Greenwich.'—Capt. Beaufort wrote in November to ask my opinion on the Preface to an edition of Groombridge's Catalogue which had been prepared by H. Taylor: Sheepshanks also wrote; he had objected to it. This was the beginning of an affair which afterwards gave me great labour.—Vernon Harcourt writes, much offended at some terms which I had used in reference to an office in the British Association.

"The Equatoreal mounting which Troughton and Simms had been preparing for Sir James South's large telescope

had not entirely succeeded. I have various letters at this
time from Sheepshanks and Simms, relating to the disposition which Sir James South shewed to resist every claim till
compelled by law to pay it.—A general election of Members
of Parliament was now coming on: Mr Lubbock was candidate for the University. On Nov. 27th I had a letter from
Sedgwick requesting me to write a letter in the newspapers
in favour of Lubbock; which I did. On Dec. 7th I have
notice of the County voting at Newmarket on Dec. 18th and
19th: I walked there to vote for Townley; he lost the
election by two or three votes in several thousands.

"The Mural Circle was now nearly ready in all respects,
and it was known that another Assistant would be required.
Mr Richardson (one of the Assistants of Greenwich Observatory) and Mr Simms recommended to me Mr Glaisher,
who was soon after appointed, and subsequently became an
Assistant at Greenwich.—On Dec. 24th I have a letter from
Bessel (the first I believe). I think that I had written to him
about a general reduction of the Greenwich Planetary Observations, using his Tabulæ Regiomontanæ as basis, and
that this was his reply approving of it.

1833

"On Jan. 4th 1833 my daughter Elizabeth was born.—I
prepared an examination paper for Smith's Prizes as usual.—
On Jan. 5th I received notice from Simms that he had
received payment (£1050) for the Mural Circle from the
Vice-Chancellor. About this time the Circle was completely
made serviceable, and I (with Mr Glaisher as Assistant) immediately began its use. A puzzling apparent defect in the
circle (exhibiting itself by the discordance of zenith points
obtained by reflection observations on opposite sides of the
zenith) shewed itself very early. On Feb. 4th I have letters
about it from Sheepshanks and Simms.—On Jan. 17th I
received notice from F. Baily that the Astronomical Society
had awarded me their Medal for my long inequality of

Venus and the Earth: on Feb. 7th I went to London, I suppose to receive the Medal.—I also inspected Sir J. South's telescope, then becoming a matter of litigation, and visited Mr Herschel at Slough: on Feb. 12th I wrote to Sir J. South about the support of the instrument, hoping to remove one of the difficulties in the litigation; but it produced no effect.—Herschel wrote to me, from Poisson, that Pontécoulant had verified my Long Inequality.

"Mar. 12th is the date of the Preface to my 1832 volume of Observations: it was of course distributed a few weeks later.—In my Report on Astronomy I had indicated the Mass of Jupiter as a subject requiring fresh investigation. During the last winter I had well employed the Equatoreal in observing elongations in R.A. of the 4th Satellite. To make these available it was necessary to work up the theory carefully, in which I discovered some remarkable errors of Laplace. Some of these, for verification, I submitted to Mr Lubbock, who entirely agreed with me. The date of my first calculations of the Mass of Jupiter is Mar. 1st: and shortly after that I gave an oral account of them to the Cambridge Philosophical Society. The date of my Paper for the Astronomical Society is April 12th. The result of my investigations (which was subsequently confirmed by Bessel) entirely removed the difficulty among Astronomers; and the mass which I obtained has ever since been received as the true one.

"On Apr. 9th my wife's two sisters, Elizabeth and Georgiana Smith, came to stay with me.—On Apr. 22nd I began lectures, and finished on May 21st: there were 54 names. During the course of the lectures I communicated a Paper to the Philosophical Society 'On the calculation of Newton's experiments on Diffraction.'—I went to London on the Visitation of the Greenwich Observatory: the dinner had been much restricted, but was now made more open.—It had been arranged that the meeting of the British Association was to be held this year at Cambridge. I invited Sir David

Brewster and Mr Herschel to lodge at the Observatory. The meeting lasted from June 24th to 30th. We gave one dinner, but had a breakfast party every day. I did not enter much into the scientific business of the meeting, except that I brought before the Committee the expediency of reducing the Greenwich Planetary Observations from 1750. They agreed to represent it to the Government, and a deputation was appointed (I among them) who were received by Lord Althorp on July 25th. On Aug. 3rd Herschel announced to me that £500 was granted.

"On Aug. 7th I started with my wife for Edensor. At Leicester we met Sedgwick and Whewell: my wife went on to Edensor, and I joined Sedgwick and Whewell in a geological expedition to Mount Sorrel and various parts of Charnwood Forest. We were received by Mr Allsop of Woodlands, who proved an estimable acquaintance. This lasted four or five days, and we then went on to Edensor.— On Aug. 15th Herschel wrote to me, communicating an offer of the Duke of Northumberland to present to the Cambridge Observatory an object-glass of about 12 inches aperture by Cauchaix. I wrote therefore to the Duke, accepting gene- rally. The Duke wrote to me from Buxton on Aug. 23rd (his letter, such was the wretched arrangement of postage, reaching Bakewell and Edensor on the 25th) and on the 26th I drove before breakfast to Buxton and had an interview with him. On Sept. 1st the Duke wrote, authorizing me to mount the telescope entirely, and he subsequently approved of Cauchaix's terms: there was much correspondence, but on Dec. 28th I instructed Cauchaix how to send the tele- scope.—On our return we paid a visit to Dr Davy, Master of Caius College, at Heacham, and reached Cambridge on Oct. 8th.

"Groombridge's Catalogue, of which the editing was for- mally entrusted to Mr Henry Taylor (son of Taylor the first- assistant of the Greenwich Observatory), had been in some measure referred to Sheepshanks: and he, in investigating the work, found reason for thinking the whole discreditable. About May he first wrote to me on his rising quarrel with H. Taylor, but on Sept. 7th he found things coming to a crisis, and denounced the whole. Capt. Beaufort the Hydrographer (in whose office this matter rested) begged me with Baily to decide upon it. We did not at first quite agree upon the terms of investigation &c., but after a time all was settled, and on Oct. 4th the Admiralty formally applied, and I formally accepted. Little or nothing had been done by Mr Baily and myself, when my work was interrupted by illness.

"Sheepshanks had thought that something might be done to advance the interests of myself or the Observatory by the favour of Lord Brougham (then Lord Chancellor), and had urged me to write an article in the Penny Cyclopædia, in which Lord Brougham took great interest. I chose the subject 'Gravitation,' and as I think wrote a good deal of it in this Autumn: when it was interrupted by my illness.

"On Dec. 9th 1833, having at first intended to attend the meeting of the Philosophical Society and then having changed my mind, I was engaged in the evening on the formulæ for effects of small errors on the computation of the Solar Eclipse of 1833. A dizziness in my head came on. I left off work, became worse, and went to bed, and in the night was in high fever with a fierce attack of scarlet fever. My wife was also attacked but very slightly. The first day of quitting my bedroom was Dec. 31st. Somewhere about the time of my illness my wife's sister, Susanna Smith, who was much reduced in the summer, died of consumption.

"Miscellaneous notes in 1833 are as follows: Henderson (at the Cape) could not endure it much longer, and on Oct. 14th Stratford writes that Maclear had just sailed to take his place: Henderson is candidate for the Edinburgh Observatory.—Stratford writes on Dec. 2nd that the Madras observations have come to England, the first whose arrangement imitates mine.—On Nov. 3rd Herschel, just going to the Cape, entrusted to me the revisal of some proof sheets, if necessary: however it was never needed.—In November I sat for my portrait to a painter named Purdon (I think): he came to the house and made a good likeness. A pencil portrait was taken for a print-seller (Mason) in Cambridge: it was begun before my illness and finished after it.—I applied through Sheepshanks for a copy of Maskelyne's Observations, to be used in the Reduction of the Planetary Observations: and on Dec. 24th (from my bedroom) I applied through Prof. Rigaud to the Delegates of the Clarendon Press for a copy of Bradley's Observations for the same. The latter request was refused. In October I applied to the Syndics of the University Press for printed forms for these Reductions: the Syndics agreed to grant me 12,000 copies.

1834

"On Jan. 11th 1834 I went with my wife to London for the recruiting of my strength. We stayed at the house of our friend Miss Sheepshanks, and returned on Feb. 13th.— I drew up a Paper of Questions for Smith's Prizes, but left the whole trouble of examination and adjudication to Professor Miller, who at my request acted for me.—While I was in London I began to look at the papers relating to Groombridge's Catalogue: and I believe that it was while in London that I agreed with Mr Baily on a Report condemnatory of H. Taylor's edition, and sent the Report to the Admiralty. The Admiralty asked for further advice, and on Feb. 28th I replied, undertaking to put the Catalogue in order. On Mar. 17th Capt. Beaufort sent me all the papers. Some time however elapsed before I could proceed with it.

"There was in this spring a furious discussion about the admission of Dissenters into the University: I took the Liberal side. On Apr. 30th there was a letter of mine in the Cambridge newspaper.—On Apr. 14th I began lectures, and finished on May 20th: there·were 87 names.—My 'Gravitation' was either finished or so nearly finished that on Jan. 24th I had some conversation with Knight the publisher about printing it. It was printed in the spring, and on Apr. 27th Sheepshanks sent a copy of it to Lord Brougham. I received from Knight £83. 17s. 1d. for this Paper.—On May 10th I went to London, I believe to attend one of the Soirées which the Duke of Sussex gave as President of the Royal Society. The Duke invited me to breakfast privately with him the next morning. He then spoke to me, on the part of the Government, about my taking the office of Astronomer Royal. On May 19th I wrote him a semi-official letter, to which reference was made in subsequent correspondence on that subject.

"On May 12th my son Arthur was born.—In June the Observatory Syndicate made a satisfied Report.—On June 7th I went to the Greenwich Visitation, and again on June 14th I went to London, I believe for the purpose of trying the mounting of South's telescope, as it had been strengthened by Mr Simms by Sheepshanks's suggestions. I was subsequently in correspondence with Sheepshanks on the subject of the Arbitration on South's telescope, and my giving evidence on it. On July 29th, as I was shortly going away, I wrote him a Report on the Telescope, to be used in case of my absence. The award, which was given in December, was entirely in favour of Simms.—On July 23rd I went out, I think to my brother's marriage at Ixworth in Suffolk.—On Aug. 1st I started for Edensor and Cumberland, with my wife, sister, and three children: Georgiana Smith joined us at Edensor. We went by Otley, Harrogate, Ripon, and Stanmoor to Keswick, from whence we made many excursions. On Aug. 11th I went with Whewell to the clouds on Skiddaw, to try hygrometers. Mr Baily called on his way to the British Association at Edinburgh. On Sept. 10th we transferred our quarters to Ambleside, and after various excursions we returned to Edensor by Skipton and Bolton. On Sept. 19th I went to Doncaster and Finningley Park to see Mr Beaumont's Observatory. On Sept. 25th we posted in one day from Edensor to Cambridge.

"On Aug. 25th Mr Spring Rice (Lord Monteagle) wrote to me to enquire whether I would accept the office of Astronomer Royal if it were vacant. I replied (from Keswick) on Aug. 30th, expressing my general willingness, stipulating for my freedom of vote, &c., and referring to my letter to the Duke of Sussex. On Oct. 8th Lord Auckland, First Lord of the Admiralty, wrote: and on Oct. 10th I provisionally accepted the office. On Oct. 30th I wrote to ask for leave to give a course of lectures at Cambridge in case that my successor at Cambridge should find difficulty in doing it in the first year: and to this Lord Auckland assented on Oct. 31st. All this arrangement was for a time upset by the change of Ministry which shortly followed.

"Amongst miscellaneous matters, in March I had some correspondence with the Duke of Northumberland about the Cauchaix Telescope. In August I had to announce to him that the flint-lens had been a little shattered in Cauchaix's shop and required regrinding: finally on Dec. 17th I announced its arrival at Cambridge.—In the Planetary Reductions, I find that I employed one computer (Glaisher) for 34 weeks.—In November the Lalande Medal was awarded to me by the French Institut, and Mr Pentland conveyed it to me in December.—On March 14th I gave the Cambridge Philosophical Society a Paper, 'Continuation of researches into the value of Jupiter's Mass.' On Apr. 14th, 'On the Latitude of Cambridge Observatory.' On June 13th, 'On the position of the Ecliptic,' and 'On the Solar Eclipse of 1833,' to the Royal Astronomical Society. On Nov. 24th, 'On Computing the Diffraction of an Object Glass,' to the Cambridge Society. And on Dec. 3rd, 'On the Calculation of Perturbations,' to the Nautical Almanac: this Paper was written at Keswick between Aug. 22nd and 29th.—I also furnished Mr Sheepshanks with investigations regarding the form of the pivots of the Cape Circle.

1835

"On Jan. 9th 1835 I was elected correspondent of the French Academy; and on Jan. 26th Mr Pentland sent me £12. 6s., the balance of the proceeds of the Lalande Medal Fund.—I prepared my Paper for Smith's Prizes, and joined in the Examination as usual.

"There had been a very sudden change of Administration, and Sir R. Peel was now Prime Minister as First Lord of the Treasury, and Lord Lyndhurst was Lord Chancellor. On Jan. 19th I wrote to Lord Lyndhurst, asking him for a Suffolk living for my brother William, which he declined to give, though he remembered my application some years later. Whether my application led to the favour which I shortly received from the Government, I do not know. But, in dining with the Duke of Sussex in the last year, I had been introduced to Sir R. Peel, and he had conversed with me a long time, and appeared to have heard favourably of me. On Feb. 17th he wrote to me an autograph letter offering a pension of £300 per annum, with no terms of any kind, and allowing it to be settled if I should think fit on my wife. I wrote on Feb. 18th accepting it for my wife. In a few days the matter went through the formal steps, and Mr Whewell and Mr Sheepshanks were nominated trustees for my wife. The subject came before Parliament, by the Whig Party vindicating their own propriety in having offered me the office of Astronomer Royal in the preceding year; and Spring Rice's letter then written to me was published in the Times, &c."

The correspondence relating to the pension above-mentioned is given below, and appears to be of interest, both as conveying in very felicitous terms the opinion of a very eminent statesman on the general subject of such pensions, and as a most convincing proof of the lofty position in Science which the subject of this Memoir had then attained.

WHITEHALL GARDENS,
Feb. 17 1835.

SIR,

You probably are aware that in a Resolution voted by the House of Commons in the last Session of Parliament, an opinion was expressed, that Pensions on the Civil List, ought not thereafter to be granted by the Crown excepting for the satisfaction of certain public claims, among which those resting on Scientific or Literary Eminence were especially mentioned.

I trust that no such Resolution would have been necessary to induce me as Minister of the Crown fully to recognize the justice of such claims, but I refer to the Resolution, as removing every impediment to a Communication of the nature of that which I am about to make to you.

In acting upon the Principle of the Resolution in so far as the Claims of Science are concerned, my *first* address is made to you, and made directly, and without previous communication with any other person, because it is dictated exclusively by public considerations, and because there can be no advantage in or any motive for indirect communication.

I consider you to have the first claim on the Royal Favour which Eminence in those high Pursuits to which your life is devoted, can give, and I fear that the Emoluments attached to your appointment in the University of Cambridge are hardly sufficient to relieve you from anxiety as to the Future on account of those in whose welfare you are deeply interested.

The state of the Civil List would enable me to advise the King to grant a pension of three hundred pounds per annum, and if the offer be acceptable to you the Pension shall be granted either to Mrs Airy or yourself as you may prefer.

I beg you distinctly to understand that your acquiescence in this Proposal, will impose upon you no obligation personal or political in the slightest degree. I make it solely upon public grounds, and I ask you, by the acceptance of it, to permit the King to give some slight encouragement to Science, by proving to those who may be disposed to follow your bright Example, that Devotion to the highest Branches of Mathematical and Astronomical Knowledge shall not necessarily involve them in constant solicitude as to the future condition of those, for whom the application of the same Talents to more lucrative Pursuits would have ensured an ample Provision.

I have the honor to be, Sir,

With true Respect and Esteem,

Your faithful Servant,

ROBERT PEEL.

Mr Professor Airy,
    &c., &c.,
        Cambridge.

OBSERVATORY, CAMBRIDGE,
1835, *Feb.* 18.

SIR,

I have the honor to acknowledge your letter of the 17th acquainting me with your intention of advising the King to grant a pension of £300 per annum from the Civil List to me or Mrs Airy.

I trust you will believe that I am sensible of the flattering terms in which this offer is made, and deeply grateful for the considerate manner in which the principal arrangement is left to my choice, as well as for the freedom from engagement in which your offer leaves me. I beg to state that I most willingly accept the offer. I should prefer that the pension be settled on Mrs Airy (by which I understand that in case of her surviving me the pension would be continued to her during her life, or in the contrary event would cease with her life).

I wish that I may have the good fortune to prove to the world that I do not accept this offer without an implied engagement on my part. I beg leave again to thank you for your attention, and to assure you that the form in which it is conveyed makes it doubly acceptable.

With sincere respect I have the honor to be, Sir,

Your very faithful Servant,

G. B. AIRY.

The Right Hon. Sir Robert Peel, Bart.,
First Lord of the Treasury, &c., &c.

WHITEHALL,
Feb. 19*th* 1835.

SIR,

   I will give immediate directions for the preparation of the Warrant settling the Pension on Mrs Airy—the effect of which will be, as you suppose, to grant the Pension to her for her life. I assure you I never gave an official order, which was accompanied with more satisfaction to myself than this.

I have the honor to be, Sir.
Your faithful Servant,
ROBERT PEEL.

Mr Professor Airy,
 &c., &c.,
  Cambridge.

———

   "On March 18th 1835 I started (meeting Sheepshanks at Kingstown) for Ireland. We visited Dublin Observatory, and then went direct to Markree near Sligo, to see Mr Cooper's telescope (our principal object). We passed on our return by Enniskillen and Ballyjamesduff, where my former pupil P. Morton was living, and returned on Apr. 3rd.—On Apr. 20th I was elected to the Royal Society, Edinburgh.—Apr. 22nd my wife wrote me from Edensor that her sister Florence was very ill: she died shortly after.—On May 4th I began lectures and finished on May 29th : there were 58 names.—My former pupil Guest asks my interest for the Recordership of Birmingham.—In June was circulated the Syndicate Report on the Observatory.—The date of the Preface to the 1834 Observations is June 16th.

   "The Ministry had been again changed in the spring, and the Whigs were again in power. On June 11th Lord Auckland, who was again First Lord of the Admiralty (as last year), again wrote to me to offer me the office of Astronomer Royal, or to request my suggestions on the filling up of the office. On June 15th I wrote my first reply, and on June 17th wrote to accept it. On June 18th Lord Auckland acknowledges, and on June 22nd the King approved. Lord

Auckland appointed to see me on Friday, June 23rd, but I was unwell. I had various correspondence with Lord Auckland, principally about buildings, and had an appointment with him for August 13th. As Lord Auckland was just quitting office, to go to India, I was introduced to Mr Charles Wood, the Secretary of the Admiralty, with whom principally the subsequent business was transacted. At this meeting Lord Auckland and Mr Wood expressed their feeling, that the Observatory had fallen into such a state of disrepute that the whole establishment ought to be cleared out. I represented that I could make it efficient with a good First Assistant; and the other Assistants were kept. But the establishment was in a queer state. The Royal Warrant under the Sign Manual was sent on August 11th. It was understood that my occupation of office would commence on October 1st, but repairs and alterations of buildings would make it impossible for me to reside at Greenwich before the end of the year. On Oct. 1st I went to the Observatory, and entered formally upon the office (though not residing for some time). Oct. 7th is the date of my Official Instructions.

" I had made it a condition of accepting the office that the then First Assistant should be removed, and accordingly I had the charge of seeking another. I determined to have a man who had taken a respectable Cambridge degree. I made enquiry first of Mr Bowstead (brother to the bishop) and Mr Steventon: at length, consulting Mr Hopkins (a well-known private tutor at Cambridge), he recommended to me Mr Robert Main, of Queens' College, with whom I corresponded in the month (principally) of August, and whom on August 30th I nominated to the Admiralty. On Oct. 21st F. W. Simms, one of the Assistants (who apparently had hoped for the office of First Assistant, for which he was quite incompetent) resigned; and on Dec. 4th I appointed in his place Mr James Glaisher, who had been at Cambridge from the beginning of 1833, and on Dec. 10th the Admiralty approved.

"During this quarter of a year I was residing at Cambridge Observatory, visiting Greenwich once a week (at least for some time), the immediate superintendence of the Observatory being placed with Mr Main. I was however engaged in reforming the system of the Greenwich Observatory, and prepared and printed 30 skeleton forms for reductions of observations and other business. On Dec. 14th I resigned my Professorship to the Vice-Chancellor. But I continued the reduction of the observations, so that not a single figure was left to my successor: the last observations were those of Halley's Comet. The Preface to my 1835 Cambridge Observations is dated Aug. 22nd, 1836.

"In regard to the Northumberland Telescope, I had for some time been speculating on plans of mounting and enclosing the instrument, and had corresponded with Simms, A. Biddell, Cubitt, and others on the subject. On Apr. 24th Tulley the younger was endeavouring to adjust the object-glass. On May 31st I plainly asked the Duke of Northumberland whether he would defray the expense of the mounting and building. On June 4th he assented, and money was placed at a banker's to my order. I then proceeded in earnest: in the autumn the building was erected, and the dome was covered before the depth of winter. I continued in 1836 to superintend the mounting of the instrument.

"In regard to the Planetary Reductions: to July 11th J. Glaisher had been employed 27 weeks, and from July 11th to Jan. 16th, 1836, 25 weeks. Mr Spring Rice, when Chancellor of the Exchequer, had promised money, but no official minute had been made, and no money had been granted. On Aug. 21st I applied to Mr Baring (Secretary of the Treasury). After another letter he answered on Oct. 15th that he found no official minute. After writing to Vernon Harcourt and to Spring Rice, the matter was arranged: my outlay was refunded, and another sum granted.—In regard to Groombridge's Observations, I find that on Dec. 16th certain trial reductions had been made under my direction by

J. Glaisher.—I had attempted some optical experiments in the summer, especially on the polarization of sky-light; but had been too busy with the Observatory to continue them.

" In August my wife was in a critical state of health.—In December I received information regarding merchant ships' chronometers, for which I had applied to Mr Charles Parker of Liverpool.—On Dec. 8th Mr Spring Rice and Lord John Russell offered me knighthood, but I declined it.—On July 23rd I went into Suffolk with my wife's sisters Elizabeth and Georgiana, and returned on August 3rd: this was all the holiday that I got in this year.—On the 14th of August I saw Mr Taylor, the Admiralty Civil Architect in London, and the extension of buildings at Greenwich Observatory was arranged.—I made various journeys to Greenwich, and on Dec. 17th, having sent off our furniture, we all quitted the Cambridge Observatory, and stayed for some days at the house of Miss Sheepshanks.

" Thus ended a busy and anxious year."

———

With reference to the offer of knighthood above-mentioned, Airy's reply is characteristic, and the short correspondence relating to it is therefore inserted.—The offer itself is an additional proof of the high estimation in which he stood at this time.

DOWNING STREET,
Dec. 8th 1835.

MY DEAR SIR,

      I have been in communication with my colleague Lord John Russell which has made me feel rather anxious to have the pleasure of seeing you, but on second thoughts it has occurred to me that the subject of my communication would render it more satisfactory to you to receive a letter than to pay a visit.

In testimony of the respect which is felt for your character and acquirements, there would be every disposition to recommend you to His Majesty to receive the distinction of Knighthood. I am quite aware that to you individually this may be a matter of small concern, but to the scientific world in general it will not be indifferent, and to foreign countries it will mark the consideration felt for you personally as well as for the position which you occupy among your learned contemporaries.

From a knowledge of the respect and esteem which I feel for you Lord John Russell has wished that the communication should be made through me rather than through any person who had not the pleasure of your acquaintance.

Pray let me hear from you and believe me my dear Sir, with compliments to Mrs Airy,

Very truly yours,

T. SPRING RICE.

P.S.—It may be right to add that when a title of honor is conferred on grounds like those which apply to your case, no fees or charges of any kind would be payable.

OBSERVATORY, CAMBRIDGE,
1835, *Dec.* 10*th.*

MY DEAR SIR,

I beg to acknowledge your letter of the 8th, which I have received at this place, conveying to me an intimation of the wish of His Majesty's Ministers to recommend me to the King for the honor of Knighthood.

I beg to assure you that I am most sensible to the liberality which I have experienced from the Government in other as well as in pecuniary matters, and that I am very highly gratified by the consideration (undeserved by me, I fear) which they have displayed in the present instance. And if I now request permission to decline the honor offered to me, I trust I may make it fully understood that it is not because I value it lightly or because I am not anxious to receive honors from such a source.

The unalterable custom of this country has attached a certain degree of light consideration to titles of honor which are not supported by considerable fortune ; or at least, it calls for the display of such an establishment as may not be conveniently supported by even a comfortable income. The provision attached to my official situation, and the liberality of the King towards one of the members of my family, have placed me in a position of great comfort. These circumstances however have bound me to consider myself as the devoted servant of the country, and to debar myself from efforts to increase my fortune which might otherwise have been open to me. I do not look forward therefore to any material increase of income, and that which I enjoy at present is hardly sufficient, in my opinion, to support respectably the honor which you and Lord John Russell have proposed to confer upon me. For this reason only I beg leave most respectfully to decline the honor of Knighthood at the present time.

I have only to add that my services will always be at the command of the Government in any scientific subject in which I can be of the smallest use.

I am, my dear Sir,

Your very faithful Servant,

G. B. AIRY.

The Right Honorable T. Spring Rice.

" In brief revision of the years from 1827 to 1835 I may confine myself to the two principal subjects—my Professorial Lectures, and my Conduct of the Cambridge Observatory.

"The Lectures as begun in 1827 included ordinary Mechanics, ordinary Hydrostatics and Pneumatics (I think that I did not touch, or touched very lightly, on the subjects connected with the Hydraulic Ram), and ordinary Optics (with a very few words on Polarization and Depolarization). In 1828 the two first were generally improved, and for the third (Optics) I introduced a few words on Circular Polarization. I believe that it was in 1829 that I made an addition to the Syllabus with a small engraving, shewing the interference of light in the best practical experiment (that of the flat prism); and I went thoroughly into the main points of the Undulatory Theory, interference, diffraction, &c. In 1830 I believe I went (in addition to what is mentioned above) into Polarization and Depolarization of all kinds. My best lecture diagrams were drawn and painted by my wife. The Lectures were universally pronounced to be valuable. The subjects underwent no material change in 1831, 2, 3, 4, 5; and I believe it was a matter of sincere regret to many persons that my removal to Greenwich terminated the series. Each lecture nominally occupied an hour. But I always encouraged students to stop and talk with me; and this supplement was usually considered a valuable part of the lecture. Practically the lecture, on most days, occupied two hours. I enjoyed the Lectures much : yet I felt that the labour (in addition to other work) made an impression on my strength, and I became at length desirous of terminating them.

" The Observatory, when I took charge of it, had only one instrument—the Transit-Instrument. The principles however which I laid down for my own direction were adapted to the expected complete equipment. Planets (totally neglected at Greenwich) were to be observed. Observations were to be reduced completely, and the reductions were to be exhibited in an orderly way: this was a novelty in Astronomy. I considered it so important that I actually proposed to omit in my publication the original observations, but was dissuaded by Herschel and others. I sometimes suspended observations for a short time, in order to obtain leisure for the reductions. I had at first no intention of correcting the places of the fundamental stars as settled at Greenwich. But I found myself compelled to do so, because they were not sufficiently accurate; and then I took the course of observing and reducing as an independent observer, without reference to any other observatory. I introduced the principle of not correcting instrumental errors, but measuring them and applying numerical corrections. I determined my longitude by chronometers, and my latitude by a repeating circle borrowed from Mr Sheepshanks, which I used so well that the result was only half a second in error. The form of my reductions in the published volume for 1828 is rather irregular, but the matter is good : it soon attracted attention. In 1829 the process was much the same : I had an assistant, Mr Baldrey. In 1830 still the same, with the additions:—that I formally gave the corrections of relative right-ascension of fundamental stars (without alteration of equinox, which I had not the means of obtaining) to be used in the year 1831; and that I reduced completely the observed occultations (with a small error, subsequently corrected). In 1831 the system of correction of broken transits was improved: the errors of assumed R. A. of Fundamental Stars were exhibited: Mean Solar Time was obtained from Sidereal Time by time of Transit of ♈ (computed by myself): the method of computing occultations was improved. In 1832 the small Equatoreal was erected, and was soon employed in observations of the elongation of the 4th Satellite of Jupiter for determining the mass of Jupiter. The Mural Circle was erected at the end of the year, but not used. The calculation of R. A. of Fundamental Stars was made homogeneously with the others: separate results of all were included in ledgers: a star-catalogue was formed: all as to the present time (1871). With the Equatoreal the difference of N. P. D. of Mars and stars was observed.

"With the beginning of 1833 the Mural Circle was established at work, a second assistant (Mr Glaisher) was appointed, and the Observatory might be considered complete. I made experiments on the graduations of the Circle. I detected and was annoyed by the R−D. I determined the latitude. I exhibited the separate results for N. P. D. of stars in ledger, and their means in Catalogue. I investigated from my observations the place of equinox and the obliquity of the ecliptic. I made another series of observations of Jupiter's 4th Satellite, for the mass of Jupiter. I observed the solar eclipse with the Equatoreal, by a method then first introduced, which I have since used several times at Cambridge and Greenwich with excellent effect. The Moon and the Planets were usually observed till near two in the morning. Correction for defective illumination applied when necessary. The volume is very complete, the only deficiency being in the observation of Moon and Planets through the severe morning hours. In 1834 the only novelties are—examination of the graduations of the declination circle of the Equatoreal (excessively bad): observations of a spot on Jupiter for rotation, and of Mars and stars. In 1835 (including January 1836) there is a more complete examination of the Equatoreal graduations : parallax and refraction for Equatoreal observations: a spot on Jupiter: a series of observations on Jupiter's 4th Satellite for the mass of Jupiter: Mars and stars: Halley's Comet (the best series of observations which could be made in the season): and a short series of meteorological observations, on a plan suggested by Sir John Herschel then at the Cape of Good Hope.

"I cannot tell precisely in which year I introduced the following useful custom. Towards the end of each year I procured a pocket-book for the following year with a space for every day, and carefully examining all the sources of elements of observations, and determining the observations to be made every day, I inserted them in the pocket-book. This system gave wonderful steadiness to the plan of observations for the next year. The system has been maintained in great perfection at the Observatory of Greenwich. (The first of these pocket-books which Prof. Adams has found is that for 1833.) Printed skeleton forms were introduced for all calculations from 1828. In the Greenwich Observatory Library there is a collection, I believe complete, of printed papers commencing with my manifesto, and containing all Syndicate Reports except for 1833 (when perhaps there was none). It seems from these that my first written Report on Observations, &c., was on May 30th, 1834. The first Syndicate Report is on May 25th, 1829."

A few remarks on Airy's private life and friends during his residence at Cambridge Observatory may be here appropriately inserted.

Amid the laborious occupations recorded in the foregoing pages, his social life and surroundings appear to have been most pleasant and congenial. At that period there were in residence in Cambridge, and particularly at Trinity, a large number of very brilliant men. Airy was essentially a Cambridge man. He had come up poor and friendless : he had gained friends and fame at the University, and his whole work had been done there. From the frequent references in after times both by him and his wife to their life at Cambridge, it is clear that they had a very pleasant recollection of it, and that the social gatherings there were remarkably attractive. He has himself recorded that with Whewell and Sedgwick, and his accomplished sisters-in-law, who were frequently on long visits at the Observatory, they formed pretty nearly one family.

His friendship with Whewell was very close. Although Whewell was at times hasty, and rough-mannered, and even extremely rude, yet he was generous and large-minded, and thoroughly upright[1]. In power of mind, in pursuits, and interests, Airy had more in common with Whewell than with any other of his friends. It was with Whewell that he undertook the experiments at Dolcoath : it was to Whewell that he first communicated the result of his remarkable investigation of the Long Inequality of Venus and the Earth ; and some of his Optical researches were conducted jointly with Whewell. Whewell took his degree in 1816, seven years before Airy, and his reputation, both for mathematical and all-round knowledge, was extremely and deservedly great, but he was always most generous in his recognition of Airy's powers. Thus in a letter of Mar. 16th, 1823 (Life of William Whewell by Mrs Stair Douglas), he says, "Airy is certainly a most extraordinary man, and deserves everything that can

---

[1] The following passage occurs in a letter from Airy to his wife, dated 1845, Sept. 17th :

"I am sorry that * * * * speaks in such terms of the 'Grand Master,' as she used to be so proud of him : it is only those who have *well* gone through the ordeal of quarrels with him and almost insults from him, like Sheepshanks and me, that thoroughly appreciate the good that is in him : I am sure he will never want a good word from me."

be said of him"; and again in the autumn of 1826 he writes to his aunt, "You mentioned a difficulty which had occurred to you in one of your late letters; how Airy should be made Professor while I was here, who, being your nephew, must of course, on that account, deserve it better than he could. Now it is a thing which you will think odd, but it is nevertheless true, that Airy is a better mathematician than your nephew, and has moreover been much more employed of late in such studies....... Seriously speaking, Airy is by very much the best person they could have chosen for the situation, and few things have given me so much pleasure as his election." How much Whewell depended upon his friends at the Observatory may be gathered from a letter which he wrote to his sister on Dec. 21st, 1833. "We have lately been in alarm here on the subject of illness. Two very near friends of mine, Prof. and Mrs Airy, have had the scarlet fever at the same time; she more slightly, he very severely. They are now, I am thankful to say, doing well and recovering rapidly. You will recollect that I was staying with them at her father's in Derbyshire in the summer. They are, I think, two of the most admirable and delightful persons that the world contains." And again on Dec. 20th, 1835, he wrote to his sister Ann, "My friends—I may almost say my dearest friends—Professor Airy and his family have left Cambridge, he being appointed Astronomer Royal at Greenwich—to me an irreparable loss; but I shall probably go and see how they look in their new abode." Their close intercourse was naturally interrupted by Airy's removal to Greenwich, but their friendly feelings and mutual respect continued without material break till Whewell's death. There was frequent correspondence between them, especially on matters connected with the conduct and teaching of the University, in which they both took a keen interest, and a warm welcome at Trinity Lodge always awaited Mr and Mrs Airy when they visited Cambridge. In a letter written to Mrs Stair Douglas on Feb. 11th, 1882, enclosing some of Whewell's letters, there occurs the following passage: "After the decease of Mrs Whewell, Whewell wrote to my wife a mournful letter, telling her of his melancholy state, and asking her to visit him at the Lodge for a few days. And she did go, and did the honours of the house for several days. You will gather from this the relation in which the families stood." Whewell died on Mar. 6th, 1866, from the effects of a fall from his horse, and the following extract is from a letter written by Airy to Whewell's niece, Mrs Sumner Gibson, on hearing of the death of his old friend:

"The Master was, I believe, my oldest surviving friend (beyond my own family), and, after an acquaintance of 46 years, I must have been one of his oldest friends. We have during that time been connected privately and officially: we travelled together and experimented together: and as opportunity served (but I need not say in very different degrees) we both laboured for our College and University. A terrible blank is left on my mind."

Sedgwick was probably 15 years older than Airy: he took his degree in 1808. But the astonishing buoyancy of spirits and bonhomie of Sedgwick fitted him for all ages alike. He was undoubtedly the most popular man in Cambridge in modern times. His ability, his brightness and wit, his fearless honesty and uprightness, his plain-speaking and good humour, rendered him a universal favourite. His close alliance with Airy was much more social than scientific. It is true that they made some geological excursions together, but, at any rate with Airy, it was far more by way of recreation than of serious study, and Sedgwick's science was entirely geological. Their friendship continued till Sedgwick's death, though it was once or twice imperilled by Sedgwick's impulsive and hasty nature.

Peacock took his degree in 1813 (Herschel's year), and was therefore probably 10 years older than Airy. He was the earliest and staunchest friend of Airy in his undergraduate years, encouraged him in every possible way, lent him books, assisted him in his studies, helped him with wise advice on many occasions, and took the greatest interest in his success. He was a good and advanced mathematician, and with a great deal of shrewdness and common-sense he united a singular kindness and gentleness of manner. It is therefore not to be wondered at that he was regarded by Airy with the greatest esteem and affection, and though they were afterwards separated, by Peacock becoming Dean of Ely and Airy Astronomer Royal, yet their warm friendship was never broken. The following letter, written by Airy to Mrs Peacock on receiving the news of the death of the Dean, well expresses his feelings towards his old friend:

TRINITY LODGE, CAMBRIDGE,
1858, *Dec.* 4.

MY DEAR MADAM,

I have desired for some time to express to you my sympathies on occasion of the sad bereavement which has come upon me perhaps as strongly as upon any one not connected by family ties with my late friend. But I can scarcely give you an idea how every disposable moment of my time has been occupied. I am now called to Cambridge on business, and I seize the first free time to write to you.

My late friend was the first person whom I knew in College (I had an introduction to him when I went up as freshman). From the first, he desired me to consider the introduction not as entitling me to a mere formal recognition from him, but as authorizing me at all times to call on him for any assistance which I might require. And this was fully carried out: I referred to him in every difficulty: I had the entire command of his rooms and library (a very important aid in following the new course of mathematics which he had been so instrumental in introducing into the University) in his occasional absences: and in all respects I looked to him as to a parent. All my debts to other friends in the University added together are not comparable to what I owe to the late Dean.

Latterly I need not say that I owed much to him and that I owe much to you for your kind notice of my two sons, even since the sad event which has put it out of his power to do more.

In the past summer, looking to my custom of making a visit to Cambridge in some part of the October Term, I had determined that a visit to Ely this year should not depend on the chance of being free to leave Cambridge, but that, if it should be found convenient to yourself and the Dean, the first journey should be made to Ely. I wish that I had formed the same resolution one or two years ago.

With many thanks for your kindness, and with deep sympathy on this occasion,

I am,

My dear Madam,

Yours very faithfully,

G. B. AIRY.

Sheepshanks was a Fellow of Trinity, in orders: he was probably seven years older than Airy (he took his degree in 1816). He was not one of Airy's earliest friends, but he had a great taste and liking for astronomy, and the friendship between them when once established became very close. He was a very staunch and fearless friend, an able and incisive writer, and remarkably energetic and diligent in astronomical investigations. He, or his sister, Miss Sheepshanks, had a house in London, and Sheepshanks was very much in London, and busied himself extremely with the work of the Royal Observatory, that of the Board of Longitude, and miscellaneous astronomical matters. He was most hospitable to his friends, and while Airy resided at Cambridge his house was always open to receive him on his frequent visits to town. In the various polemical discussions on scientific matters in which Airy was engaged, Sheepshanks was an invaluable ally, and after Airy's removal to Greenwich had more or less separated him from his Cambridge friends, Sheepshanks was still associated with him and took a keen interest in his Greenwich work. And this continued till Sheepshanks's death. The warmest friendship always subsisted between the family at the Observatory and Mr and Miss Sheepshanks.

There were many other friends, able and talented men, but these four were the chief, and it is curious to note that they were all much older than Airy. It would seem as if Airy's knowledge had matured in so remarkable a manner, and the original work that he produced was so brilliant and copious, that by common consent he ranked with men who were much his seniors : and the natural gravity and decorum of his manners when quite a young man well supported the idea of an age considerably greater than was actually the case.

CHAPTER V.

1836

"Through the last quarter of 1835 I had kept everything going on at the Greenwich Observatory in the same manner in which Mr Pond had carried it on. With the beginning of 1836 my new system began. I had already prepared 30 printed skeleton forms (a system totally unknown to Mr Pond) which were now brought into use. And, having seen the utility of the Copying Press in merchants' offices, I procured one. From this time my correspondence, public and private, is exceedingly perfect.

"At this time the dwelling house was still unconnected with the Observatory. It had no staircase to the Octagon Room. Four new rooms had been built for me on the western side of the dwelling house, but they were not yet habitable. The North-east Dome ground floor was still a passage room. The North Terrace was the official passage to the North-west Dome, where there was a miserable Equatoreal, and to the 25-foot Zenith Tube (in a square tower like a steeple, which connected the N.W. Dome with Flamsteed's house). The southern boundary of the garden ran down a hollow which divides the peninsula from the site of the present Magnetic Observatory, in such a manner that the principal part of the garden was fully exposed to the public. The Computing Room was a most pitiful little room. There was so little room for me that I transported the principal table to a room in my house, where I conducted much of my own official business. A large useless reflecting telescope (Ramage's), on the plan and nearly of the size of Sir W. Herschel's principal telescope, encumbered the centre of the Front Court.

"On Jan. 11th I addressed Mr Buck, agent of the Princess Sophia of Gloucester, Ranger of Greenwich Park, for leave to enclose a portion of the ground overlooking my garden. This was soon granted, and I was partially delivered from the inconvenience of the public gaze. The liberation was not complete till the Magnetic ground was enclosed in 1837.

"In the inferior departments of the Admiralty, especially in the Hydrographic Office (then represented by Captain Beaufort) with which I was principally connected, the Observatory was considered rather as a place for managing Government chronometers than as a place of science. The preceding First Assistant (Taylor) had kept a book of letter references, and I found that out of 840 letters, 820 related to Government chronometers only. On Jan. 17th I mentally sketched my regulations for my own share in chronometer business. I had some correspondence with Captain Beaufort, but we could not agree, and the matter was referred to the Admiralty. Finally arrangements were made which put the chronometer business in proper subordination to the scientific charge of the Observatory.

"In my first negociations with the Admiralty referring to acceptance of the office of Astronomer Royal, in 1834, Lord Auckland being then First Lord of the Admiralty, I had stipulated that, as my successor at Cambridge would be unprepared to carry on my Lectures, I should have permission to give a final course of Lectures there. At the end of 1835 Lord Auckland was succeeded by Lord Minto: I claimed the permission from him and he refused it. When this was known in Cambridge a petition was presented by many Cambridge residents, and Lord Minto yielded. On April 18th I went to Cambridge with my wife, residing at the Bull Inn, and began Lectures on April 21st: they continued (apparently) to May 27th. My lecture-room was crowded (the number of names was 110) and the lectures gave great satisfaction. I offered to the Admiralty to put all the profits in their hands, and transmitted a cheque to the Accountant General of the Navy: but the Admiralty declined to receive them.

"On June 4th the Annual Visitation of the Observatory was held, Mr F. Baily in the Chair. I presented a written Report on the Observatory (a custom which I had introduced at Cambridge) in which I did not suppress the expression of my feelings about chronometer business. The Hydrographer, Captain Beaufort, who was one of the Official Visitors, was irritated: and by his influence the Report was not printed. I kept it and succeeding Reports safe for three years, and then the Board of Visitors agreed to print them; and four Reports were printed together, and bound with the Greenwich Observations of 1838.

"In the course of this year I completed the volume of Observations made at Cambridge Observatory in 1835 and on Nov. 10th the printed copies were distributed. About the end of 1835 the Dome for the Northumberland Telescope was erected: but apparently the polar frame was not erected."

The following account of an accident which occurred during the construction of the dome is extracted from a letter by Airy to his wife dated 1836 Jan. 31st. "The workmen's account of the dome blowing off is very curi-ous: it must have been a strange gust. It started suddenly when the men were all inside and Beaumont was looking up at it: the cannon balls were thrown in with great violence (one of them going between the spokes of Ransomes' large casting), and instantly after the dome had started, the boards of the outside scaffolding which had been tossed up by the same gust dropped down into the gap which the dome had left. It is a wonder that none of the men were hurt and that the iron was not broken. The dome is quite covered and I think does not look so well as when the hooping was visible."

"Previous to 1836 I had begun to contemplate the attachment of Magnetic Observations to the Observatory, and had corresponded with Prof. Christie, Prof. Lloyd, Prof. J. D. Forbes, and Mr Gauss on the subject. On Jan. 12th 1836 I addressed a formal letter to the Admiralty, and on Jan. 18th received their answer that they had referred it to the Board of Visitors. On March 25th I received authority for the expenditure of £30, and I believe that I then ordered Merz's 2-foot magnet. The Visitors met on Feb. 26th and after some discussion the site was chosen and the extent of ground generally defined, and on Dec. 22nd Mr Spring Rice (Lord Monteagle) as Chancellor of the Exchequer virtually effected the transfer of the ground. But no further steps were taken in 1836. A letter on a systematic course of magnetic observations in various parts of the world was addressed by Baron Alexander Humboldt to the Duke of Sussex, President of the Royal Society: and was referred to Prof. Christie and me. We reported on it on June 9th 1836, strongly recommending the adoption of the scheme.

"A plan had been proposed by the Promoters of the London and Gravesend Railway (Col. Landman, Engineer) for carrying a railway at high level across the bottom of the Park. On Jan. 9th I received orders from the Admiralty to examine into its possible effect in producing vibrations in the Observatory. After much correspondence, examination of ground, &c., I fixed upon a part of the Greenwich Railway (not yet opened for traffic) near the place where the Croydon trunk line now joins it, as the place for trains to run upon, while I made observations with a telescope viewing a collimator by reflection in mercury at the distance of 500 feet. The experiments were made on Jan. 25th, and I reported on Feb. 4th. It was shewn that there would be some danger to the Observatory. On Nov. 2nd Mr James Walker, Engineer, brought a model of a railway to pass by tunnel under the lower part of the Park: apparently this scheme was not pressed.

"In addition to the routine work of the Observatory, a special set of observations were made to determine the mass of Jupiter.—Also the Solar Eclipse of May 15th was observed at Greenwich in the manner which I had introduced at Cambridge.—The Ordnance Zenith Sector, and the instruments for the St Helena Observatory were brought for examination.— Much attention was given to chronometers, and various steps were taken for their improvement.—I had some important correspondence with Mr (Sir John) Lubbock, upon the Lunar Theory generally and his proposed empirical lunar tables. This was the first germ of the great reduction of Lunar Observations which I subsequently carried out.—In October I was nominated on the Council of the Royal Society, having been admitted a Fellow on Feb. 18th 1836.   I was President of the Astronomical Society during this and the preceding year (1836 and 1835).

"My connection with Groombridge's Catalogue of Stars began in 1832, and the examination, in concert with Mr Baily, of the edition printed by Mr Henry Taylor, resulted in its condemnation.   In 1834 I volunteered to the Admiralty to prepare a new edition, and received their thanks and their authority for proceeding.   It required a great deal of examination of details, and much time was spent on it in 1836: but it was not brought to the state of readiness for press.

"My predecessor, Mr Pond, died on Sept. 7th 1836, and was interred in Halley's tomb in Lee churchyard."

The following letter was written by Airy in support of the application for a pension to Mrs Pond, who had been left in great distress:

To HENRY WARBURTON, Esq.

"The points upon which in my opinion Mr Pond's claims to the gratitude of Astronomers are founded, are principally the following. *First* and chief, the accuracy which he introduced into all the principal observations. This is a thing which from its nature it is extremely difficult to estimate now, so long after the change has been made, and I can only say that so far as I can ascertain from books the change is one of very great extent: for certainty and accuracy, Astronomy is quite a different thing from what it was, and this is mainly due to Mr Pond. The most striking exemplification of this is in his laborious working out of every conceivable cause or indication of error in the Circle and the two Circles : but very great praise is also due for the new system which he introduced in working the Transit. In comparing Mr Pond's systems of observation with Dr Maskelyne's, no one can avoid being impressed with the inferiority of Dr Maskelyne's. It is very important to notice that the continental observatories which have since attracted so much attention did not at that time exist or did not exist in vigour. *Secondly*, the attention bestowed by Mr Pond on those points (chiefly of sidereal astronomy) which he regarded as fundamental : to which such masses of observations were directed as entirely to remove the doubts from probable error of individual observations or chance circumstances which have injured many other determinations. *Thirdly*, the regularity of observation. The effect of all these has been that, since the commencement of Mr Pond's residence at Greenwich, Astronomy considered as an accurate representation of the state of the heavens in the most material points has acquired a certainty and an extent which it never had before. There is no period in the history of the science so clean. On some matters (in regard to the choice of observations) I might say that my own judgment would have differed in some degree from Mr Pond's, but one thing could have been gained only by giving up another, and upon the general accuracy no improvement could have been made. Mr Pond understood nothing of physical astronomy ; but neither did anybody else, in England.

The supposed decrease of general efficiency in the last few years is to be ascribed to the following causes :

1.　Mr Pond's ill health.
2.　The inefficiency of his first assistant.
3.　The oppression of business connected with chronometers.

The last of these, as I have reason to think, operated very far. Business of this nature which (necessarily) is *daily* and *peremptory* will always prevail over that which is *general* and *confidential*. I will not trouble you with an account of the various ways in which the chronometer business teazed the Astronomer Royal (several alterations having been made at my representation), but shall merely remark that much of the business had no connection whatever with astronomy.

"I beg to submit these remarks to your perusal, requesting you to point out to me *what part* of them should be laid before any of the King's Ministers, *at what time, in what shape*, and to whom addressed. I am quite sure that Mrs Pond's claims require nothing to ensure favourable consideration but the impression of such a feeling of Mr Pond's astronomical merits as must be entertained by any reasonable astronomer; and I am most anxious to assist in conveying this impression."

"Of private history: I went to Suffolk for a week on Mar. 25th. On Sept. 19th my son Wilfrid (my fourth child) was born. In October I made an excursion for a week round the coast of Kent. In November I went to my brother's house at Keysoe in Bedfordshire: I was much exposed to cold on the return-journey, which probably aggravated the illness that soon followed. From Nov. 27th I was ill; made the last journal entry of the year on Dec. 6th; the next was on Jan. 14th, 1837. I find that in this year I had introduced Arthur Biddell to the Tithe Commutation Office, where he was soon favourably received, and from which connection he obtained very profitable employment as a valuer."

1837

"My connection with Cambridge Observatory was not yet finished. I had determined that I would not leave a figure to be computed by my successor. In October I had (at my private expense) set Mr Glaisher to work on reducing the observations of Sun, Moon, and Planets made in 1833, 1834, 1835; and subsequently had the calculations examined by Mr Hartnup. This employed me at times through 1837. I state here, once for all, that every calculation or other work in reference to the Cambridge Observatory, in this and subsequent years, was done at my private expense. The work of the Northumberland Telescope was going on through the year: from Nov. 24th to 29th I was at Cambridge on these works.

"An object-glass of 6¾ inches aperture (a most unusual size at this time, when it was difficult to find a 4-inch or 5-inch glass) had been presented to the Greenwich Observatory by my friend Mr Sheepshanks, and on Mar. 29th I received from the Admiralty authority for mounting it equatoreally in the empty South Dome, which had been intended for a copy of the Palermo Circle.—In the month of July the Admiralty wished for my political assistance in a Greenwich election, but I refused to give any.—On Jan. 3rd I gave notice to the Admiralty that I had finished the computations of Groombridge's Catalogue, and was ready to print. The printing was authorized and proceeded (the introduction was finished on Nov. 22nd), but the book was not quite ready till the beginning of 1838.—In connection with the Cavendish experiment: on June 10th I wrote to Spring Rice (Chancellor of the Exchequer) for £500, which was soon granted: and from this time there is a great deal of correspondence (mainly with Mr Baily) upon the details of the experiment and the theory of the calculation.—On July 24th I saw the descent of the parachute by which Mr Cocking was killed. I attended the coroner's inquest and gave evidence a few days later.

"The Planetary Reductions from 1750 to 1830 had been going on: the computers (Glaisher, Hartnup, and Thomas) worked in the Octagon Room, and considerable advance was made.—In consequence of the agitation of the proposal by Mr Lubbock to form empirical tables of the Moon, for which I proposed to substitute complete reduction of the observations of the Moon from 1750, the British Association at York (Oct. 23rd, 1837) appointed a deputation (including myself) to place the matter before the Government. I wrote on the matter to Mr Wood (Lord Halifax) stating that it would be proper to raise the First Assistant's salary, and to give me more indefinite power about employing computers. In all these things I received cordial assistance from Mr Wood. The Chancellor of the Exchequer (Mr Spring Rice) received us on Dec. 20th: statements were furnished by me, and the business was sanctioned immediately.—During this year I was very much engaged in correspondence with Lubbock and others on improvements of the Lunar Theory.

"In the operations of 1836 and 1837 a great quantity of papers had been accumulated. I had kept them in reasonably good order, tied up in bundles: but this method began to fail in convenience, as the number increased. The great lines of classification were however now well understood. I believe it was in the latter part of the year 1837 that I finally settled on the principle of arranging papers in packets and subordinate packets, every paper being flat, by the use of four punched holes in every paper. I have never seen any principle of arrangement comparable to this. It has been adopted with the greatest ease by every assistant, and is used to the present time (1871) without alteration.

"On Jan. 3rd I was informed unofficially by Mr Wood (Admiralty Secretary) that the addition of the Magnetic Ground was sanctioned. On Feb. 16th Mr Rhodes (an officer of the Department of Woods and Works) came to put me formally in possession of the ground. Between Apr. 26th and May 13th the ground was enclosed, and my garden was completely protected from the public. The plan of the building was settled, and numerous experiments were made on various kinds of concrete: at last it was decided to build with wood.

"After a dinner given by Lord Burlington, Chancellor, the first meeting of the London University was held on Mar. 4th, and others followed. On Apr. 18th I handed to the Chancellor a written protest against a vote of a salary of £1000 to the Registrar: which salary, in fact, the Government refused to sanction. Dissensions on the question of religious examination were already beginning, but I took little part in them.

"In 1833 Mr Henderson had resigned the superintendance of the Cape of Good Hope Observatory, and Mr Maclear was appointed. I recommended the same Official Instructions for him (they had included an allusion to La Caille's Arc of Meridian) with an addition on the probability of Trigonometrical Survey, on Aug. 8th, 1837. On Feb. 24th, 1837, I wrote to Beaufort suggesting that Bradley's Sector should be used for verifying the astronomical determinations, and subsequently received the approval of the Admiralty. In June Sir J. Herschel and I had an interview with Mr Wood on the Cape equipment generally. The Sector was erected with its new mounting, careful drawings were made of every part, instructions were prepared for its use, and on Aug. 10th it was sent to Woolwich Dockyard and shipped for the Cape.

"Of private history: On Aug. 23rd I started with my wife for an excursion in South Wales, &c. On Sept. 9th I gave a lecture in the Town Hall of Neath. While at Swansea we received news of the death of my wife's father, the Rev. Richard Smith, and returned at once.—In this year Arthur Biddell bought the little Eye estate for me."

1838

"Cambridge Observatory:—On Dec. 29th, 1837, I had set Mr Glaisher to work in collecting the annual results for star-places from the Cambridge Observations, to form one catalogue: I examined the calculations and the deduced catalogue, and on Dec. 14, 1838, presented it to the Royal Astronomical Society, under the title of 'The First Cambridge Catalogue.'—For the Northumberland telescope I was engaged with Simms about the clockwork from time to time up to Apr. 30th, and went to Cambridge about it. The instrument was brought to a useable state, but some small parts were still wanting.

"At Greenwich:—In April I drew up a little history of the Observatory for the Penny Cyclopædia.—On June 30th the Lords of the Admiralty paid a short visit to the Observatory : on this occasion Mr Wood suggested a passage connecting the Observatory with the dwelling-house, and I subsequently prepared sketches for it; it was made in the next year.—In the course of the year the Sheepshanks Equatoreal was mounted, and Encke's Comet was observed with it from Oct. 26th to Nov. 13th.—On Mar. 31st, &c. I reported to the Admiralty on the selection of chronometers for purchase, from a long list : this was an important beginning of a new system.—The Magnetic Observatory was built, in the form originally planned for it (a four-armed cross with equal arms, one axis being in the magnetic meridian) in the beginning of this year. (No alteration has since been made in form up to the present time, 1871, except that the north arm has been lengthened 8 feet a few years ago.) On May 21st a magnet was suspended for the first time, Mr Baily and Lieut. (afterwards Sir William) Denison being present.—Groombridge's Catalogue was finished, and on Mar. 3rd I arranged for sending out copies.—The Planetary Reductions were carried on vigorously. On May 31st, 1838, the Treasury assented to the undertaking of the Lunar Reductions and allotted £2,000 for it : preparations were made, and in the autumn 7 computers were employed upon it. It will easily be seen that this undertaking added much to my labours and cares.—The geodetic affairs of the Cape of Good Hope began to be actively pressed, and in February Beaufort wrote to me in consequence of an application from Maclear, asking about a standard of length for Maclear (as foundation for a geodetic survey). I made enquiries, and on Mar. 13th wrote to Mr Wood, alluding also generally to the want of a National English standard after the destruction of the Houses of Parliament. On Apr. 24th the Admiralty sanctioned my procuring proper Standard Bars.—In connection with the Cavendish Experiment, I have an immense quantity of correspondence with Mr Baily, and all the mathematics were furnished by me : the experiment was not finished at the end of the year.—The Perturbations of Uranus were now attracting attention. I had had some correspondence on this subject with Dr Hussey in 1834, and in 1837 with Eugène Bouvard. On Feb. 24th, of 1838, I wrote to Schumacher regarding the error in the tabular radius-vector of Uranus, which my mode of reducing the observations enabled me to see.

"The National Standards of Length and Weight had been destroyed in the fire of the Houses of Parliament. On May 11th I received a letter from Mr Spring Rice, requesting me to act (as chairman) with a committee consisting of F. Baily, J. E. Drinkwater Bethune, Davies Gilbert, J. G. S. Lefevre, J. W. Lubbock, G. Peacock, and R. Sheepshanks, to report on the steps now to be taken. I accepted the charge, and the first meeting was held at the Observatory on May 22nd; all subsequent meetings in London, usually in the apartments of the Royal Astronomical Society. I acted both as chairman and as working secretary. Our enquiries went into a very wide field, and I had much correspondence.

"On Jan. 4th Mr Wood wrote to me, mentioning that Capt. Johnson had made some observations on the magnetism of iron ships, and asking whether they ought to be continued; a steamer being offered at £50 per week. I applied to Beaufort for a copy of Johnson's Observations, and on Jan. 7th replied very fully, discouraging such observations; but recommending a train of observations expressly directed to theoretical points. On Feb. 17th I reported that I had examined the Deptford Basin, and found that it would do fairly well for experiments. On July 14th, 1838, Capt. Beaufort wrote to me that the Admiralty wished for experiments on the ship, the 'Rainbow,' then in the river, and enquired whether I would undertake them and what assistance I desired, as for instance that of Christie or Barlow. I replied that one person should undertake it, either Christie, Barlow, or myself, and that a basin was desirable. On July 16th and 17th I looked at the basins of Woolwich and Deptford, approving the latter. On July 21st the Admiralty gave me full powers. From July 23rd I was almost entirely employed on preparations. The course of operations is described in my printed Paper: the original maps, curves, and graphical projections, are in the bound MSS.: 'Correction of Compass in Iron Ships—"Rainbow,"' at the Greenwich Observatory. The angular disturbances were found on July 26th and 30th, requiring some further work on a raft, so that they were finally worked out on Aug. 11th. I struggled hard with the numbers, but should not have succeeded if it had not occurred to me to examine the horizontal magnetic intensities. This was done on Aug. 14th, and the explanation of the whole was suggested at once: graphical projections were made on Aug. 16th and 17th for comparison of my explanation with observations, and the business was complete. On Aug. 17th and 18th I measured the intensity of some magnets, to be used in the ship for correction. It is to be remarked that, besides the effect of polar magnetism, there was no doubt of the existence of an effect of induced magnetism requiring correction by other induced magnetism: and experiments for this were made in the Magnetic Observatory. All was ready for trial: and on Aug. 20th I carried my magnets and iron correctors to Deptford, mounted them in the proper places, tried the ship, and the compass, which had been disturbed 50 degrees to the right and 50 degrees to the left, was now sensibly correct. On Aug. 21st I reported this to the Admiralty, and on Aug. 24th I tried the ship to Gravesend. On Aug. 30th I had the loan of her for an expedition with a party of friends to Sheerness, and on Sept. 9th I accompanied her to Gravesend, on her first voyage to Antwerp.—On Oct. 5th application was made to me by the owner of the 'Ironsides' to correct her compasses. In consequence of this I went to Liverpool on Oct. 25th, and on this occasion made a very important improvement in the practical mode of performing the correction.—On Nov. 16th I reported to the Admiralty in considerable detail. On Dec. 4th I had an interview with Lord Minto (First Lord of the Admiralty) and Mr Wood. They refused to sanction any reward to me.— The following is a copy of the report of the Captain of the

'Rainbow' after her voyage to Antwerp: 'Having had the command of the Rainbow steamer the two voyages between London and Antwerp, I have the pleasure to inform you that I am perfectly satisfied as to the correctness of the compasses, and feel quite certain they will continue so. I took particular notice from land to land from our departure and found the bearings by compass to be exact.'"—The following extracts from letters to his wife refer to the "Ironsides": on Oct. 28th 1838 he writes, " I worked up the observations so much as to see that the compass disturbance is not so great as in the 'Rainbow' (35° instead of 50°), but quite enough to make the vessel worthless; and that it is quite different in direction from that in the 'Rainbow'—so that if they had stolen one of the 'Rainbow' correctors and put it into this ship it would have been much worse than before." And on Nov. 1st he writes, "On Wednesday I again went to the ship and tried small alterations in the correctors: I am confident now that the thing is very near, but we were most abominably baffled by the sluggishness of the compass."

"The University of London:—On Jan. 6th I attended a sub-committee meeting on the minimum of acquirements for B.A. degree, and various meetings of the Senate. On July 14th I intimated to Mr Spring Rice my wish to resign. I had various correspondence, especially with Mr Lubbock, and on Dec. 13th I wrote to him on the necessity of stipends to Members of Senate. The dissensions on religious examination became very strong. I took a middle course, demanding examination in the languages and books, but absolutely refusing to claim any religious assent. I expressed this to Dr Jerrard, the principal representative on the religious side, by calling on him to substitute the words 'Recognition of Christian Literature' for 'Recognition of Christian Religion': I addressed a printed letter to Lord Burlington (Chancellor) and the Members of the Senate, on this subject.

"Of private history: In January I made a short excursion in Norfolk and Suffolk, and visited Prof. Sedgwick at

Norwich. In April I paid a short visit to Mr Courtney at Sanderstead, with my wife. On June 14th my son Hubert was born. In September I went with my sister by Cambridge, &c., to Luddington, where I made much enquiry concerning my father and the family of Airy who had long been settled there. We then visited various places in Yorkshire, and arrived at Brampton, near Chesterfield, where Mrs Smith, my wife's mother, now resided. And returned by Rugby. I had much correspondence with my brother and for him about private pupils and a better church living. I complained to the Bishop of Norwich about the mutilation of a celebrated monument in Playford Church by the incumbent and curate."

The following extracts are from letters to his wife relating to the above-mentioned journeys:

CLOSE, NORWICH.
1838, *Jan.* 21.

I do not know what degree of cold you may have had last night, but here it was (I believe) colder than before—thermometer close to the house at 3°. I have not suffered at all. However I do not intend to go to Lowestoft.

BRAMPTON.
1838, *Sept.* 30th.

We began to think that we had seen enough of Scarborough, so we took a chaise in the afternoon to Pickering, a small agricultural town, and lodged in a comfortable inn there. On Wednesday morning at 8 we started by the railroad for Whitby, in a huge carriage denominated the Lady Hilda capable of containing 40 persons or more drawn by one horse, or in the steep parts of the railway by two horses. The road goes through a set of defiles of the eastern moorlands of Yorkshire which are extremely pretty: at first woody and rich, then gradually poorer, and at last opening on a black moor with higher moors in sight: descending in one part by a long crooked inclined plane, the carriage drawing up another load by its weight: through a little tunnel: and then along a valley to Whitby. The rate of travelling was about 10 miles an hour. Betsy declares that it was the most agreeable travelling that she ever had.

Yesterday (Saturday) Caroline drove Betsy and Miss Barnes drove me to Clay Cross to see the works at the great railroad tunnel there. Coming from the north, the railroad passes up the Chesterfield valley close by the town and continues up the same valley, till it is necessary for it to enter the valley which runs the opposite way towards Buttersley : the tunnel passes under the high ground between these two vallies : so that it is in reality at the water-shed : it is to be I think more than a mile long, and when finished 27 feet clear in height, so it is a grand place. We saw the preparations for a blast, and heard it fired : the ladies stopping their ears in due form.

1839

" Cambridge Observatory :—On Mar. 7th I went to Cambridge on the business of the Northumberland Telescope : I was subsequently engaged on the accounts, and on Aug. 16th I finally resigned it to Prof. Challis, who accepted it on Aug. 19th. On Sept. 11th I communicated its completion and the settlement of accounts to the Duke of Northumberland. The total expense was £1938. 9s. 2d. + 15000 francs for the object-glass.

" At Greenwich Observatory :—On Jan. 3rd I received the last revise of the 1837 Observations, and on Jan. 8th the first sheet for 1838.—In July I report on selection from a long list of chronometers which had been on trial, and on Sept. 2nd I pointed out to Capt. Beaufort that the system of offering only one price would be ruinous to the manufacture of chronometers, and to the character of those supplied to the Admiralty : and that I would undertake any trouble of classifying the chronometers tried. This letter introduced the system still in use (1871), which has been most beneficial to the manufacture. On Sept. 11th I proposed that all trials begin in the first week of January : this also has been in use as an established system to the present time.—It was pointed out to me that a certain chronometer was affected by external magnetic power. I remedied this by placing under it a free compass magnet : a stand was specially prepared for it. I have never found another chronometer sensibly affected by magnetism.—In November and December I tried my new double-image micrometer.—Between May 16th and Oct. 13th a fireproof room was constructed in the southern part of the quadrant room ; and in November a small shed was erected over the entrance to the North Terrace.—The position of the free Meridional Magnet (now mounted in the Magnetic Observatory) was observed at every 5 m. through 24 hours on Feb. 22nd and 23rd, May 24th and 25th, Aug. 30th and 31st, and Nov. 29th and 30th. This was done in co-operation with the system of the Magnetic Union established by Gauss in Germany.—The Reduction of the Greenwich Planetary and Lunar Observations, 1750 to 1830, went on steadily. I had six and sometimes seven computers constantly at work, in the Octagon Room.—As in 1838 I had a great amount of correspondence with Mr Baily on the Cavendish Experiment.—I attended as regularly as I could to the business of the University of London. The religious question did not rise very prominently. I took a very active part, and have a great deal of correspondence, on the nature of the intended examinations in Hydrography and Civil Engineering.—On the Standards Commission the chief work was in external enquiries.—On June 6th I had enquiries from John Quincey Adams (U. S. A.) on the expense, &c., of observatories : an observatory was contemplated in America.—I had correspondence about the proposed establishment of observatories at Durham, Glasgow, and Liverpool.

"I had in this year a great deal of troublesome and on the whole unpleasant correspondence with the Admiralty about the correction of the compass in iron ships. I naturally expected some acknowledgment of an important service rendered to Navigation : but the Admiralty peremptorily refused it. My account of the Experiments &c. for the Royal Society is dated April 9th. The general success of the undertaking soon became notorious, and (as I understood) led immediately to extensive building of iron ships : and it led also to applications to me for correction of compasses. On Jan. 9th I was addressed in reference to the Royal Sovereign and Royal George at Liverpool ; July 18th the Orwell ; May 11th two Russian ships built on the Thames ; Sept. 4th the ships of the Lancaster Company.

"I had much work in connection with the Cape of Good Hope Observatory, chiefly relating to the instrumental equipment and to the geodetical work. As it was considered advisable that any base measured in the Cape Colony should be measured with compensation bars, I applied to Major Jervis for the loan of those belonging to the East Indian Survey, but he positively refused to lend them. On Jan. 20th I applied to Col. Colby for the compensation bars of the British Survey, and he immediately assented to lending them. Col. Colby had suggested to the Ordnance Department that Capt. Henderson and several sappers should be sent to use the measuring bars, and it was so arranged. It still appeared desirable to have the command of some soldiers from the Garrison of Cape Town, and this matter was soon arranged with the military authorities by the Admiralty.

" The following are the principal points of my private history: it was a very sad year. On Jan. 24th I went with my wife to Norwich, on a visit to Prof. Sedgwick, and in June I visited Sir J. Herschel at Slough. On June 13th my dear boy Arthur was taken ill: his malady soon proved to be scarlet fever, of which he died on June 24th at 7 in the morning. It was arranged that he should be buried in Playford churchyard on the 28th, and on that day I proceeded to Playford with my wife and my eldest son George Richard. At Chelmsford my son was attacked with slight sickness, and being a little unwell did not attend his brother's funeral. On July 1st at 4 h. 15 m. in the morning he also died : he had some time before suffered severely from an attack of measles, and it seemed probable that his brain had suffered. On July

5th he was buried by the side of his brother Arthur in Playford churchyard.—On July 23rd I went to Colchester on my
way to Walton-on-the-Naze, with my wife and all my family;
all my children had been touched, though very lightly, with
the scarlet fever.—It was near the end of this year that my
mother quitted the house (Luck's) at Playford, and came to
live with me at Greenwich Observatory, where she lived till
her death; having her own attendant, and living in perfect
confidence with my wife and myself, and being I trust as
happy as her years and widowhood permitted. My sister
also lived with me at the Observatory."

1840

" In the latter part of 1839, and through 1840, I had much
correspondence with the Admiralty, in which I obtained a
complete account of the transfer of the Observatory from the
Ordnance Department to the Admiralty, and the transfer of
the Visitation of the Observatory from the Royal Society to
the present Board of Visitors. In 1840 I found that the
papers of the Board of Longitude were divided between the
Royal Society and the Admiralty: I obtained the consent of
both to bring them to the Observatory.

" In this year I began to arrange about an annual dinner
to be held at the Visitation.—My double-image micrometer
was much used for observations of circumpolar double stars.
—In Magnetism and Meteorology, certain quarterly observations were kept up; but in November the system of incessant
eye-observations was commenced. I refused to commence
this until I had secured a 'Watchman's Clock' for mechanical verification of the regular attendance of the Assistants.—
With regard to chronometers: In this year, for the first time,
I took the very important step of publishing the rates obtained by comparisons at the Observatory. I confined myself
on this occasion to the chronometers purchased by the
Admiralty. In March a pigeon-house was made for expo-

sure of chronometers to cold.—The Lunar and Planetary Reductions were going on steadily.—I was consulted about an Observatory at Oxford, where I supported the introduction of the Heliometer.—The stipend of the Bakerian Lecture was paid to me for my explanation of Brewster's new prismatic fringes.—The business of the Cape Observatory and Survey occupied much of my time.—In 1838 the Rev. H. J. Rose (Editor of the Encyclopædia Metropolitana) had proposed my writing a Paper on Tides, &c.; in Oct. 1840 I gave him notice that I must connect Tides with Waves, and in that way I will take up the subject. Much correspondence on Tides, &c., with Whewell and others followed.

"With regard to the Magnetical and Meteorological Establishment. On June 18th Mr Lubbock reported from the Committee of Physics of the Royal Society to the Council in favour of a Magnetic and Meteorological Observatory near London. After correspondence with Sheepshanks, Lord Northampton, and Herschel, I wrote to the Council on July 9th, pointing out what the Admiralty had done at Greenwich, and offering to cooperate. In a letter to Lord Minto I stated that my estimate was £550, including £100 to the First Assistant: Lubbock's was £3,000. On Aug. 11th the Treasury assented, limiting it to the duration of Ross's voyage. On Aug. 17th Wheatstone looked at our buildings and was satisfied. My estimate was sent to the Admiralty, viz. £150 outfit, £520 annual expense; and Glaisher to be Superintendent. I believe this was allowed for the present; for the following year it was placed on the Estimates. Most of the contemplated observations were begun before the end of 1840: as much as possible in conformity with the Royal Society's plan. Mr Hind (subsequently the Superintendent of the Nautical Almanac) and Mr Paul were the first extra assistants.

"Of private history. On Feb. 29th I went to Cambridge with my Paper on the Going Fusee. On Mar. 27th I went to visit Mrs Smith, my wife's mother, at Brampton near

Chesterfield. I made a short visit to Playford in April and a short expedition to Winchester, Portsmouth, &c., in June. From Sept. 5th to Oct. 3rd I was travelling in the North of England and South of Scotland." [This was an extremely active and interesting journey, in the course of which a great number of places were visited by Airy, especially places on the Border mentioned in Scott's Poems, which always had a great attraction for him. He also attended a Meeting of the British Association at Glasgow and made a statement regarding the Planetary and Lunar Reductions : and looked at a site for the Glasgow Observatory.] " In November I went for a short time to Cambridge and to Keysoe (my brother's residence). On Dec. 26th my daughter Hilda was born (subsequently married to E. J. Routh). In this year I had a loss of £350 by a fire on my Eye estate."

The following extracts are from letters to his wife. Some of them relate to matters of general interest. They are all of them characteristic, and serve to shew the keen interest which he took in matters around him, and especially in architecture and scenery. The first letter relates to his journey from Chesterfield on the previous day.

FLAMSTEED HOUSE,
1840, *April* 2.

I was obliged to put up with an outside place to Derby yesterday, much against my will, for I was apprehensive that the cold would bring on the pain in my face. Of that I had not much ; but I have caught something of sore throat and catarrh. The coach came up at about 22 minutes past 8. It arrived in Derby at 20 minutes or less past 11 (same guard and coachman who brought us), and drew up in the street opposite the inn at which we got no dinner, abreast of an omnibus. I had to go to a coach office opposite the inn to pay and be booked for London, and was duly set down in a way-bill with *name ;* and then entered the omnibus : was transferred to the Railway Station, and then received the Railway Ticket by shouting out my name. If you should come the same way, you would find it convenient to book your place at Chesterfield to London by your name (paying for the whole, namely, coach fare, omnibus fare -/6, and railway fare £1. 15s. od. first class). Then you will only have to step out of the coach into the omnibus, and to scream out once or twice to the guard to make sure that you are entered in the way-bill and that your luggage is put on the omnibus.

FLAMSTEED HOUSE, GREENWICH,
1840, *April* 15.

I forgot to tell you that at Lord Northampton's I saw some specimens of the Daguerrotype, pictures made by the Camera Obscura, and they surpass in beauty of execution anything that I could have imagined. Baily who has two or three has promised to lend them for your inspection when you return. Also I saw some post-office stamps and stamped envelopes: I do not much admire the latter.

The following relates to the fire on his Eye farm, referred to above:

PLAYFORD,
1840, *April* 23.

On Wednesday (yesterday) went with my uncle to the Eye Estate, to see the effects of the fire. The farming buildings of every kind are as completely cleared away as if they had been mown down: not a bit of anything but one or two short brick walls and the brick foundations of the barns and stacks. The aspect of the place is much changed, because in approaching the house you do not see it upon a back-ground of barns, &c., but standing alone. The house is in particularly neat and good order. I did not think it at all worth while to make troublesome enquiries of the people who reside there, but took Mr Case's account. There seems no doubt that the fire was caused by the maid-servant throwing cinders into a sort of muck-place into which they had been commonly thrown. I suppose there was after all this dry weather straw or muck drier than usual, and the cinders were hotter than usual. The whole was on fire in an exceedingly short time; and everything was down in less than an hour. Two engines came from Eye, and all the population of the town (as the fire began shortly after two o'clock in the afternoon). It is entirely owing to these that my house, and the farm (Sewell's) on the opposite side of the road, were not burned down. At the beginning of the fire the wind was N.E. which blew directly towards the opposite farm (Sewell's): although the nearest part of it (tiled dwelling house) was 100 yards off or near it, and the great barn (thatched roof) considerably further, yet both were set on fire several times. All this while, the tail of my house was growing very hot: and shortly after the buildings fell in burning ruins, the wind changed to N.W., blowing directly to my house. If this change had happened while the buildings were standing and burning, there would have been no possibility of saving the house. As it was, the solder is melted from the window next the farm-yard, and the roof was set on fire in three or four places. One engine was kept working on my house and one on the opposite farm. A large pond was pretty nearly emptied. Mr Case's horses and bullocks were got out, not without great difficulty, as the progress of the fire was fearfully rapid. A sow and nine pigs were burnt, and a large hog ran out burnt so much that the people killed it immediately.

GEORGE INN, WINCHESTER,
1840, *June* 21.

At Winchester we established ourselves at the George and then without delay proceeded to St Cross. I did not know before the nature of its hospital establishment, but I find that it is a veritable set of alms-houses. The church is a most curious specimen of the latest Norman. I never saw one so well marked before—Norman ornaments on pointed arches, pilasters detached with cushion capitals, and various signs: and it is clearly an instance of that state of the style when people had been forced by the difficulties and inelegancies of the round arch in groining to adopt pointed arches for groining but had not learnt to use them for windows.......This morning after breakfast went to the Cathedral (looking by the way at a curious old cross in the street). I thought that its inside was wholly Norman, and was most agreeably surprised by finding the whole inside groined in every part with excellent late decorated or perpendicular work. Yet there are several signs about it which lead me to think that the whole inside has been Norman, and even that the pilasters now worked up into the perpendicular are Norman. The transepts are most massive old Norman, with side-aisles running round their ends (which I never saw before). The groining of the side aisles of the nave very effective from the strength of the cross ribs. The clerestory windows of the quire very large. The organ is on one side. But the best thing about the quire is the wooden stall-work, of early decorated, very beautiful. A superb Lady Chapel, of early English.

PORTSMOUTH,
1840, *June* 23.

We left Winchester by evening train to the Dolphin, Southampton, and slept there. At nine in the morning we went by steamboat down the river to Ryde in the Isle of Wight : our steamer was going on to Portsmouth, but we thought it better to land at Ryde and take a boat for ourselves. We then sailed out (rather a blowing day) to the vessel attending Col. Pasley's operations, and after a good deal of going from one boat to another (the sea being so rough that our boat could not be got up to the ships) and a good deal of waiting, we got on board the barge or lump in which Col. Pasley was. Here we had the satisfaction of seeing the barrel of gunpowder lowered (there was more than a ton of gunpowder), and seeing the divers go down to fix it, dressed in their diving helmets and supplied with air from the great air-pump above. When all was ready and the divers had ascended again, the barge in which we were was warped away, and by a galvanic battery in another barge (which we had seen carried there, and whose connection with the barrel we had seen), upon signal given by sound of trumpet, the gunpowder was fired. The effect was most wonderful. The firing followed the signal instantaneously. We were at between 100 and 200 yards from the place (as I judge), and the effects were as follows. As soon as the signal was given, there was a report, louder than a musket but not so loud as a small cannon, and a severe shock was felt at our feet, just as if our barge had struck on a rock. Almost immediately, a very slight swell was perceived over the place of the explosion, and the water looked rather foamy : then in about a second it began to rise, and there was the most enormous outbreak of spray that you can conceive. It rose in one column of 60 or 70 feet high, and broad at the base, resembling a stumpy sheaf with jagged masses of spray spreading out at the sides, and seemed to grow outwards till I almost feared that it was coming to us. It sunk, I suppose, in separate parts, for it did not make any grand squash down, and then there were seen logs of wood rising, and a dense mass of black mud, which spread gradually round till it occupied a very large space. Fish were stunned by it: our boatmen picked up some. It was said by all present that this was the best explosion which had been seen: it was truly wonderful. Then we sailed to Portsmouth.......The explosion was a thing worth going many miles to see. There were many yachts and sailing boats out to see it (I counted 26 before they were at the fullest), so that the scene was very gay.

Here are some notes on York Cathedral after the fire :

RED LION HOTEL, REDCAR,
1840, *Sept.* 7.

My first letter was closed after service at York Cathedral. As soon as I had posted it, I walked sedately twice round the cathedral, and then I found the sexton at the door, who commiserating me of my former vain applications, and having the hope of lucre before his eyes, let me in. I saw the burnt part, which looks not melancholy but unfinished. Every bit of wood is carried away clean, with scarcely a smoke-daub to mark where it has been: the building looks as if the walls were just prepared for a roof, but there are some deep dints in the pavement, shewing where large masses have fallen. The lower parts of some of the columns (to the height of 8 or 10 feet) are much scaled and cracked. The windows are scarcely touched. I also refreshed my memory of the chapter-house, which is most beautiful, and which has much of its old gilding reasonably bright, and some of its old paint quite conspicuous. And I looked again at the old crypt with its late Norman work, and at the still older crypt of the pre-existing church.

1841

" The routine work of the Observatory in its several departments was carried on steadily during this year.—The Camera Obscura was removed from the N.W. Turret of the Great Room, to make way for the Anemometer.—In Magnetism and Meteorology the most important thing was the great magnetic storm of Sept. 25th, which revealed a new class of magnetic phenomena. It was very well observed by Mr Glaisher, and I immediately printed and circulated an account of it.—In April I reported that the Planetary Reductions were completed, and furnished estimates for the printing.—In August I applied for 18,000 copies of the great skeleton form for computing Lunar Tabular Places, which were granted.—I reported, as usual, on various Papers for the Royal Society, and was still engaged on the Cavendish Experiment.—In the University of London I attended the meeting of Dec. 8th, on the reduction of Examiners' salaries, which were extravagant.—I furnished Col. Colby with a plan of a new Sector, still used in the British Survey.—I appealed to Colby about the injury to the cistern on the Great Gable in Cumberland, by the pile raised for the Survey Signal.—On Jan. 3rd occurred a most remarkable tidal disturbance: the tide in the Thames was 5 feet too low. I endeavoured to trace it on the coasts, and had a vast amount of correspondence: but it elicited little.

"Of private history: I was a short time in Suffolk in March.—On Mar. 31st I started with my wife (whose health had suffered much) for a trip to Bath, Bristol, Cardiff, Swansea, &c. While at Swansea we received news on Apr. 24th of the deadly illness of my dear mother. We travelled by Neath and Cardiff to Bath, where I solicited a rest for my wife from my kind friend Miss Sutcliffe, and returned alone to Greenwich. My dear mother had died on the morning of the 24th. The funeral took place at Little Whelnetham (near Bury) on May 1st, where my mother was buried by the side of my father. We went to Cambridge, where my wife consulted Dr Haviland to her great advantage, and returned to Greenwich on May 7th.—On May 14th to 16th I was at Sanderstead (Rev. J. Courtney) with Whewell as one sponsor, at the christening of my daughter Hilda.—In September I went for a trip with my sister to Yorkshire and Cumberland, in the course of which we visited Dent (Sedgwick's birth-place), and paid visits to Mr Wordsworth, Miss Southey, and Miss Bristow, returning to Greenwich on the 30th Sept.— From June 15th to 19th I visited my brother at Keysoe."

The following extracts are from letters written to his wife while on the above trip in Yorkshire and Cumberland:

RED LION INN, REDCAR,
1841, *Sept.* 11.

We stopped at York: went to the Tavern Hotel. In the morn-ing (Friday) went into the Cathedral. I think that it improves on acquaintance. The nave is now almost filled with scaffolding for the repair of the roof, so that it has not the bare unfinished appear-ance that it had when I was there last year. The tower in which the fire began seems to be a good deal repaired: there are new mul-lions in its windows, &c. We stopped to hear part of the service, which was not very effective.

Here are notes of his visit to Dentdale in Yorkshire, the birthplace of his friend Sedgwick:

KING'S HEAD, KENDAL,
1841, *Sept.* 15.

The day was quite fine, and the hills quite clear. The ascent out of Hawes is dull; the little branch dale is simple and monoto-nous, and so are the hills about the great dale which are in sight. The only thing which interested us was the sort of bird's-eye view of Hardraw dell, which appeared a most petty and insignificant opening in the great hill side. But when we got to the top of the pass there was a magnificent view of Ingleborough. The dale which was most nearly in front of us is that which goes down to Ingleton, past the side of Ingleborough. The mountain was about nine miles distant. We turned to the right and immediately descended Dent-dale. The three dales (to Hawes, to Ingleton, and to Dent) lay their heads together in a most amicable way, so that, when at the top, it is equally easy to descend down either of them. We found very soon that Dent-dale is much more beautiful than that by which we had ascended. The sides of the hills are steeper, and perhaps higher: the bottom is richer. The road is also better. The river is a con-

tinued succession of very pretty falls, almost all of which have
scooped out the lower strata of the rock, so that the water shoots
clear over.   For several miles (perhaps 10) it runs upon bare lime-
stone without a particle of earth.   From the head of the dale to the
village of Dent is eight miles.   At about half-way is a new chapel,
very neat, with a transept at its west end.   The village of Dent is
one of the strangest places that I ever saw.   Narrow street, up and
down, with no possibility of two carriages bigger than children's
carts passing each other.   We stopped at the head inn and enquired
about the Geolog: but he is not in the country.   We then called
on his brother, who was much surprised and pleased to see us.   His
wife came in soon after (his daughter having gone with a party to see
some waterfall) and they urged us to stop and dine with them.   So
we walked about and saw every place about the house, church, and
school, connected with the history of the Geolog: and then dined.
I promised that you should call there some time when we are in the
north together and spend a day or two with them.   Mr Sedgwick
says it is reported that Whewell will take Sedbergh living (which is
now vacant: Trinity College is patron).   Then we had our chaise
and went to Sedbergh.   The very mouth of Dent-dale is more con-
tracted than its higher parts.   Sedbergh is embosomed among lump-
ing hills.   Then we had another carriage to drive to Kendal.

Here is a recollection of Wordsworth :

SALUTATION, AMBLESIDE,
1841, *Sept.* 19.

We then got our dinner at Lowwood, and walked straight to
Ambleside, changed our shoes, and walked on to Rydal to catch
Wordsworth at tea.   Miss Wordsworth was being drawn about in a
chair just as she was seven years ago.   I do not recollect her appearance then so as to say whether she is much altered, but I think not.
Mr Wordsworth is as full of good talk as ever, and seems quite
strong and well.   Mrs Wordsworth looks older.   Their son William
was at tea, but he had come over only for the day or evening.
There was also a little girl, who I think is Mrs Wordsworth's niece.

1842

"In this year I commenced a troublesome work, the Description of the Northumberland Telescope. On Sept. 9th I wrote to the Duke of Northumberland suggesting this, sending him a list of Plates, and submitting an estimate of expense £120. On Sept. 19th I received the Duke's assent. I applied to Prof. Challis (at the Cambridge Observatory) requesting him to receive the draughtsman, Sly, in his house, which he kindly consented to do.

"With regard to Estimates. I now began to point out to the Admiralty the inconvenience of furnishing separate estimates, viz. to the Admiralty for the Astronomical Establishment, and to the Treasury for the Magnetical and Meteorological Establishment.—The great work of the Lunar Reductions proceeded steadily: 14 computers were employed on them.—With regard to the Magnetical and Meteorological Establishment: I suppose that James Ross's expedition had returned: and with this, according to the terms of the original grant, the Magnetical and Meteorological Establishments expired. There was much correspondence with the Royal Society and the Treasury, and ultimately Sir R. Peel consented to the continuation of the establishments to the end of 1845.—In this year began my correspondence with Mr Mitchell about the Cincinnati Observatory. On Aug. 25 Mr Mitchell settled himself at Greenwich, and worked for a long time in the Computing Room.—And in this year Mr Aiken of Liverpool first wrote to me about the Liverpool Observatory, and a great deal of correspondence followed: the plans were in fact entirely entrusted to me.—July 7th was the day of the Total Eclipse of the Sun, which I observed with my wife at the Superga, near Turin. I wrote an account of my observations for the Royal Astronomical Society.—On Jan. 10th I notified to Mr Goulburn that our Report on the Restoration of the Standards was ready, and on Jan. 12th I presented it.

After this followed a great deal of correspondence, principally concerning the collection of authenticated copies of the Old Standards from all sides.—In some discussions with Capt. Shirreff, then Captain Superintendent of the Chatham Dockyard, I suggested that machinery might be made which would saw ship-timbers to their proper form, and I sent him some plans on Nov. 8th. This was the beginning of a correspondence which lasted long, but which led to nothing, as will appear hereafter.—On Dec. 15th, being on a visit to Dean Peacock at Ely, I examined the Drainage Scoop Wheel at Prickwillow, and made a Report to him by letter, which obtained circulation and was well known.—On May 26th the manuscript of my article, 'Tides and Waves,' for the Encyclopædia Metropolitana was sent to the printer. I had extensive correspondence, principally on local tides, with Whewell and others. Tides were observed for me by Colby's officers at Southampton, by myself at Christchurch and Poole, at Ipswich by Ransome's man; and a great series of observations of Irish Tides were made on my plan under Colby's direction in June, July and August.—On Sept. 15th Mr Goulburn, Chancellor of the Exchequer, asked my opinion on the utility of Babbage's calculating machine, and the propriety of expending further sums of money on it. I replied, entering fully into the matter, and giving my opinion that it was worthless.—I was elected an Honorary Member of the Institution of Civil Engineers, London.

"The reduction and printing of the astronomical observations had been getting into arrear: the last revise of the 1840 observations went to press on May 18th, 1842. On Aug. 18th came into operation a new organization of Assistants' hours of attendance, &c., required for bringing up reductions. I worked hard myself and my example had good effect." His reference to this subject in his Report to the Visitors is as follows: "I have in one of the preceding articles alluded to the backwardness of our reductions. In those which follow it I trust that I have sufficiently explained it. To say nothing of the loss, from ill health, of the services of most efficient assistants, I am certain that the quantity of current work will amply explain any backwardness. Perhaps I may particularly mention that in the observations of 1840 there was an unusual quantity of equatoreal observations, and the reductions attending these occupied a very great time. But, as regards myself, there has been another cause. The reduction of the Ancient Lunar and Planetary Observations, the attention to chronometer constructions, the proposed management of the printing of papers relating to important operations at the Cape of Good Hope; these and similar operations have taken up much of my time. I trust that I am doing well in rendering Greenwich, even more distinctly than it has been heretofore, the place of reference to all the world for the important observations, and results of observations, on which the system of the universe is founded. As regards myself, I have been accustomed, in these matters, to lay aside private considerations; to consider that I am not a mere Superintendent of current observations, but a Trustee for the honour of Greenwich Observatory generally, and for its utility generally to the world; nay, to consider myself not as mere Director of Greenwich Observatory, but (however unworthy personally) as British Astronomer, required sometimes by my office to interfere (when no personal offence is given) in the concerns of other establishments of the State. If the Board supports me in this view there can be little doubt that the present delay of computations, relating to current observations, will be considered by them as a very small sacrifice to the important advantage that may be gained by proper attention to the observations of other times and other places."

"Of private history: In February I went for a week to Playford and Norwich, visiting Prof. Sedgwick at the latter place. On Mar. 1st my third daughter Christabel was born. In March I paid a short visit to Sir John Herschel at Hawkhurst. From June 12th to Aug. 11th I was travelling with my wife on the Continent, being partly occupied with the

observation of the Total Eclipse of the Sun on July 7th.
The journey was in Switzerland and North Italy. In December I went to Cambridge and Ely, visiting Dr Peacock at the
latter place."

From Feb. 23rd to 28th Airy was engaged on Observations of Tides at Southampton, Christchurch, Poole, and
Weymouth. During this expedition he wrote frequently (as
he always did) to his wife on the incidents of his journey,
and the following letters appear characteristic:

KING'S ARMS, CHRISTCHURCH,
OR XCHURCH,
1842, <i>Feb.</i> 24.

The lower of the above descriptions of my present place of
abode is the correct one, as I fearlessly assert on the authority of
divers direction-posts on the roads leading to it (by the bye this
supports my doctrine that x in Latin was not pronounced eks but
khi, because the latter is the first letter of Christ, for which x is here
traditionally put). Finding this morning that Yolland (who called
on me as soon as I had closed the letter to you) was perfectly
inclined to go on with the tide observations at Southampton, and
that his corporals of sappers were conducting them in the most
exemplary manner, I determined on starting at once. However we
first went to look at the New Docks (mud up to the knees) and truly
it is a very great work. There is to be enclosed a good number of
acres of water 22 feet deep: one dock locked in, the other a tidal
dock or basin with that depth at low water. They are surrounded
by brick walls eight feet thick at top, 10 or more at bottom; and all
the parts that ever can be exposed are faced with granite. The
people reckon that this work when finished will attract a good deal
of the London commerce, and I should not be surprised at it. For
it is very much easier for ships to get into Southampton than into
London, and the railway carriage will make them almost one. A
very large steamer is lying in Southampton Water: the Oriental,
which goes to Alexandria. The Lady Mary Wood, a large steamer
for Lisbon and Gibraltar, was lying at the pier. The said pier is a
very pleasant place of promenade, the water and banks are so pretty,
and there is so much liveliness of ships about it. Well I started in
a gig, in a swashing rain, which continued off and on for a good
while. Of the 21 miles, I should think that 15 were across the New

Forest. I do not much admire it. As for Norman William's destruction of houses and churches to make it hunting ground, that is utter nonsense which never could have been written by anybody that ever saw it: but as to hunting, except his horses wore something like mud-pattens or snow-shoes, it is difficult to conceive it. Almost the whole Forest is like a great sponge, water standing in every part. In the part nearer to Xchurch forest trees, especially beeches, seem to grow well. We stopped to bait at Lyndhurst, a small place high up in the Forest: a good view, such as it is, from the churchyard. The hills of the Isle of Wight occasionally in sight. On approaching Xchurch the chalk cliffs of the west end of the Isle of Wight (leading to the Needles) were partly visible; and, as the sun was shining on them, they fairly blazed. Xchurch is a small place with a magnificent-looking church (with lofty clerestory, double transept, &c., but with much irregularity) which I propose to visit to-morrow. Also a ruin which looks like an abbey, but the people call it a castle. There is a good deal of low land about it, and the part between the town and the sea reminded me a good deal of the estuary above Cardigan, flat ill-looking bogs (generally islands) among the water. I walked to the mouth of the river (more than two miles) passing a nice little place called Sandford, with a hotel and a lot of lodgings for summer sea-people. At the entrance of the river is a coastguard station, and this I find is the place to which I must go in the morning to observe the tide. I had some talk with the coastguard people, and they assure me that the tide is really double as reported. As I came away the great full moon was rising, and I could read in her unusually broad face (indicating her nearness to the earth) that there will be a powerful tide. I came in and have had dinner and tea, and am now going to bed, endeavouring to negociate for a breakfast at six o'clock to-morrow morning. It is raining cats and dogs.

LUCE'S HOTEL, WEYMOUTH,
1842, *Feb.* 27.

This morning when I got up I found that it was blowing fresh from S.W. and the sea was bursting over the wall of the eastern extremity of the Esplanade very magnanimously. So (the swell not being favourable for tide-observations) I gave them up and determined to go to see the surf on the Chesil Bank. I started with my great-coat on, more for defence against the wind than against rain;

but in a short time it began to rain, and just when I was approaching the bridge which connects the mainland with the point where the Chesil Bank ends at Portland (there being an arm of the sea behind the Chesil Bank) it rained and blew most dreadfully. However I kept on and mounted the bank and descended a little way towards the sea, and there was the surf in all its glory. I cannot give you an idea of its majestic appearance. It was evidently very high, but that was not the most striking part of it, for there was no such thing as going within a considerable distance of it (the occasional outbreaks of the water advancing so far) so that its magnitude could not be well seen. My impression is that the height of the surf was from 10 to 20 feet. But the striking part was the clouds of solid spray which formed immediately and which completely concealed all the other operations of the water. They rose a good deal higher than the top of the surf, so the state of things was this. A great swell is seen coming, growing steeper and steeper; then it all turns over and you see a face just like the pictures of falls of Niagara; but in a little more than one second this is totally lost and there is nothing before you but an enormous impenetrable cloud of white spray. In about another second there comes from the bottom of this cloud the foaming current of water up the bank, and it returns grating the pebbles together till their jar penetrates the very brain. I stood in the face of the wind and rain watching this a good while, and should have stood longer but that I was so miserably wet. It appeared to me that the surf was higher farther along the bank, but the air was so thickened by the rain and the spray that I could not tell. When I returned the bad weather abated. I have now borrowed somebody else's trowsers while mine are drying (having got little wet in other parts, thanks to my great-coat, which successfully brought home a hundredweight of water), and do not intend to stir out again except perhaps to post this letter.

FLAMSTEED HOUSE,
1842, *May* 15.

Yesterday after posting the letter for you I went per steamboat to Hungerford. I then found Mr Vignoles, and we trundled off together, with another engineer named Smith, picking up Stratford by the way, to Wormwood Scrubs. There was a party to see the Atmospheric Railway in action: including (among others) Sir John

Burgoyne, whom I met in Ireland several years ago, and Mr Pym, the Engineer of the Dublin and Kingstown Railway, whom I have seen several times, and who is very sanguine about this construction; and Mr Clegg, the proposer of the scheme (the man that invented gas in its present arrangements), and Messrs Samuda, two Jews who are the owners of the experiment now going on; and Sir James South! With the latter hero and mechanician we did not come in contact. Unfortunately the stationary engine (for working the air-pump which draws the air out of the pipes and thus sucks the carriages along) broke down during the experiment, but not till we had seen the carriage have one right good run. And to be sure it is very funny to see a carriage running all alone "as if the Devil drove it" without any visible cause whatever. The mechanical arrangements we were able to examine as well after the engine had broken down as at any time. And they are very simple and apparently very satisfactory, and there is no doubt of the mechanical practicability of the thing even in places where locomotives can hardly be used: whether it will pay or not is doubtful. I dare say that the Commissioners' Report has taken a very good line of discrimination.

1843

"In March I wrote to Dr Wynter (Vice-Chancellor) at Oxford, requesting permission to see Bradley's and Bliss's manuscript Observations, with the view of taking a copy of them. This was granted, and the books of Transits were subsequently copied under Mr Breen's superintendence.— The following paragraph is extracted from the Report to the Visitors: ' In the Report of last year, I stated that our reductions had dropped considerably in arrear. I have the satisfaction now of stating that this arrear and very much more have been completely recovered, and that the reductions are now in as forward a state as at any time since my connection with the Observatory.' In fact the observations of 1842 were sent to press on Mar. 1st, 1843.—About this year the Annual Dinner at the Visitation began to be more important, principally under the management of Capt. W. H. Smyth, R.N.— In November I was enquiring about an 8-inch object-glass.

I had already in mind the furnishing of our meridional instruments with greater optical powers.—On July 14th the Admiralty referred to me a Memorial of Mr J. G. Ulrich, a chronometer maker, claiming a reward for improvements in chronometers. I took a great deal of trouble in the investigation of this matter, by books, witnesses, &c., and finally reported on Nov. 4th that there was no ground for claim.—In April I received the first application of the Royal Exchange Committee, for assistance in the construction of the Clock: this led to a great deal of correspondence, especially with Dent.— The Lunar Reductions were going on in full vigour.—I had much work in connection with the Cape Observatory: partly about an equatoreal required for the Observatory, but chiefly in getting Maclear's work through the press.—In this year I began to think seriously of determining the longitude of Valencia in Ireland, as a most important basis for the scale of longitude in these latitudes, by the transmission of chronometers; and in August I went to Valencia and examined the localities. In September I submitted a plan to the Admiralty, but it was deferred.—The new Commission for restoring the Standards was appointed on June 20th, I being Chairman. The work of collecting standards and arranging plans was going on; Mr Baily attending to Standards of Length, and Prof. W. H. Miller to Standards of Weight. We held two meetings.—A small assistance was rendered to me by Mr Charles May (of the firm of Ransomes and May), which has contributed much to the good order of papers in the Observatory. Mr Robert Ransome had remarked my method of punching holes in the paper by a hand-punch, the places of the holes being guided by holes in a piece of card, and said that they could furnish me with something better. Accordingly, on Aug. 28th Mr May sent me the punching machine, the prototype of all now used in the Observatory.

"On Sept. 25th was made my proposal for an Altazimuth Instrument for making observations of the Moon's place more frequently and through parts of her orbit where she

could never be observed with meridional instruments; the
most important addition to the Observatory since its foundation. The Board of Visitors recommended it to the Admiralty, and the Admiralty sanctioned the construction of the
instrument and the building to contain it." The following
passage is quoted from the Address of the Astronomer Royal
to the Board of Visitors at the Special Meeting of Nov. 10th,
1843 : " The most important object in the institution and
maintenance of the Royal Observatory has always been the
Observations of the Moon. In this term I include the determination of the places of fixed stars which are necessary for
ascertaining the instrumental errors applicable to the instrumental observations of the Moon. These, as regards the
objects of the institution, were merely auxiliaries : the history
of the circumstances which led the Government of the day to
supply the funds for the construction of the Observatory
shews that, but for the demands of accurate Lunar Determinations as aids to navigation, the erection of a National
Observatory would never have been thought of. And this
object has been steadily kept in view when others (necessary
as fundamental auxiliaries) were passed by. Thus, during
the latter part of Bradley's time, and Bliss's time (which two
periods are the least efficient in the modern history of the
Observatory), and during the latter part of Maskelyne's
presidency (when, for years together, there is scarcely a
single observation of the declination of a star), the Observations of the Moon were kept up with the utmost regularity.
And the effect of this regularity, as regards its peculiar object,
has been most honourable to the institution. The existing
Theories and Tables of the Moon are founded entirely upon
the Greenwich Observations; the Observatory of Greenwich
has been looked to as that from which alone adequate observations can be expected, and from which they will not be
expected in vain : and it is not perhaps venturing too much
to predict that, unless some gross dereliction of duty by the
managers of the Observatory should occur, the Lunar Tables

will always be founded on Greenwich Observations. With
this impression it has long been to me a matter of consideration whether means should not be taken for rendering the
series of Observations of the Moon more complete than it
can be made by the means at present recognized in our
observatories."—In illustration of the foregoing remarks, the
original inscription still remaining on the outside of the wall
of the Octagon Room of the Observatory may be quoted.
It runs thus: 'Carolus IIˢ Rex Optimus Astronomiæ et
Nauticæ Artis Patronus Maximus Speculam hanc in utri-
usque commodum fecit Anno Dⁿⁱ MDCLXXVI Regni sui
XXVIII curante Iona Moore milite RTSG.'

   "The Ashburton Treaty had been settled with the United
States, for the boundary between Canada and the State of
Maine, and one of its conditions was, that a straight line
about 65 miles in length should be drawn through dense
woods, connecting definite points. It soon appeared that
this could scarcely be done except by astronomical operations. Lord Canning, Under Secretary of the Foreign Office,
requested me to nominate two astronomers to undertake the
work. I strongly recommended that Military Officers should
carry out the work, and Capt. Robinson and Lieut. Pipon
were detached for this service. On Mar. 1st they took
lodgings at Greenwich, and worked at the Observatory every
day and night through the month. My detailed astronomical
instructions to them were drawn out on Mar. 29th. I prepared all the necessary skeleton forms, &c., and looked to
their scientific equipment in every way. The result will be
given in 1844.

   "Of private history: In January I went to Dover with
my wife to see the blasting of a cliff there: we also visited
Sir J. Herschel at Hawkhurst. In April I was at Playford,
on a visit to Arthur Biddell. On Apr. 9th my daughter
Annot was born. From July 22nd to August 25th I was
travelling in the South of Ireland, chiefly to see Valencia
and consider the question of determining its longitude:

during this journey I visited Lord Rosse at Birr Castle, and returned to Weymouth, where my family were staying at the time. In October I visited Cambridge, and in December I was again at Playford."

The journey to Cambridge (Oct. 24th to 27th) was apparently in order to be present on the occasion of the Queen's visit there on the 25th: the following letter relating to it was written to his wife:

SEDGWICK'S ROOMS,
TRINITY COLLEGE, CAMBRIDGE.
1843, *Oct.* 26, *Thursday.*

I have this morning received your letter: I had no time to write yesterday. There are more things to tell of than I can possibly remember. The Dean of Ely yesterday was in a most ludicrous state of misery because his servant had sent his portmanteau (containing his scarlet academicals as well as everything else) to London, and it went to Watford before it was recovered: but he got it in time to shew himself to-day. Yesterday morning I came early to breakfast with Sedgwick. Then I walked about the streets to look at the flags. Cambridge never had such an appearance before. In looking along Trinity Street or Trumpington Street there were arches and flags as close as they could stand, and a cord stretched from King's Entrance to Mr Deck's or the next house with flags on all its length: a flag on St Mary's, and a huge royal standard ready to hoist on Trinity Gateway: laurels without end. I applied at the Registrar's office for a ticket which was to admit me to Trinity Court, the Senate House, &c., and received from Peacock one for King's Chapel. Then there was an infinity of standing about, and very much I was fatigued, till I got some luncheon at Blakesley's rooms at 1 o'clock. This was necessary because there was to be no dinner in hall on account of the Address presentation. The Queen was expected at 2, and arrived about 10 minutes after 2. When she drove up to Trinity Gate, the Vice-Chancellor, masters, and beadles went to meet her, and the beadles laid down their staves, which she desired them to take again. Then she came towards the Lodge as far as the Sundial, where Whewell as master took the college keys (a bundle of rusty keys tied together by a particularly greasy strap) from the bursar Martin, and handed them to the Queen, who returned them. Then she drove round by the turret-corner of the court to the Lodge door. Almost every member of the University was in the court, and there was a great hurraing except when the ceremonies were going forward. Presently the Queen appeared at a window and bowed, and was loudly cheered. Then notice was given that the Queen and Prince would receive the Addresses of the University in Trinity hall, and a procession was formed, in which I had a good place, as I claimed rank with the Professors. A throne and canopy were erected at the top of the hall, but the Queen did not sit, which was her own determination, because if she had sat it would have been proper that everybody should back out before presenting the Address to the Prince : which operation would have suffocated at least 100 people. The Queen wore a blue gown and a brown shawl with an immense quantity of gold embroidery, and a bonnet. Then it was known that the Queen was going to service at King's Chapel at half past three : so everybody went there. I saw the Queen walk up the antechapel and she looked at nothing but the roof. I was not able to see her in chapel or to see the throne erected for her with its back to the Table, which has given great offence to many people. (I should have said that before the Queen came I called on Dr Haviland, also on Scholefield, also on the Master of Christ's.) After this she returned to Trinity, and took into her head to look at the chapel. The cloth laid on the pavement was not long enough and the undergraduates laid down their gowns. Several of the undergraduate noblemen carried candles to illuminate Newton's statue. After this the Prince went by torchlight to the library. Then I suppose came dinner, and then it was made known that at half-past nine the Queen would receive some Members of the University. So I rigged myself up and went to the levée at the Lodge and was presented in my turn by the Vice-Chancellor as " Ex-Professor Airy, your Majesty's Astronomer Royal." The Queen and the Prince stood together, and a bow was made to and received from each. The Prince recognised me and said " I am glad to see you," or something like that. Next to him stood Goulburn, and next Lord Lyndhurst, who to my great surprise spoke very civilly to me (as I will tell you afterwards). The Queen had her head bare and a sort of French white gown and looked very well. She had the ribbon of the Garter on her breast ; but like a ninny I forgot to look whether she had the Garter upon her arm. The Prince wore his Garter. I went to bed dead tired and got up with a headache.—About the degree to the Prince and the other movements I will write again.

Here is a note from Cubitt relating to the blasting of the Round Down Cliff at Dover referred to above:

GREAT GEORGE STREET,
Jan. 20th, 1843.

MY DEAR SIR,

Thursday next the 26th at 12 is the time fixed for the attempt to blow out the foot of the "Round Down" Cliff near Dover.

The Galvanic apparatus has been repeatedly tried in place—that is by exploding cartridges in the very chambers of the rock prepared for the powder—with the batteries at 1200 feet distance they are in full form and act admirably so that I see but little fear of failure on that head.

They have been rehearsing the explosions on the plan I most strongly recommended, that is—to fire each chamber by an independent battery and circuit and to discharge the three batteries simultaneously by signal or word of command which answers well and "no mistake."

I shall write to Sir John Herschel to-day, and remain

My dear Sir,

Very truly yours,

W. CUBITT.

G. B. Airy, Esq.

The following extracts are from letters to his wife written in Ireland when on his journey to consider the determination of the longitude of Valencia.

SKIBBEREEN,
1843, *July* 28.

By the bye, to shew the quiet of Ireland now, I saw in a newspaper at Cork this account. At some place through which a repeal-association was to pass (I forget its name) the repealers of the place set up a triumphal arch. The police pulled it down, and were pelted by the repealers, and one of the policemen was much bruised. O'Connell has denounced this place as a disgrace to the cause of repeal, and has moved in the full meeting that the inhabitants of this place be struck off the repeal list, with no exception but that of the

parish priest who was proved to be absent. And O'Connell declares
that he will not pass through this place. Now for my journey. It
is a sort of half-mountain country all the way, with some bogs to
refresh my eyes.

VALENCIA HOTEL,
1843, *August* 6.

It seems that my coming here has caused infinite alarm. The
common people do not know what to conjecture, but have some
notion that the "sappers and miners" are to build a bridge to admit
the charge of cavalry into the island. An attendant of Mrs Fitz-
gerald expressed how strange it was that a man looking so mild and
gentle could meditate such things "but never fear, Maam, those that
look so mild are always the worst": then she narrated how that her
husband was building some stables, but that she was demanding of
him "Pat, you broth of a boy, what is the use of your building stables
when these people are coming to destroy everything." I suspect that
the people who saw me walking up through the storm yesterday must
have thought me the prince of the powers of the air at least.

HIBERNIAN HOTEL, TRALEE,
1843, *August* 7.

I sailed from Valencia to Cahersiveen town in a sail-boat up the
water (not crossing at the ferry). I had accommodated my time to
the wish of the boatman, who desired to be there in time for prayers :
so that I had a long waiting at Cahersiveen for the mail car. In
walking through the little town, I passed the chapel (a convent
chapel) to which the people were going: and really the scene was
very curious. The chapel appeared to be overflowing full, and the
court in front of it was full of people, some sitting on the ground,
some kneeling, and some prostrate. There were also people in the
street, kneeling with their faces towards the gate pillars, &c. It
seemed to me that the priest and the chapel were of less use here
than even in the continental churches, and I do not see why both
parties should not have stopped at home. When the chapel broke
up, it seemed as if the streets were crammed with people. The turn-
out that even a small village in Ireland produces is perfectly amazing.

1844

" In the course of 1843 I had put in hand the engraving of the drawings of the Northumberland Telescope at Cambridge Observatory, and wrote the description for letterpress. In the course of 1844 the work was completed, and the books were bound and distributed.

" The building to receive the Altazimuth Instrum nt was erected in the course of the year; during the construction a foreman fell into the foundation pit and broke his leg, of which accident he died. This is the only accident that I have known at the Observatory.—The Electrometer Mast and sliding frame were erected near the Magnetic Observatory.—The six-year Catalogue of 1439 stars was finished; this work had been in progress during the last few years.—In May I went to Woolwich to correct the compasses of the ' Dover,' a small iron steamer carrying mails between Dover and Ostend : this I believe was the first iron ship possessed by the Admiralty.—The Lunar Reductions were making good progress; 16 computers were employed upon them. I made application for printing them and the required sum (£1000) was granted by the Treasury.—In this year commenced that remarkable movement which led to the discovery of Neptune. On Feb. 13th Prof. Challis introduced Mr Adams to me by letter. On Feb. 15th I sent my observed places of Uranus, which were wanted. On June 19th I also sent places to Mr E. Bouvard.—As regards the National Standards, Mr Baily (who undertook the comparisons relating to standards of length) died soon, and Mr Sheepshanks then undertook the work.—I attended the meeting of the British Association held at York (principally in compliment to the President, Dr Peacock), and gave an oral account of my work on Irish Tides.—At the Oxford Commemoration in June, the honorary degree of D.C.L. was conferred on M. Struve and on me, and then a demand was made on each of us for £6. 6s. for fees. We were much disgusted and refused to pay it, and I wrote angrily to Dr Wynter, the Vice-Chancellor. The fees were ultimately paid out of the University Chest.

"In this year the longitude of Altona was determined by M. Struve for the Russian Government. For this purpose it was essential that facilities should be given for landing chronometers at Greenwich. But the consent of the custom-house authorities had first to be obtained, and this required a good deal of negotiation. Ultimately the determination was completed in the most satisfactory manner. The chronometers, forty-two in number, crossed the German Sea sixteen times. The transit observers were twice interchanged, in order to eliminate not only their Personal Equation, but also the gradual change of Personal Equation. On Sept. 30th Otto Struve formally wrote his thanks for assistance rendered.

"For the determination of the longitude of Valencia, which was carried out in this year, various methods were discussed, but the plan of sending chronometers by mail conveyance was finally approved. From London to Liverpool the chronometers were conveyed by the railways, from Liverpool to Kingstown by steamer, from Dublin to Tralee by the Mail Coaches, from Tralee to Cahersiveen by car, from Cahersiveen to Knightstown by boat, and from Knightstown to the station on the hill the box was carried like a sedan-chair. There were numerous other arrangements, and all succeeded perfectly without a failure of any kind. Thirty pocket chronometers traversed the line between Greenwich and Kingstown about twenty-two times, and that between Kingstown and Valencia twenty times. The chronometrical longitudes of Liverpool Observatory, Kingstown Station, and Valencia Station are 12^m 0.05^s, 24^m 31.17^s, 41^m 23.25^s; the geodetic longitudes, computed from elements which I published long ago in the Encyclopædia Metropolitana, are 12^m 0.34^s, 24^m 31.47^s, 41^m 23.06^s. It appears from this that the elements to which I have alluded represent the form of the Earth here as nearly as is possible. On the whole, I think it probable that this is the best arc of parallel that has ever been measured.

"With regard to the Maine Boundary: on May 7th Col. Estcourt, the British Commissioner, wrote to me describing the perfect success of following out my plan: the line of 64 miles was cut by directions laid out at the two ends, and the cuttings met within 341 feet. The country through which this line was to pass is described as surpassing in its difficulties the conception of any European. It consists of impervious forests, steep ravines, and dismal swamps. A survey for the line was impossible, and a tentative process would have broken the spirit of the best men. I therefore arranged a plan of operations founded on a determination of the absolute latitudes and the difference of longitudes of the two extremities. The difference of longitudes was determined by the transfer of chronometers by the very circuitous route from one extremity to the other; and it was necessary to divide the whole arc into four parts, and to add a small part by measure and bearing. When this was finished, the azimuths of the line for the two ends were computed, and marks were laid off for starting with the line from both ends. One party, after cutting more than forty-two miles through the woods, were agreeably surprised, on the brow of a hill, at seeing directly before them a gap in the woods on the next line of hill; it opened gradually, and proved to be the line of the opposite party. On continuing the lines till they passed abreast of each other, their distance was found to be 341 feet. To form an estimate of the magnitude of this error, it is to be observed that it implies an error of only a quarter of a second of time in the difference of longitudes; and that it is only one-third (or nearly so) of the error which would have been committed if the spheroidal form of the Earth had been neglected. I must point out the extraordinary merit of the officers who effected this operation. Transits were observed and chronometers were interchanged when the temperature was lower than 19° below zero : and when the native assistants, though paid highly, deserted on account of the severity of the weather, the British officers still continued the observations upon whose delicacy everything depended.

"Of private history : From July 3rd to Aug. 13th I was in Ireland with my wife. This was partly a business journey in connection with the determination of the longitude of Valencia. On Jan. 4th I asked Lord Lyndhurst (Lord Chancellor) to present my brother to the living of Helmingham, which he declined to do : but on Dec. 12th he offered Binbrooke, which I accepted for my brother."

1845

" A map of the Buildings and Grounds of the Observatory was commenced in 1844, and was still in progress.—On Mar. 19th I was employed on a matter which had for some time occupied my thoughts, viz., the re-arrangement of current manuscripts. I had prepared a sloping box (still in use) to hold 24 portfolios : and at this time I arranged papers A, and went on with B, C, &c. Very little change has been made in these.—In reference to the time given to the weekly report on Meteorology to the Registrar General, the Report to the Board of Visitors contains the following paragraph : ' The devotion of some of my assistants' time and labour to the preparation of the Meteorological Report attached to the weekly report of the Registrar General, is, in my opinion, justified by the bearing of the meteorological facts upon the medical facts, and by the attention which I understand that Report to have excited.'—On Dec. 13th the sleep of Astronomy was broken by the announcement that a new planet, Astræa, was discovered by Mr Hencke. I immediately circulated notices.—But in this year began a more remarkable planetary

discussion. On Sept. 22nd Challis wrote to me to say that
Mr Adams would leave with me his results on the explanation
of the irregularities of Uranus by the action of an exterior
planet. In October Adams called, in my absence. On
Nov. 5th I wrote to him, enquiring whether his theory explained the irregularity of radius-vector (as well as that of
longitude). I waited for an answer, but received none.
(See the Papers printed in the Royal Astronomical Society's
Memoirs and Monthly Notices).—In the Royal Society, the
Royal Medal was awarded to me for my Paper on the Irish
Tides.—In the Royal Astronomical Society I was President ;
and, with a speech, delivered the Medal to Capt. Smyth for
the Bedford Catalogue of Double Stars.—On Jan. 21st I was
appointed (with Schumacher) one of the Referees for the King
of Denmark's Comet Medal : I have the King's Warrant under
his sign manual.—The Tidal Harbour Commission commenced on Apr. 5th : on July 21st my Report on Wexford
Harbour (in which I think I introduced important principles)
was communicated. One Report was made this year to the
Government.—In the matter of Saw Mills (which had begun
in 1842), I had prepared a second set of plans in 1844, and in
this year Mr Nasmyth made a very favourable report on my
plan. A machinist of the Chatham Dock Yard, Sylvester,
was set to work (but not under my immediate command) to
make a model : and this produced so much delay as ultimately to ruin the design.—On Jan. 1st I was engaged on my
Paper ' On the flexure of a uniform bar, supported by equal
pressures at equidistant points.' " (This was probably in connection with the support of Standards of Length, for the
Commission. Ed.).—In June I attended the Meeting of the
British Association at Cambridge, and on the 20th I gave a
Lecture on Magnetism in the Senate House. The following
quotation relating to this Lecture is taken from a letter by
Whewell to his wife (see Life of William Whewell by Mrs
Stair Douglas): " I did not go to the Senate House yester-
day evening. Airy was the performer, and appears to have

outdone himself in his art of giving clearness and simplicity
to the hardest and most complex subjects. He kept the
attention of his audience quite enchained for above two
hours, talking about terrestrial magnetism."—On Nov. 29th
I gave evidence before a Committee of the House of Commons on Dover Harbour Pier.

.." With respect to the Magnetical and Meteorological Establishment, the transactions in this year were most important.
It had been understood that the Government establishments
had been sanctioned twice for three-year periods, of which the
second would expire at the end of 1845 : and it was a question with the scientific public whether they should be continued. My own opinion was in favour of stopping the
observations and carefully discussing them. And I am
convinced that this would have been best, except for the
subsequent introduction of self-registering systems, in which
I had so large a share. There was much discussion and correspondence, and on June 7th the Board of Visitors resolved
that ' In the opinion of the Visitors it is of the utmost importance that these observations should continue to be made
on the most extensive scale which the interests of those
sciences may require.' The meeting of the British Association was held at Cambridge in June : and one of the most
important matters there was the Congress of Magnetic
Philosophers, many of them foreigners. It was resolved
that the Magnetic Observatory at Greenwich be continued
permanently. At this meeting I proposed a resolution which
has proved to be exceedingly important. I had remarked
the distress which the continuous two-hourly observations
through the night produced to my Assistants, and determined
if possible to remove it. I therefore proposed ' That it is
highly desirable to encourage by specific pecuniary reward
the improvement of self-recording magnetical and meteorological apparatus : and that the President of the British
Association and the President of the Royal Society be
requested to solicit the favourable consideration of Her

Majesty's Government to this subject,' which was adopted. In October the Admiralty expressed their willingness to grant a reward up to £500. Mr Charles Brooke had written to me proposing a plan on Sept. 23rd, and he sent me his first register on Nov. 24th. On Nov. 1st the Treasury informed the Admiralty that the Magnetic Observatories will be continued for a further period.

"The Railway Gauge Commission in this year was an important employment. The Railways, which had begun with the Manchester and Liverpool Railway (followed by the London and Birmingham) had advanced over the country with some variation in their breadth of gauge. The gauge of the Colchester Railway had been altered to suit that of the Cambridge Railway. And finally there remained but two gauges: the broad gauge (principally in the system allied with the Great Western Railway); and the narrow gauge (through the rest of England). These came in contact at Gloucester, and were likely to come in contact at many other points—to the enormous inconvenience of the public. The Government determined to interfere, beginning with a Commission. On July 3rd Mr Laing (then on the Board of Trade) rode to Greenwich, bearing a letter of introduction from Sir John Lefevre and a request from Lord Dalhousie (President of the Board of Trade) that I would act as second of a Royal Commission (Col. Sir Frederick Smith, Airy, Prof. Barlow). I assented to this: and very soon began a vigorous course of business. On July 23rd and 24th I went with Prof. Barlow and our Secretary to Bristol, Gloucester, and Birmingham: on Dec. 17th I went on railway experiments to Didcot: and on Dec. 29th to Jan. 2nd I went to York, with Prof. Barlow and George Arthur Biddell, for railway experiments. On Nov. 21st I finished a draft Report of the Railway Gauge Commission, which served in great measure as a basis for that adopted next year.

"Of private history: I wrote to Lord Lyndhurst on Feb. 20th, requesting an exchange of the living to which he had presented my brother in Dec. 1844 for that of Swineshead : to which he consented.—On Jan. 29th I went with my wife on a visit to my uncle George Biddell, at Bradfield St George, near Bury.—On June 9th I went into the mining district of Cornwall with George Arthur Biddell.—From Aug. 25th to Sept. 26th I was travelling in France with my sister and my wife's sister, Georgiana Smith. I was well introduced, and the journey was interesting.—On Oct. 29th my son Osmund was born.— Mr F. Baily bequeathed to me £500, which realized £450."

Here are some extracts from letters written to his wife relating to the visit to the Cornish mines, &c.—

PEARCE'S HOTEL, FALMOUTH,
1845, *June 12th, Thursday.*

Then we walked to the United Mines in Gwennap. The day was very fine and now it was perfectly broiling : and the hills here are long and steep. At the United Mines we found the Captain, and he invited us to join in a rough dinner, to which he and the other captains were going to sit down. Then we examined one of the great pumping engines, which is considered the best in the country: and some other engines. Between 3 and 4 there was to be a setting out of some work to the men by a sort of Dutch Auction (the usual way of setting out the work here): some refuse ores were to be broken up and made marketable, and the subject of competition was, for how little in the pound on the gross produce the men would work them up. While we were here a man was brought up who was hurt in blasting: a piece of rock had fallen on him. At this mine besides the ladder ways, they have buckets sliding in guides by which the men are brought up: and they are just preparing for work another apparatus which they say is tried successfully at another mine (Tresavean): there are two wooden rods A and B reaching from the top to the bottom, moved by cranks from the same wheel, so that one goes up when the other goes down, and vice versâ: each of these rods has small stages, at such a distance that when the rod A is down and the rod B is up, the first stage of A is level with the first stage of B: but when the rod A is up and the rod B is down, the second stage of A is level with the first stage of B: so a man who wants to descend steps on the first stage of A and waits till it goes down: then he steps sideways on the first stage of B and waits till it goes down: then he steps sideways to the second stage of *A* and waits till it goes down, and so on: or if a man is coming up he does just the same. While we were here Mr R. Taylor came. We walked home (a long step, perhaps seven miles) in a very hot sun. Went to tea to Mr Alfred Fox, who has a house in a beautiful position looking to the outside of Falmouth Harbour.

PENZANCE,
1845, *June* 14, *Saturday*.

Yesterday morning we breakfasted early at Falmouth, and before 9 started towards Gwennap. I had ascertained on Thursday that John Williams (the senior of a very wealthy and influential family in this country) was probably returned from London. So we drove first to his house Burntcoose or Barncoose, and found him and his wife at home. (They are Quakers, the rest of the family are not.) Sedgwick, and Whewell, and I, or some of our party including me, had slept once at their house. They received George and me most cordially, and pressed us to come and dine with them after our visit to Tresavean mine, of which intention I spoke in my last letter: so I named 4 o'clock as hour for dinner. After a little stay we drove to Tresavean, where I found the Captain of the mine prepared to send an Underground Captain and a Pit-man to descend with us. So we changed our clothes and descended by the ladders in the pumpshaft. Pretty work to descend with the huge pump-rods (garnished with large iron bolts) working violently, making strokes of 12 feet, close to our elbows; and with a nearly bottomless pit at the foot of every ladder, where we had to turn round the foot of the ladder walking on only a narrow board. However we got down to the bottom of the mine with great safety and credit, seeing all the mighty machinery on the way, to a greater depth than I ever reached before, namely 1900 feet. From the bottom of the pump we went aside a short distance into the lowest workings where two men nearly naked were driving a level towards the lode or vein of ore. Here I felt a most intolerable heat: and upon moving to get out of the place, I had a dreadful feeling of feebleness and fainting, such as I never had in my life before. The men urged me to climb the ladders to a level where the air was better, but they might as well have urged me to lift up the rock. I could do nothing but sit down and lean fainting against the rocks. This arose entirely from the badness of the air. After a time I felt a trifle better, and then I climbed one short ladder, and sat down very faint again. When I recovered, two men tied a rope round me, and went up the ladder before me, supporting a part of my weight, and in this way I ascended four or five ladders (with long rests between) till we came to a level, 260 fathoms below the adit or nearly 300 fathoms below the surface, where there was a tolerable current of pretty good air. Here I speedily recovered, though I was a little weak for a short time afterwards. George also felt the bad air a good deal, but not so much as I. He descended to some workings equally low in another place (towards which the party that I spoke of were directing their works), but said that the air there was by no means so bad. We all met at the bottom of the man-engine 260 fathoms below the adit. We sat still a little while, and I acquired sufficient strength and nerve, so that I did not feel the slightest alarm in the operation of ascending by the man-engine. This is the funniest operation that I ever saw: it is the only absolute novelty that I have seen since I was in the country before: it has been introduced 2½ years in Tresavean, and one day in the United Mines. In my last letter I described the principle. In the actual use there is no other motion to be made by the person who is ascending or descending than that of stepping sideways each time (there being proper hand-holds) with no exertion at all, except that of stepping exactly at the proper instant: and not the shadow of unpleasant feeling in the motion. Any woman may go with the most perfect comfort, if she will but attend to the rules of stepping, and forget that there is an open pit down to the very bottom of the mine. In this way we were pumped up to the surface, and came up as cool as cucumbers, instead of being drenched with perspiration. In my description in last letter I forgot to mention that between the stages on the moving rods which I have there described there are intermediate stages on the moving rods (for which there is ample room, inasmuch as the interval between the stages on each rod used by one person is 24 feet), and these intermediate stages are used by persons *descending*: so that there are persons *ascending* and persons *descending* at the same time, who never interfere with each other and never step on the same stages, but merely see each other passing on the other rods.—It is a most valuable invention. We then changed our clothes and washed, and drove to Barncoose, arriving in good time for the dinner. I found myself much restored by some superb Sauterne with water. When we were proposing to go on to Camborne, Mr and Mrs Williams pressed us so affectionately to stop that we at length decided on stopping for the night, only bargaining for an early breakfast this morning. This morning after breakfast, we started for Redruth and Camborne. The population between them has increased immensely since I was here before. &c. &c.

Here is a letter written to his wife while he was engaged on the business of the Railway Gauge Commission. It contains reminiscences of some people who made a great figure in the railway world at that time, and was preceded by a letter which was playfully addressed " From the Palace of King Hudson, York."

GEORGE INN, YORK,
1845, *Dec.* 30.

I wrote yesterday from Mr Hudson's in time for the late post, and hope that my letter might be posted by the servant to whom it was given. Our affairs yesterday were simple : we reached Euston Station properly, found Watson there, found a carriage reserved for us, eat pork-pie at Wolverton (not so good as formerly), dined at Derby, and arrived in York at 5.20. On the way Watson informed me that the Government have awarded us £500 each. Sir F. Smith had talked over the matter with us, and I laid it down as a principle that we considered the business as an important one and one of very great responsibility, and that we wished either that the Government should treat us handsomely or should consider us as servants of the State acting gratuitously, to which they assented. I think the Government have done very well. Mr Hudson, as I have said, met us on the platform and pressed us to dine with him (though I had dined twice). Then we found the rival parties quarrelling, and had to arrange between them. This prevented me from writing for the early post. (I forgot to mention that Saunders, the Great Western Secretary, rode with us all the way). At Hudson's we had really a very pleasant dinner : I sat between Vernon Harcourt and Mrs Malcolm (his sister Georgiana) and near to Mr Hudson. This morn-ing we were prepared at 9 at the Station for some runs. Brunel and other people had arrived in the night. And we have been to Darlington and back, with a large party in our experimental train. George Arthur Biddell rode on the engine as representing me. But the side wind was so dreadfully heavy that, as regards the wants of the case, this day is quite thrown away. We have since been to lunch with Vernon Harcourt (Mrs Harcourt not at home) and then went with him to look at the Cathedral. The Chapter-house, which was a little injured, has been pretty well restored: all other things in good order. The Cathedral looks smaller and lower than French cathedrals. Now that we have come in, the Lord Mayor of York has just called to invite us to dinner to-morrow.—I propose to George Arthur Biddell that he go to Newcastle this evening, in order to see glass works and other things there to-morrow, and to return when he can.

I think that I can persuade Barlow to stop to see the experiments out, and if so I shall endeavour to return as soon as possible. The earliest day would be the day after to-morrow.

The following extract is from a letter written to Mr Murray for insertion in his Handbook of France, relating to the Breakwater at Cherbourg, which Airy had visited during his journey in France in the autumn of this year.

ROYAL OBSERVATORY, GREENWICH,
1845, Oct. 8th.

My opinion on the construction I need not say ought not to be quoted: but you are quite welcome to found any general statement on it; or perhaps it may guide you in further enquiries. To make it clear, I must speak rather generally upon the subject. There are three ways in which a breakwater may be constructed. 1. By building a strong wall with perpendicular face from the bottom of the sea. 2. By making a bank with nothing but slopes towards the sea. 3. By making a sloping bank to a certain height and then building a perpendicular wall upon it.—Now if the 1st of these constructions could be arranged, I have no doubt that it would be the best of all, because a sea does not *break* against a perpendicular face, but recoils in an unbroken swell, merely making a slow quiet push at the wall, and not making a violent impact. But practically it is nearly impossible. The 2nd construction makes the sea to break tremendously, but if the sloping surface be made of square stone put together with reasonable care there is not the smallest tendency to unseat these stones. This is the principle of construction of Plymouth Breakwater. In the 3rd construction, the slope makes the sea to break tremendously, and then it strikes the perpendicular face with the force of a battering ram : and therefore in my opinion this is the worst construction of all. A few face-stones may easily be dislodged, and then the sea entering with this enormous force will speedily destroy the whole. This is the form of the Cherbourg Digue.

From this you will gather that I have a full belief that Plymouth Breakwater will last very long, and that the Digue of Cherbourg, at least its upper wall, will not last long. The great bank will last a good while, gradually suffering degradation, but still protecting the Road pretty well.

I was assured by the officers residing on the Digue that the sea which on breaking is thrown vertically upwards and then falls down upon the pavement does sometimes push the stones about which are lying there and which weigh three or four tons.

I saw some preparations for the foundations of the fort at the eastern extremity of the Digue. One artificial stone of concrete measured $12'\cdot9'' \times 6'\cdot7'' \times 5'\cdot7''$, and was estimated to weigh 25000 kilogrammes.

CHAPTER VI.

1846

" ON Nov. 7th I proposed a change in the form of Estimates for the Observatory. The original astronomical part was provided by the Admiralty, and the new magnetical and meteorological part was provided by the Treasury: and the whole Estimates and Accounts of the Observatory never appeared in one public paper. I proposed that the whole should be placed on the Navy Estimates, but the Admiralty refused. I repeated this in subsequent years, with no success. Meantime I always sent to the Admiralty a duplicate of my Treasury Estimate with the proper Admiralty Estimate.—Stephenson's Railway through the lower part of the Park, in tunnel about 850 feet from the Observatory, was again brought forward. On Feb. 20th it was put before me by the Government, and on March 9th I made experiments at Kensal Green, specially on the effect of a tunnel: which I found to be considerable in suppressing the tremors. On May 6th I made my Report, generally favourable, supposing the railway to be in tunnel. On May 13th I, with Mr Stephenson, had an interview at the Admiralty with Lord Ellenborough and Sir George Cockburn. The Earl appeared willing to relax in his scruples about allowing a railway through the Park, when Sir George Cockburn made a most solemn protest against it, on the ground of danger to an institution of such importance as the Observatory. I have no doubt that this protest of Sir George Cockburn's really determined the Government. On June 10th I was informed that the Government refused their consent. After this the South Eastern Railway Company adopted the line through Tranquil Vale.—In consequence of the defective state of Paramatta Observatory I had written to Sir Robert Peel on April 16th raising the question of a General Superintending Board for Colonial Observatories : and on June 27th I saw Mr Gladstone at the Colonial Office to enquire about the possibility of establishing local Boards. On June 29th a general plan was settled, but it never came to anything.—Forty volumes of the Observatory MSS. were bound—an important beginning.—Deep-sunk thermometers were prepared by Prof. Forbes.—On June 22nd Sir Robert Inglis procured an Order of the House of Commons for printing a paper of Sir James South's, ostensibly on the effects of a railway passing through Greenwich Park, but really attacking almost everything that I did in the Observatory. I replied to this on July 21st by a letter in the Athenæum addressed to Sir Robert Inglis, in terms so strong and so well supported that Sir James South was effectually silenced." The following extract from a letter of Airy's to the Earl of Rosse, dated Dec. 15th 1846, will shew how pronounced the quarrel between Airy and South had become in consequence of the above-mentioned attack and previous differences: " After the public exposure which his conduct in the last summer compelled me to make, I certainly cannot meet him on equal terms, and desire not to meet him at all." (Ed.).—" In the Mag. and Met. Department, I was constantly engaged with Mr Charles Brooke in the preparation and mounting of the self-registering instruments, and the chemical arrangements for their use, to the end of the year. With Mr Ronalds I was similarly engaged: but I had the greatest difficulty in transacting business with him, from his unpractical habits.—The equipment of the Liverpool Observatory, under me, was still going on : I introduced the use of Siemens's Chronometric Governor for giving horary motion to an Equatoreal there. I have since introduced the same prin-

180 GEORGE BIDDELL AIRY.

ciple in the Chronograph Barrel and the Great Equatoreal at Greenwich: I consider it important.—On Feb. 13th I received the Astronomical Society's Medal for the Planetary Reductions.—In the University of London: At this time seriously began the discussion whether there should be a compulsory examination in matters bearing on religious subjects. After this there was no peace.—For discovery of Comets three medals were awarded by Schumacher and me: one to Peters, two to De Vico. A comet was seen by Hind, and by no other observer: after correspondence, principally in 1848, the medal was refused to him.—With respect to the Railway Gauge Commission: On Jan. 1st, in our experiments near York, the engine ran off the rails. On Jan. 29th the Commissioners signed the Report, and the business was concluded by the end of April. Our recommendation was that the narrow gauge should be carried throughout. This was opposed most violently by partisans of the broad gauge, and they had sufficient influence in Parliament to prevent our recommendation from being carried into effect. But the policy, even of the Great Western Railway (in which the broad gauge originated), has supported our views: the narrow gauge has been gradually substituted for the broad: and the broad now (1872) scarcely exists.—On June 20th Lord Canning enquired of me about makers for the clock in the Clock Tower of Westminster Palace. I suggested Vulliamy, Dent, Whitehurst; and made other suggestions: I had some correspondence with E. B. Denison, about clocks.—I had much correspondence with Stephenson about the Tubular Bridge over the Menai Straits. Stephenson afterwards spoke of my assistance as having much supported him in this anxious work: on Dec. 11th I was requested to make a Report, and to charge a fee as a Civil Engineer ; but I declined to do so. In January I went, with George Arthur Biddell, to Portsmouth, to examine Lord Dundonald's rotary engine as mounted in the ' Janus,' and made a Report on the same to the Admiralty: and I made several subsequent Reports on the same matter.

The scheme was abandoned in the course of next year; the real cause of failure, as I believe, was in the bad mounting in the ship.

"The engrossing subject of this year was the discovery of Neptune. As I have said (1845) I obtained no answer from Adams to a letter of enquiry. Beginning with June 26th of 1846 I had correspondence of a satisfactory character with Le Verrier, who had taken up the subject of the disturbance of Uranus, and arrived at conclusions not very different from those of Adams. I wrote from Ely on July 9th to Challis, begging him, as in possession of the largest telescope in England, to sweep for the planet, and suggesting a plan. I received information of its recognition by Galle, when I was visiting Hansen at Gotha. For further official history, see my communications to the Royal Astronomical Society, and for private history see the papers in the Royal Observatory. I was abused most savagely both by English and French."

The Report to the Visitors contains an interesting account of the Great Lunar Reductions, from which the following passage is extracted : " Of the Third Section, containing the comparison of Observed Places with Tabular Places, three sheets are printed, from 1750 to 1756. This comparison, it is to be observed, does not contain a simple comparison of places, but contains also the coefficients of the various changes in the moon's place depending on changes in the elements. . . . The process for the correction of the elements by means of these comparisons is now going on : and the extent of this work, even after so much has been prepared, almost exceeds belief. For the longitude, ten columns are added in groups, formed in thirteen different ways, each different way having on the average about nine hundred groups. For the ecliptic polar distance, five columns are added in groups, formed in seven different ways, each different way having on the average about nine hundred groups. Thus it will appear that there are not fewer than 150,000 additions of columns of figures. This part of the work is not only completed but is verified, so that the books of comparison of Observed and Tabular Places are, as regards this work, completely cleared out. The next step is to take the means of these groups, a process which is now in hand: it will be followed by the formation and solution of the equations on which the corrections of the elements depend."

The following remarks, extracted from the Report to the Visitors, with respect to the instrumental equipment of the Observatory, embody the views of the Astronomer Royal at this time: "The utmost change, which I contemplate as likely to occur in many years, in regard to our meridional instruments, is the substitution of instruments of the same class carrying telescopes of larger aperture. The only instrument which, as I think, may possibly be called for by the demands of the astronomer or the astronomical public, is a telescope of the largest size, for the observation of faint nebulæ and minute double stars. Whether the addition of such an instrument to our apparatus would be an advantage, is, in my opinion, not free from doubt. The line of conduct for the Observatory is sufficiently well traced; there can be no doubt that our primary objects ought to be the accurate determination of places of the fundamental Stars, the Sun, the Planets, and, above all, the Moon. Any addition whatever to our powers or our instrumental luxuries, which should tend to withdraw our energies from these objects, would be a misfortune to the Observatory."

Of private history: "In March I visited Prof. Sedgwick at Norwich.—On Mar. 28th the 'Sir Henry Pottinger' was launched from Fairbairn's Yard on the Isle of Dogs, where I was thrown down and dislocated my right thumb.—From Apr. 10th to 15th I was at Playford.—On June 10th Prof. Hansen arrived, and stayed with me to July 4th.—From July 6th to 10th I was visiting Dean Peacock at Ely.—From July 23rd to 29th I was at Playford, where for the first time I lodged in my own cottage. I had bought it some time before, and my sister had superintended alterations and the addition of a room.   I was much pleased thus to be connected with the happy scenes of my youth.—From Aug. 10th to Oct. 11th I was with my wife and her sister Elizabeth Smith on the Continent.   We stayed for some time at Wiesbaden, as my nerves were shaken by the work on the Railway Gauge Commission, and I wanted the Wiesbaden waters.   We visited various places in Germany, and made a 10-days' excursion among the Swiss Mountains.   At Gotha we lodged with Prof. Hansen for three days; and it was while staying here that I heard from Prof. Encke (on Sept. 29th) that Galle had discovered the expected planet.   We visited Gauss at Göttingen and Miss Caroline Herschel at Hannover. We had a very bad passage from Hamburgh to London, lasting five days: a crank-pin broke and had to be repaired: after four days our sea-sickness had gone off, during the gale—a valuable discovery for me, as I never afterwards feared sea-sickness.—On Dec. 22nd I attended the celebration of the 300th anniversary of Trinity College."

The following extracts relating to the engines of the " Janus" are taken from letters to his wife dated from Portsmouth, Jan. 6th and 7th, 1846:

As soon as possible we repaired to the Dock Yard and presented ourselves to the Admiral Superintendant—Admiral Hyde Parker (not Sir Hyde Parker). Found that the " Janus" had not arrived: the Admiral Superintendant (who does not spare a hard word) expressing himself curiously thereon.   But he had got the proper orders from the Admiralty relating to me : so he immediately sent for Mr Taplin, the superintendant of machinery: and we went off to see the small engine of Lord D—d's construction which is working some pumps and other machinery in the yard.   It was kept at work a little longer than usual for us to see it.   And I have no hesitation in saying that it was working extremely well.   It had not been opened in any way for half a year, and not for repair or packing for a much longer time. . . . This morning we went to the Dock Yard, and on entering the engine house there was Shirreff, and Lord D—d soon appeared.   The

"Janus" had come to anchor at Spithead late last night, and had entered the harbour this morning. Blowing weather on Saturday night. We had the engine pretty well pulled to pieces, and sat contemplating her a long time. Before this Denison had come to us. We then went on board the "Janus" with Shirreff but not with Lord D—d. The engines were still hot, and so they were turned backwards a little for my edification. (This was convenient because, the vessel being moored by her head, she could thus strain backwards without doing mischief.) The vacuum not good. Then, after a luncheon on board, it was agreed to run out a little way. But the engines absolutely stuck fast, and would not stir a bit. This I considered a perfect Godsend. So the paddle-wheels (at my desire) were lashed fast, and we are to see her opened to-morrow morning.

This morning (Jan. 7th) we all went off to the "Janus," where we expected to find the end of the cylinder (where we believe yesterday's block to have taken place) withdrawn. But it was not near it. After a great many bolts were drawn, it was discovered that one bolt could not be drawn, and in order to get room for working at it, it was necessary to take off the end of the other cylinder. And such a job! Three pulley hooks were broken in my sight, and I believe some out of my sight. However this auxiliary end was at last got off: and the people began to act on the refractory bolt. But by this time it was getting dark and the men were leaving the dockyard, so I left, arranging that what they could do in preparation for me might be done in good time to-morrow morning.

1847

"On Nov. 13th I circulated an Address, proposing to discontinue the use of the Zenith Tube, because it had been found by a long course of comparative trials that the Zenith Tube was not more accurate than the Mural Circle. The Address stated that 'This want of superior efficiency of the Zenith Tube (which, considered in reference to the expectations that had been formed of its accuracy, must be estimated as a positive failure) is probably due to two circumstances. One is, the use of a plumb-line; which appears to be affected with various ill-understood causes of unsteadiness. The other is, the insuperable difficulty of ventilating the room in which the instrument is mounted.'—On December 20th I circulated an Address, proposing a Transit Circle, with telescope of 8 inches aperture.   The Address states as follows: 'The clear aperture of the Object-Glass of our Transit Instrument is very nearly 5 inches, that of our Mural Circle is very nearly 4 inches.'—I had been requested by the Master-General of Ordnance (I think) to examine Candidates for a Mastership in Woolwich Academy, and I was employed on it in February and March, in conjunction with Prof. Christie.—In January I applied to Lord Auckland for money-assistance to make an astronomical journey on the Continent, but he refused.—On Mar. 19th Sir James South addressed to the Admiralty a formal complaint against me for not observing with the astronomical instruments: on Mar. 31st I was triumphantly acquitted by the Admiralty.— In June I was requested by the Commissioners of Railways to act as President of a Commission on Iron Bridges (suggested by the fall of the bridge at Chester).   Lord Auckland objected to it, and I was not sorry to be spared the trouble of it.—In December I was requested, and undertook to prepare the Astronomical part of the Scientific Manual for Naval Officers.—On Sept. 24th occurred a very remarkable Magnetic Storm, to which there had been nothing comparable before.   Mr Glaisher had it observed by eye extremely well, and I printed and circulated a paper concerning it.—Hansen, stimulated by the Lunar Reductions, discovered two long inequalities in the motion of the Moon, produced by the action of Venus.   In the Report to the Visitors this matter is thus referred to: 'In the last summer I had the pleasure of visiting Prof. Hansen at Gotha, and I was so fortunate as to exhibit to him the corrections of the elements from these Reductions, and strongly to call his attention to their certainty, the peculiarity of their fluctuations, and the necessity of seeking for some physical explanation.   I have much pleasure in indulging in the thought, that it was mainly owing to this representation that Prof. Hansen undertook that quest, which has terminated in the discovery of his two new lunar inequalities, the most remarkable discovery, I think, in Physical Astronomy.'—In discussing points relating to the discovery of Neptune, I made an unfortunate blunder. In a paper hastily sent to the Athenæum (Feb. 18th) I said that Arago's conduct had been indelicate. I perceived instantly that I had used a wrong expression, and by the very next post I sent an altered expression. This altered expression was not received in time, and the original expression was printed, to my great sorrow. I could not then apologize. But at what appeared to be the first opportunity, in December, I did apologize; and my apology was accepted. But I think that Arago was never again so cordial as before.—On July 4th Hebe was discovered. After this Iris and Flora. Now commenced that train of discoveries which has added more than 100 planets to the Solar System.—On Oct. 8th was an Annular Eclipse of the Sun, of which the limit of annularity passed near to Greenwich. To determine the exact place, I equipped observatories at Hayes, Lewisham South End, Lewisham Village, Blackwall, Stratford, Walthamstow, and Chingford. The weather was bad and no observation was obtained.—In the Royal Astronomical Society: In 1846, the dispute between the partisans of Adams and Le Verrier was so violent that no medal could be awarded to either. In 1847 I (with other Fellows of the Society) promoted a special Meeting for considering such a modification of the bye-laws that for this occasion only it might be permissible to give two medals. After two days' stormy discussion, it was rejected.—In the University of London: At a meeting in July, where the religious question was discussed, it was proposed to receive some testimonial from affiliated bodies, or to consider that or some other plan for introducing religious literature. As the propriety of this was doubtful, there was a general feeling for taking legal advice: and it was set aside solely on purpose to raise the question about legal consultation. *That* was negatived by vote : and I then claimed the consideration of the question which we had put aside for it. By the influence of H. Warburton, M.P., this was denied. I wrote a letter to be laid before the Meeting on July 28th, when I was necessarily absent, urging my claim : my letter was put aside. I determined never to sit with Warburton again : on Aug. 2nd I intimated to Lord Burlington my wish to retire, and on Aug. 29th he transmitted to the Home Secretary my resignation. He (Lord Burlington) fully expressed his opinion that my claim ought to have been allowed.—On June 9th, on the occasion of Prince Albert's state visit to Cambridge, knighthood was offered to me through his Secretary, Prof. Sedgwick, but I declined it.—In September, the Russian Order of St Stanislas was offered to me, Mr De Berg, the Secretary of Embassy, coming to Greenwich personally to announce it: but I was compelled by our Government Rules to decline it.—I invited Le Verrier to England, and escorted him to the Meeting of the British Association at Oxford in June.—As regards the Westminster Clock on the Parliamentary Building: in May I examined and reported on Dent's and Whitehurst's clock factories. Vulliamy was excessively angry with me. On May 31st a great Parliamentary Paper was prepared in return to an Order of the House of Lords for correspondence relating to the Clock. —With respect to the Saw Mills for Ship Timber: work was going on under the direction of Sylvester to Mar. 18th. It was, I believe, at that time, that the fire occurred in Chatham Dock Yard which burnt the whole of the saw-machinery. I was tired of my machinery: and, from the extending use of iron ships, the probable value of it was much diminished ; and I made no effort to restore it."

Of private history : " In February I went to Derby to see Whitehurst's clock factory ; and went on with my wife to Brampton near Chesterfield, where her mother was living.— From Apr. 1st to 5th I was at Playford.—On Holy Thursday, I walked the Parish Bounds (of Greenwich) with the Parish officers and others. From Apr. 19th to 24th I was at Birmingham (on a visit to Guest, my former pupil, and afterwards Master of Caius College) and its neighbourhood, with George Arthur Biddell.—From June 23rd to 28th I was at Oxford and Malvern : my sister was at Malvern, for water-cure : the meeting of the British Association was at Oxford and I escorted Le Verrier thither.—July 28th to 30th I was at Brampton.—From August 10th to September 18th I was engaged on an expedition to St Petersburg, chiefly with the object of inspecting the Pulkowa Observatory. I went by Hamburg to Altona, where I met Struve, and started with him in an open waggon for Lübeck, where we arrived on Aug. 14th. We proceeded by steamer to Cronstadt and Petersburg, and so to Pulkowa, where I lodged with O. Struve. I was here engaged till Sept. 4th, in the Observatory, in expeditions in the neighbourhood and at St Petersburg, and at dinner-parties, &c. I met Count Colloredo, Count Ouvaroff, Count Stroganoff, Lord Bloomfield (British Ambassador), and others. On Sept. 4th I went in a small steamer to Cronstadt, and then in the Vladimir to Swinemünde: we were then towed in a passage boat to Stettin, and I proceeded by railway to Berlin. On Sept. 9th I found Galle and saw the Observatory. On Sept. 10th I went to Potzdam and saw Humboldt. On the 12th I went to Hamburg and lodged with Schumacher : I here visited Repsold and Rümker. On Sept. 14th I embarked in the John Bull for London, and arrived there on the evening of the 18th : on the 16th it was blowing 'a whole gale,' reported to be the heaviest gale known for so many hours ; 4 bullocks and 24 sheep were thrown overboard.—From Dec. 3rd to 8th I was at Cambridge, and from the 22nd to 31st at Playford."

Here is a letter to his wife written from Birmingham, containing a note of the progress of the ironwork for the Menai Bridge :

EDGBASTON, BIRMINGHAM,
1847, *Apr.* 22.

Yesterday morning we started between 10 and 11 for Stourbridge, first to see some clay which is celebrated all over the world as the only clay which is fit to make pots for melting glass, &c.   You know that in all these fiery regions, fire-clay is a thing of very great importance, as no furnace will stand if made of any ordinary bricks (and even with the fire-clay, the small furnaces are examined every week), but this Stourbridge clay is as superior to fire-clay as fire-clay is to common brick-earth.   Then we went to Fosters' puddling and rolling works near Stourbridge.   These are on a very large scale: of course much that we saw was a repetition of what we had seen before, but there were slitting mills, machines for rolling the puddled blooms instead of hammering them, &c., and we had the satisfaction of handling the puddling irons ourselves.   Then we went to another work of the Fosters not far from Dudley, where part of the work of the Tube Bridge for the Menai is going on.   The Fosters are, I believe, the largest iron masters in the country, and the two principal partners, the elder Mr Foster and his Nephew, accompanied us in all our inspections and steppings from one set of works to another. The length of Tube Bridge which they have in hand here is only 120 feet, about $\frac{1}{4}$ of the whole length: and at present they are only busy on the bottom part of it: but it is a prodigious thing.   I shall be anxious about it.   Then we went to other works of the Fosters' at King's Wynford, where they have blast furnaces: and here after seeing all other usual things we saw the furnaces tapped.   In this district the Fosters work the 10-yard coal in a way different from any body else: they work out the upper half of its thickness and then leave the ground to fall in: after a year or two this ground becomes so hard as to make a good safe roof, and then they work away the other half: thus they avoid much of the danger and difficulty of working the thick bed all at once.   The ventilation of these mines scarcely ever requires fires, and then only what they call "lamps," those little fire-places which are used for giving light at night.   (In the Northumberland and Durham pits, they constantly have immense roaring fires to make a draught.)   Then we came home through Dudley.

During his stay in Russia, there was a great desire manifested by the astronomers and scientific men of Russia that he should be presented to the Emperor. This would no doubt have taken place had not the movements of the Court, and his own want of time prevented it. The following letter to the British Ambassador, Lord Bloomfield, relates to this matter:

PULKOWA,
1847, *August 25th.*
Wednesday evening.

MY LORD,

   I had the honour yesterday to receive your Lordship's note of Sunday last, which by some irregularity in the communications with this place reached me, I believe, later than it ought. From this circumstance, and also from my being made acquainted only this afternoon with some official arrangements, I am compelled to trouble you at a time which I fear is less convenient than I could have desired.

   The object of my present communication is, to ask whether (if the movements of the Court permit it) it would be agreeable to your Lordship to present me to the Emperor. In explanation of this enquiry, I beg leave to state that this is an honour to which, personally, I could not think of aspiring. My presence however at Pulkowa at this time is in an official character. As Astronomer Royal of England, I have thought it my duty to make myself perfectly acquainted with the Observatory of Pulkowa, and this is the sole object of my journey to Russia. It is understood that the Emperor takes great interest in the reputation of the Observatory, and I am confident that the remarks upon it which I am able to make would be agreeable to him.

   I place these reasons before you, awaiting entirely Your Lordship's decision on the propriety of the step to which I have alluded. I am to leave St Petersburg on Saturday the 4th of September.

I have the honor to be
My Lord,
Your Lordship's very faithful servant,

G. B. AIRY.

Lord Bloomfield, &c., &c.

It was probably in acknowledgment of this letter that in due time he received the following letter with the offer of the Russian Order of St Stanislas :

MONSIEUR L'ASTRONOME ROYAL,

Sa Majesté l'Empereur en appréciant les travaux assidus qui vous ont donné une place distinguée au rang des plus illustres Astronomes de l'Europe, et la coopération bienveillante, que vous n'avez cessé de témoigner aux Astronomes Russes dans les expéditions, dont ils étaient chargés, et en dernier lieu par votre visite à l'Observatoire central de Poulkova, a daigné sur mon rapport, vous nommer Chevalier de la seconde classe de l'Ordre Impérial et Royal de St Stanislas. Je ne manquerai pas de vous faire parvenir par l'entremise de Lord Bloomfield les insignes et la patente de l'ordre.

Veuillez en attendant, Monsieur, recevoir mes sincères félicitations et l'assurance de ma parfaite considération.

Le Ministre de l'instruction publique,

CTE OUVAROFF.

ST PÉTERSBOURG,

ce $\dfrac{24\ Août}{5\ Septbr.}$, 1847.

à Mr G. B. Airy, Esq.,
Astronome Royal de S. M. Britannique à
Greenwich.

Airy provisionally accepted the Order, but wrote at once to Lord John Russell the following letter of enquiry :

ROYAL OBSERVATORY, GREENWICH,
1847, Oct. 15.

MY LORD,

In respect of the office of Astronomer Royal, I refer to the first Lord of the Treasury as Official Patron. In virtue of this relation I have the honour to lay before your Lordship the following statement, and to solicit your instructions thereon.

For conducting with efficiency and with credit to the nation the institution which is entrusted to me, I have judged it proper to cultivate intimate relations with the principal Observatories of

Europe, and in particular with the great Observatory founded by the Emperor of Russia at Pulkowa near St Petersburg. I have several times received Mr Struve, the Director of that Observatory, at Greenwich: and in the past summer I made a journey to St Petersburg for the purpose of seeing the Observatory of Pulkowa.

Since my return from Russia, I have received a communication from Count Ouvaroff, Minister of Public Instruction in the Russian Empire, informing me that the Emperor of Russia desires to confer on me the decoration of Knight Commander in the second rank of the Order of St Stanislas.

And I have the honour now to enquire of your Lordship whether it is permitted to me to accept from the Emperor of Russia this decoration.

I have the honour to be,

My Lord,

Your Lordship's very obedient servant,

G. B. AIRY.

The Rt Honble Lord John Russell,
&c.   &c.   &c.
First Lord of the Treasury.

The answer was as follows:

DOWNING STREET,
October 19, 1847.

SIR,

I am desired by Lord John Russell to acknowledge the receipt of your letter, of the 14th inst. and to transmit to you the enclosed paper respecting Foreign Orders by which you will perceive that it would be contrary to the regulations to grant you the permission you desire.

I am, Sir,

Your obedient servant,

C. A. GREY.

G. B. Airy, Esq.

The passage in the Regulations referred to above is quoted in the following letter to Count Ouvaroff:

ROYAL OBSERVATORY, GREENWICH,
1847, *Oct.* 22.

SIR,

Referring to your Excellency's letter of the $\dfrac{24 \text{ August}}{5 \text{ September}}$, and to my answer of the 25th September, in which I expressed my sense of the high honor conferred on me by His Majesty the Emperor of Russia in offering me, through your Excellency, the Order of St Stanislas, and my pride in accepting it:—I beg leave further to acquaint you that I have thought it necessary to make enquiry of Lord John Russell, First Lord of Her Majesty's Treasury, as to my competency to accept this decoration from His Majesty the Emperor of Russia: and that his Lordship in reply has referred me to the following Regulation of the British Court;

"5th. That no Subject of Her Majesty could be allowed to accept the Insignia of a Foreign Order from any Sovereign of a Foreign State, except they shall be so conferred in consequence of active and distinguished services before the Enemy, either at Sea, or in the Field; or unless he shall have been actually employed in the Service of the Foreign Sovereign."

In consequence of the stringency of this Regulation, it is my duty now to state to your Excellency that I am unable to accept the decoration which His Majesty the Emperor of Russia was pleased, through your Excellency, to offer to me.

I beg leave to repeat the expression of my profound reverence to His Majesty and of my deep sense of the honor which he has done me.

I have the honor to be,

Sir,

Your Excellency's very faithful

and obedient servant,

G. B. AIRY.

To His Excellency
Count Ouvaroff,
&c.   &c.

In the course of the following year a very handsome gold medal, specially struck, was transmitted by Count Ouvaroff on the part of the Emperor of Russia, to Mr Airy.

1848

"In April I received authority to purchase of Simms an 8-inch object-glass for the new Transit Circle for £300. The glass was tested and found satisfactory. While at Playford in January I drew the first plans of the Transit Circle: and C. May sketched some parts. Definite plans were soon sent to Ransomes and May, and to Simms in March. The instrument and the building were proceeded with during the year. The New Transit Circle was to be erected in the Circle Room, and considerable arrangement was necessary for continuing the Circle Observations with the existing instruments, whilst the new instrument was under erection. When the new Transit is completely mounted, the old Transit Instrument may be removed, and the Transit Room will be free for any other purpose. I propose to take it as Private Room for the Astronomer Royal.—On May 12th I made my first proposal of the Reflex Zenith Tube. The principle of it is as follows: Let the micrometer be placed close to the object-glass, the frame of the micrometer being firmly connected with the object-glass cell, and a reflecting eye-piece being used with no material tube passing over the object-glass: and let a basin of quicksilver be placed below the object-glass, but in no mechanical connection with it, at a distance equal to half the focal length of the object-glass. Such an instrument would at least be free from all uncertainties of twist of plumb-line, viscosity of water, attachment of upper plumb-line microscope, attachment of lower plumb-line microscope, and the observations connected with them: and might be expected, as a result of this extreme simplicity, to give accurate results.—A considerable error was discovered in the graduation of Troughton's Circle, amounting in one part to six seconds, which is referred to as follows: 'This instance has strongly confirmed me in an opinion which I have long held—that no independent division is comparable in general accuracy to engine-division, where the fundamental divisions of the engine have been made by Troughton's method, and where in any case the determination by the astronomer of errors of a few divisions will suffice, in consequence of the uniformity of law of error, to give the errors of the intermediate divisions.'—The method of observing with the Altazimuth is carefully described, and the effect of it, in increasing the number of observations of the Moon, is thus given for the thirteen lunations between 1847, May 15, and 1848, May 30. 'Number of days of complete observations with the Meridional Instruments, 111 ; number of days of complete observations with Altitude and Azimuth Instrument, 203. The results of the observations appear very good ; perhaps a little, and but a little, inferior to those of the Meridional Instruments. I consider that the object for which this instrument was erected is successfully attained.'— Being satisfied with the general efficiency of the system arranged by Mr Brooke for our photographic records (of magnetical observations) I wrote to the Admiralty in his favour, and on Aug. 25th the Admiralty ordered the payment of £500 to him. A Committee of the Royal Society also recommended a reward of £250 to Mr Ronalds, which I believe was paid to him.—On May 1st the last revise of the Lunar Reductions was passed, and on May 5th, 500 copies were sent for binding.—In this year Schumacher and I refused a medal to Miss Mitchell for a Comet discovered, because the rules of correspondence had not been strictly followed: the King of Denmark gave one by special favour.— In this year occurred the discovery of Saturn's 8th Satellite by Mr Lassell : upon which I have various correspondence.— On the 18th of December the degree of LL.D. was conferred upon me by the University of Edinburgh.—The Ipswich Lectures: A wish had been expressed that I would give a series of Astronomical Lectures to the people of Ipswich. I therefore arranged with great care the necessary apparatus, and lectured six evenings in a room (I forget its name—it

might be Temperance Hall — high above St Matthew's
Street), from Mar. 13th to the end of the week. A short-
hand writer took them down : and these formed the ' Ipswich
Lectures,' which were afterwards published by the Ipswich
Museum (for whose benefit the lectures were given) and by
myself, in several editions, and afterwards by Messrs Macmillan
in repeated editions under the title of ' Airy's Popular Astronomy.'—It had been found necessary to include under
one body all the unconnected Commissions of Sewers for
the Metropolis, and Lord Morpeth requested me to be a
member. Its operations began on Oct. 28th. In constitution it was the most foolish that I ever knew : consisting of,
I think, some 200 persons, who could not possibly attend to
it. It came to an end in the next year."

Of private history : " I was at Playford from Jan. 1st to
11th, and again from Jan. 17th to 25th: also at Playford from
June 21st to July 12th.—From Aug. 23rd to Sept. 12th I was
in Ireland on a visit to Lord Rosse at Parsonstown, chiefly
engaged on trials of his large telescope. I returned by Liverpool, where I inspected the Liverpool Equatoreal and Clockwork, and examined Mr Lassell's telescopes and grinding
apparatus.—From Dec. 6th to 20th I was at Edinburgh with
my wife, on a visit to Prof. J. D. Forbes. We made various
excursions, and I attended lectures by Prof. Wilson and Sir
W. Hamilton : on the 18th I gave a lecture in Prof. Forbes's
room. I received the Honorary Degree of LL.D., and made
a statement on the Telescopes of Lord Rosse and Mr Lassell
to the Royal Society of Edinburgh. Returned to Greenwich
by Brampton."

Here is a reminiscence of the "Ipswich Lectures," in a
letter to his wife, dated Playford, 1848 Mar. 14, " At the
proper time I went to the hall : found a chairman installed
(Mr Western) : was presented to him, and by him presented
to the audience : made my bow and commenced. The room

was quite full: I have rarely seen such a sea of faces; about 700 I believe.   Everything went off extremely well, except that the rollers of the moving piece of sky would squeak: but people did not mind it: and when first a star passed the meridian, then Jupiter, then some stars, and then Saturn, he was much applauded.   Before beginning I gave notice that I should wait to answer questions: and as soon as the lecture was finished the Chairman repeated this and begged people to ask.   So several people did ask very pertinent questions (from the benches) shewing that they had attended well. Others came up and asked questions."

The following extracts are from letters written to his wife while on his visit to Lord Rosse at Parsonstown in Ireland.   On the way he stopped at Bangor and looked at the Tubular Bridge Works, which are thus referred to: " Stopped at Bangor, settled *pro tem.* at the Castle, and then walked past the Suspension Bridge towards the Tube Works, which are about 1½ mile south-west of the Suspension Bridge. The way was by a path through fields near the water side: and from one or two points in this, the appearance of the Suspension Bridge was most majestic.   The Tube Bridge consists of four spans, two over water and two over sloping land.   The parts for the double tube over the water spans (four lengths of tube) are building on a platform as at Conway, to be floated by barges as there: the parts over the sloping banks are to be built in their place, on an immense scaffolding.   I suspect that, in regard to these parts, Stephenson is sacrificing a great deal of money to uniformity of plan: and that it would have been much cheaper to build out stone arches to the piers touching the water. ...... The Tube Works are evidently the grand promenade of the idlers about Bangor: I saw many scores of ladies and gentlemen walking that way with their baskets of provision, evidently going to gipsy in the fields close by."

THE CASTLE, PARSONSTOWN,
1848, *Aug.* 29.

After tea it was voted that the night was likely to be fine, so we all turned out. The night was uncertain : sometimes entirely clouded, sometimes partially, but objects were pretty well seen when the sky was clear : the latter part was much steadier.   From the interruption by clouds, the slowness of finding with and managing a large instrument (especially as their finding apparatus is not perfectly arranged) and the desire of looking well at an object when we had got it, we did not look at many objects.   The principal were, Saturn and the Annular Nebula of Lyra with the 3-feet; Saturn, a remarkable cluster of stars, and a remarkable planetary nebula, with the 6-feet. With the large telescope, the evidence of the quantity of light is prodigious. And the light of an object is seen in the field without any colour or any spreading of stray light : and it is easy to see that the vision with a reflecting telescope may be much more perfect than with a refractor.   With these large apertures, the rings round the stars are insensible.   The planetary nebula looked a mass of living and intensely brilliant light : this is an object which I do not suppose can be seen at all in our ordinary telescopes.   The definition of the stars near the zenith is extremely good : with a high power (as 800) they are points or very nearly so—indeed I believe quite so—so that it is clear that the whole light from the great 6-feet mirror is collected into a space not bigger than the point of a needle.   But in other positions of the telescope the definition is not good : and we must look to-day to see what is the cause of this fault.   It is not a fault in the telescope, properly so-called, but it is either a tilt of the mirror, or an edge-pressure upon the mirror when the telescope points lower down which distorts its figure, or something of that kind.   So I could not see Saturn at all well, for which I was sorry, as I could so well have compared his appearance with what I have seen before.   I shall be very much pleased if we can make out what is the fault of adjustment, and so correct it as to get good images everywhere.   It is evident that the figuring of the mirror, the polishing, and the general arrangement, are perfectly managed.

THE CASTLE, PARSONSTOWN,
1848, *Aug.* 30.

Yesterday we were employed entirely about the Great Telescope, beginning rather late.   The principal objects had relation to the fault of definition when the telescope is pointed low (which I had remarked on the preceding night), and were, to make ourselves acquainted with the mechanism of the mirror's mounting generally, and to measure in various ways whether the mirror actually does shift its place when the telescope is set to different angles of elevation. For the latter we found that the mirror actually does tilt $\frac{1}{4}$ of an inch when the tube points low.   This of itself will not account for the fault but it indicates that the lower part is held fast in a way that may cause a strain which would produce the fault.   These operations and reasonings took a good deal of time.   Lord Rosse is disposed to make an alteration in the mounting for the purpose of correcting this possible strain.

THE CASTLE, PARSONSTOWN,
1848, *Aug.* 31.

The weather here is still vexatious : but not absolutely repulsive. Yesterday morning Lord Rosse arranged a new method of suspending the great mirror, so as to take its edgewise pressure in a manner that allowed the springy supports of its flat back to act.   This employed his workmen all day, so that the proposed finish of polishing the new mirror could not go on.   I took one Camera Lucida sketch of the instrument in the morning, dodging the heavy showers as well as I could; then, as the afternoon was extremely fine, I took another, with my head almost roasted by the sun.   This last view is extremely pretty and characteristic, embracing parts of the mounting not shewn well in the others, and also shewing the Castle, the Observatory, and the 3-feet telescope.   The night promised exceedingly well : but when we got actually to the telescope it began to cloud and at length became hopeless.   However I saw that the fault which I had remarked on the two preceding nights was gone.   There is now a slight exhibition of another fault to a much smaller extent. We shall probably be looking at the telescope to-day in reference to it.

THE CASTLE, PARSONSTOWN,
1848, *Sept.* 1.

Yesterday we made some alterations in the mounting of the great mirror. We found that sundry levers were loose which ought to be firm, and we conjectured with great probability the cause of this, for correction of which a change in other parts was necessary. The mirror was then found to preserve its position much more fixedly than before. . . . At night, upon trying the telescope, we found it very faulty for stars near the zenith, where it had been free from fault before. The screws which we had driven hard were then loosened, and immediately it was made very good. Then we tried with some lower objects, and it was good, almost equally good, there. For Saturn it was very greatly superior to what it had been before. Still it is not satisfactory to us, and at this time a strong chain is in preparation, to support the mirror edgeways instead of the posts that there were at first or the iron hoop which we had on it yesterday.

Nobody would have conceived that an edgewise gripe of such a mass of metal could derange its form in this way.

Last night was the finest night we have had as regards clouds, though perhaps not the best for definition of objects.

THE CASTLE, PARSONSTOWN,
1848, *Sept.* 2.

I cannot learn that the fault in the mirror had been noticed before, but I fancy that the observations had been very much confined to the Zenith and its neighbourhood.

1849

" In July the new constant-service water-pipes to the Observatory were laid from Blackheath. Before this time the supply of water to the Observatory had been made by a pipe leading up from the lower part of the Park, and was not constant.—In May the new staircase from my dwelling-house to the Octagon Room was commenced.—In the Report to the Visitors there is a curious account of Mr Breen's (one of the Assistants) personal equation, which was found to be different in quantity for observations of the Moon and observations of the Stars.—The most important set of observations (of planets) was a series of measures of Saturn in four directions, at the time when his ring had disappeared. They appear completely to negative the idea that Saturn's form differs sensibly from an ellipsoid.—Among the General Remarks of the Report the following appears: 'Another change (in prospect) will depend on the use of galvanism; and as a probable instance of the application of this agent, I may mention that, although no positive step has hitherto been taken, I fully expect in no long time to make the going of all the clocks in the Observatory depend on one original regulator. The same means will probably be employed to increase the general utility of the Observatory, by the extensive dissemination throughout the kingdom of accurate time-signals, moved by an original clock at the Royal Observatory; and I have already entered into correspondence with the authorities of the South Eastern Railway (whose line of galvanic communication will shortly pass within nine furlongs of the Observatory) in reference to this subject.'—I agreed with Schumacher in giving no medal to Mr G. P. Bond; his comet was found to be Petersen's. Five medals were awarded for comets in 1847 (Hind, Colla, Mauvais, Brorsen, Schweizer). —The Liverpool Observatory was finished this year: and the thanks of the Town Council were presented to me.—Respecting Fallows's Observations at the Cape of Good Hope: I had received the Admiralty sanction for proceeding with calculations in 1846, and I employed computers as was convenient. On July 20th of this year I was ready with final results, and began to make enquiries about Fallows's personal history, and the early history of the Cape Observatory. On Oct. 23rd I applied for sanction for printing, which was given, and the work was soon finished off, in the Astronomical Society's Memoirs.—In the month of March I had commenced correspondence with various persons on the imperfect

state of publication of the British Survey. Sheets of the
Map were issued by scores, but not one of them had an indication of latitude or longitude engraved. I knew that great
pains had been taken in giving to the principal triangulation
a degree of accuracy never before reached, and in fixing the
astronomical latitudes of many stations with unequalled pre-
cision. Finally I prepared for the Council of the Royal
Society a very strong representation on these subjects, which
was adopted and presented to the Government. It was
entirely successful, and the Maps were in future furnished
with latitude and longitude lines.—I was elected President
of the Royal Astronomical Society on Feb. 9th.—In June I
went with Sheepshanks to see some of the operation of
measuring a Bàse on Salisbury Plain. The following extract
from a letter to his wife dated 1849, June 27th, relates to this
expedition : ' In the morning we started before eight in an
open carriage to the Plain : looking into Old Sarum on our
way. The Base is measured on what I should think a most
unfavourable line, its north end (from which they have begun
now, in verification of the old measure) being the very highest
point in the whole plain, called Beacon Hill. The soldiers
measure only 252 feet in a day, so it will take them a good
while to measure the whole seven miles. While we were
there Col. Hall (Colby's successor) and Yolland and Gosset
came.' "

Of private history : " I made short visits to Playford in
January, April and July. From July 28th to Sept. 12th I
made an expedition with my wife to Orkney and Shetland.—
From Dec. 24th to 26th I was at Hawkhurst, on a visit to
Sir John Herschel."

1850

" The Report to the Board of Visitors opens with the
following paragraph : ' In recording the proceedings at the
Royal Observatory during the last year, I have less of novelty

to communicate to the Visitors than in the Reports of several years past. Still I trust that the present Report will not be uninteresting; as exhibiting, I hope, a steady and vigorous adherence to a general plan long since matured, accompanied with a reasonable watchfulness for the introduction of new instruments and new methods when they may seem desirable.' —Since the introduction of the self-registering instruments a good many experiments had been made to obtain the most suitable light, and the Report states that 'No change whatever has been made in these instruments, except by the introduction of the light of coal-gas charged with the vapour of coal-naptha, for photographic self-registration both of the magnetic and of the meteorological instruments....The chemical treatment of the paper is now so well understood by the Assistants that a failure is almost unknown. And, generally speaking, the photographs are most beautiful, and give conceptions of the continual disturbances in terrestrial magnetism which it would be impossible to acquire from eye-observation.'—Amongst the General Remarks of the Report it is stated that 'There are two points which have distinctly engaged my attention. The first of these is, the introduction of the American method of observing transits, by completing a galvanic circuit by means of a touch of the finger at the instant of appulse of the transiting body to the wire of the instrument, which circuit will then animate a magnet that will make an impression upon a moving paper. After careful consideration of this method, I am inclined to believe that, in Prof. Mitchell's form, it does possess the advantages which have been ascribed to it, and that it may possess peculiar advantages in this Observatory, where the time-connection of transits made with two different instruments (the Transit and the Altazimuth) is of the highest importance....The second point is, the connection of the Observatory with the galvanic telegraph of the South Eastern Railway, and with other lines of galvanic wire with which that telegraph communicates. I had formerly in mind only the connection of this Observatory with different parts of the great British island: but I now think it possible that our communications may be extended far beyond its shores. The promoters of the submarine telegraph are very confident of the practicability of completing a galvanic connection between England and France: and I now begin to think it more than possible that, within a few years, observations at Paris and Brussels may be registered on the recording surfaces at Greenwich, and vice versâ.'—Prof. Hansen was engaged in forming Lunar Tables from his Lunar Theory, but was stopped for want of money. On Mar. 7th I represented this privately to Mr Baring, First Lord of the Admiralty; and on Mar. 30th I wrote officially to the Admiralty, soliciting £150 with the prospect, if necessary, of making it £200. On Apr. 10th the Admiralty gave their assent. The existence of Hansen's Lunar Tables is due to this grant.—The King of Denmark's Medal for Comets was discontinued, owing to the difficulties produced by the hostility of Prussia.—On Aug. 1st I gave to the Treasury my opinion on the first proposal for a large reflector in Australia: it was not strongly favourable.—In August, being (with my wife and Otto Struve) on a visit to Lady Breadalbane at Taymouth Castle, I examined the mountain Schehallien.—As in other years, I reported on several Papers for the Royal Society, and took part in various business for them.—In the Royal Astronomical Society I had much official business, as President.—In March I communicated to the Athenæum my views on the Exodus of the Israelites: this brought me into correspondence with Miss Corbaux, Robert Stephenson, Capt. Vetch, and Prof. J. D. Forbes.—In December I went to the London Custom House, to see Sir T. Freemantle (Chairman of Customs), and to see how far decimal subdivisions were used in the Custom House."

Of private history: "From Mar. 19th to 22nd I was on an expedition to Folkestone, Dover, Dungeness, &c.—From Apr. 3rd to 8th at Playford, and again for short periods in June and July.—From Aug. 1st to Sept. 5th I was travelling

in Scotland with my wife and Otto Struve (for part of the
time).   At Edinburgh I attended the Meeting of the British
Association, and spoke a little in Section A.   I was nominated President for 1851 at Ipswich.   We travelled to Cape
Wrath and returned by Inverness and the Caledonian Canal.
—I was at Playford for a short time in October and December."

1851

" In this year the great shed was built (first erected on the
Magnetic Ground, and about the year 1868 transferred to the
South Ground).—The chronometers were taken from the old
Chronometer Room (a room on the upper story fronting the
south, now, 1872, called Library 2) and were put in the room
above the Computing Room (where they remained for 10 or
12 years, I think) : it had a chronometer-oven with gas-heat,
erected in 1850.—The following passage is quoted from the
Report to the Visitors :—' As regards Meridional Astronomy
our equipment may now be considered complete.   As I have
stated above, an improvement might yet be made in our
Transit Circle ; nevertheless I do not hesitate to express my
belief that no other existing meridional instrument can be
compared with it.   This presumed excellence has not been
obtained without much thought on my part and much anxiety
on the part of the constructors of the instrument (Messrs
Ransomes and May, and Mr Simms).   But it would be very
unjust to omit the further statement that the expense of the
construction has considerably exceeded the original estimate,
and that this excess has been most liberally defrayed by the
Government.'—In December Sir John Herschel gave his
opinion (to the Admiralty, I believe) in favour of procuring
for the Cape Observatory a Transit Circle similar to that at
Greenwich.—I had much correspondence about sending Pierce
Morton (formerly a pupil of mine at Cambridge, a clever
gentlemanly man, and a high wrangler, but somewhat flighty)

as Magnetic Assistant to the Cape Observatory : he was with me from May to October, and arrived at the Cape on Nov. 27th.—I was much engaged with the clock with conical motion of pendulum, for uniform movement of the Chronographic Barrel.—Regarding galvanic communications : On Sept. 19th I had prepared a Draft of Agreement with the South Eastern Railway Company, to which they agreed. In November I wrote to Sir T. Baring (First Lord of the Admiralty) and to the Admiralty for sanction, which was given on Dec. 18th. In December I had various communications about laying wires through the Park, &c., &c., and correspondence about the possibility of using sympathetic clocks : in June, apparently, I had seen Shepherd's sympathetic clock at the Great Exhibition, and had seen the system of sympathetic clocks at Pawson's, St Paul's Churchyard.—In the last quarter of this year I was engaged in a series of calculations of chronological eclipses. On Sept. 30th Mr Bosanquet wrote to me about the Eclipse of Thales, and I urged on the computations related to it, through Mr Breen. In October the eclipse of Agathocles (the critical eclipse for the motion of the Moon's node) was going on. In October Hansteen referred me to the darkness at Stiklastad.—I went to Sweden to observe the total eclipse of July 28th, having received assistance from the Admiralty for the journeys of myself, Mr Dunkin, Mr Humphreys and his friend, and Capt. Blackwood. I had prepared a map of its track, in which an important error of the *Berliner Jahrbuch* (arising from neglect of the earth's oblateness) was corrected. I gave a lecture at the Royal Institution, in preparation for the eclipse, and drew up suggestions for observations, and I prepared a scheme of observations for Greenwich, but the weather was bad. The official account of the Observations of the Eclipse, with diagrams and conclusions, is given in full in a paper published in the Royal Astr. Society's Memoirs.—This year I was President of the British Association, at the Ipswich Meeting : it necessarily produced a great deal of business. I lectured one evening on the coming eclipse. Prince Albert was present, as guest of Sir William Middleton : I was engaged to meet him at dinner, but when I found that the dinner day was one of the principal soirée days, I broke off the engagement.—On May 26th I had the first letter from E. Hamilton (whom I had known at Cambridge) regarding the selection of professors for the University of Sydney. Herschel, Maldon, and H. Denison were named as my coadjutors. Plenty of work was done, but it was not finished till 1852.—In connection with the clock for Westminster Palace, in February there were considerations about providing other clocks for the various buildings; and this probably was one reason for my examining Shepherd's Clocks at the Great Exhibition and at Pawson's. In November I first proposed that Mr E. B. Denison should be associated with me. About the end of the year, the plan of the tower was supplied to me, with reference to the suspension of the weights and other particulars.—In 1850 Admiral Dundas (M.P. for Greenwich and one of the Board of Admiralty) had requested me to aid the Trustees of the Dee Navigation against an attack; and on Mar. 19th 1851 I went to Chester to see the state of the river. On Jan. 1st 1852 I went to give evidence at the Official Enquiry.—At a discussion on the construction of the Great Exhibition building in the Institution of Civil Engineers, I expressed myself strongly on the faulty principles of its construction.—In this year I wrote my first Paper on the landing of Julius Cæsar in Britain, and was engaged in investigations of the geography, tides, sands, &c., relating to the subject."

Of private history : " I was several times at Playford during January, and went there again on Dec. 23rd.—In this year a very heavy misfortune fell on us. My daughter, Elizabeth, had been on a visit to Lady Herschel at Hawkhurst, and on Apr. 2nd Sir J. Herschel wrote to me, saying that she was so well in health. She returned a few days later, and from her appearance I was sure that she was suffering under deadly disease. After some time, an able physician was consulted, who at once pronounced it to be pulmonary. A sea voyage was thought desirable, and my wife took her to Shetland, where there was again a kind welcome from Mr Edmonston. But this, and the care taken on her return, availed nothing: and it was determined to take her to Madeira. My wife and daughter sailed in the brig 'Eclipse' from Southampton on Dec. 11th. The termination came in 1852.—On Nov. 23rd I went to Bradfield, near Bury: my uncle, George Biddell, died, and I attended the funeral on Nov. 29th.—From July 18th to Aug. 24th I was in Sweden for the Observation of the Eclipse, and returned through Holland.—In October I was about a week at Ventnor and Torquay, and from Dec. 7th to 11th at Southampton, on matters connected with my daughter's illness."

The following extracts are from letters to his wife, relating to the Observation of the eclipse, his interview with the King of Sweden, &c., and his visit to the pumping engines at Haarlem:

July 28, *half-past* 10, *morning.*

The weather is at present most perfectly doubtful. Nearly the whole sky is closely covered, yet there is now and then a momentary gleam of sun. The chances are greatly against much of the eclipse being seen. All is arranged to carry off the telescope, &c., at 11: they can be carted to the foot of the hill, and we have made out a walking-pass then to the top. We are to dine with Mr Dickson afterwards.

July 28, 10 *at night.*

Well we have had a glorious day. As soon as we started, the weather began to look better. We went up the hill and planted my telescope, and the sky shewed a large proportion of blue. At first I placed the telescope on the highest rock, but the wind blew almost a gale, and shook it slightly: so I descended about 8 feet to one side. (The power of doing this was one of the elements in my choice of this station, which made me prefer it to the high hill beyond the river.) The view of scenery was inexpressibly beautiful.

The beginning of eclipse was well seen.   The sky gradually thickened from that time, so that the sun was in whitish cloud at the totality, and barely visible in dense cloud at the end of the eclipse.   The progress of the eclipse brought on the wonderful changes that you know : just before the totality I saw a large piece of blue sky become pitch black ; the horror of totality was very great ; and then flashed into existence (I do not know how) a broad irregular corona with red flames *instantly seen* of the most fantastic kind.   The darkness was such that my assistant had very great trouble in reading his box chronometer.   (A free-hand explanatory diagram is here given.) Some important points are made out from this.   1st the red flames certainly belong to the sun.   2nd they certainly are in some instances detached.   3rd they are sometimes quite crooked.   4th they seem to be connected with spots.   The corona was brilliant white.   One star brilliant : I believe Venus.   I had no time to make observations of polarization, &c., although prepared.   When the totality was more than half over I looked to N. and N.W., and in these regions there was the fullest rosy day-break light.   After the sun-light reappeared, the black shadow went travelling away to the S.E. exactly like the thunder-storm from the Main.   The day then grew worse, and we came home here (after dinner) in pouring rain.

STOCKHOLM,
1851, Aug. 5.

I then by appointment with Sir Edmund Lyons went with him to the Minister for Foreign Affairs, Baron Stjerneld, who received me most civilly.   My business was to thank him for the orders which had been given to facilitate the landing of our telescopes, &c., &c. He was quite familiar with the names of my party, Humphreys Milaud, &c., so that I trust they have been well received (I have had no letter).   He intimated, I suppose at Sir E. Lyons's suggestion, that perhaps King Oscar might wish to see me, but that it would not be on Tuesday.   So I replied that I was infinitely flattered and he said that he would send a message to Sir E. Lyons by Tuesday evening.   Now all this put me in a quandary : because I wanted to see Upsala, 47 miles off : and the steamboats on the Mälar only go in the morning and return in the morning : and this was irreconcile-able with waiting for his Majesty's appointment which might be for Wednesday morning.   So after consultation Sir E. Lyons put me in the hands of a sort of courier attached to the Embassy, and he procured a calèche, and I posted to Upsala yesterday afternoon (knocking the people up at 11 at night) and posted back this afternoon. And sure enough a message has come that the king expects me at 11 to-morrow morning. Posting of course is much dearer than steam-boat travelling, but it is cheap in comparison with England: two horses cost 1s. for nearly 7 miles. At Upsala there is a very good old cathedral, I suppose the only one in Sweden: and many things about the University which interested me. I sent my card to Professor Fries, and he entirely devoted himself to me: but imagine our conversation—he spoke in *Latin* and I in French: however we understood each other very well. It is on the whole a dreary country except where enlivened by lakes: some parts are pine forests and birch forests, but others are featureless ground with boulder stones, like the worst part of the Highlands.

August 6, *Wednesday*, 3 *o'clock.*

I rigged myself in black trowsers and white waistcoat and neck-cloth this morning. Sir Edmund Lyons called. Baron Wrede called on me: he had observed the Eclipse at Calmar and brought his drawing, much like mine. He conducted me to the Palace. The Minister for Foreign Affairs came to me. In the waiting-room I was introduced to the Lieutenant-Governor of Christianstad, who had had the charge of Humphreys and Milaud. He had placed a *guard of soldiers* round them while they were observing. They saw the eclipse well. Captain Blackwood went to Helsingborg instead of Bornholm, and saw well. I am sorry to hear that it was cloudy at Christiania, Mr Dunkin's station. I heard some days ago that Hind had lost his telescope, but I now heard a very different story: that he landed at Ystad, and found a very bad hotel there: that he learnt from Murray that the hotels at Carlscrona (or wherever he meant to go) were much worse; and so he grew faint at heart and turned back. I was summoned in to the King and presented by the Minister (Stjerneld), and had a long conversation with him: on the eclipse, the arc of meridian, the languages, and the Universities. We spoke in French. Then Baron Wrede went with me to the Rittershus (House of Lords or Nobles) in Session, and to the Gallery of Scandinavian Antiquities, which is very remarkable: the collection of stone axes and chisels, bronze do., iron do., ornaments, &c. is quite amazing. I was struck with seeing specimens from a very distant age of the Maid of Norway's brooch : the use of which I explained to the Director.

I dined and drove out with Sir E. Lyons, and called at the houses of the Baron Stjerneld and of the Norwegian Minister Baron Duë, and had tea at the latter. Most of these people speak English well, and they seem to live in a very domestic family style. I should soon be quite at home here : for I perceive that my reception at Court, &c., make people think that I am a very proper sort of person.

The extract concerning his visit to the Pumping-Engines at Haarlem is as follows :

LEYDEN,
1851, August 20, Wednesday.

I went to see the great North Holland Canal, and went a mile or two in a horse-drawn-boat upon it : a very comfortable conveyance. Saw windmills used for sawing timber and other purposes, as well as some for grinding and many for draining. Yesterday at half-past one I went by railway to Haarlem. I did not look at anything in the town except going through it and seeing that it is a curious fantastic place, but I drove at once to the burgomaster to ask permission to visit one of the three great pumping engines for draining the immense Haarlem lake, and then drove to it. Imagine a round tower with a steam-cylinder in its center ; and the piston which works up-and-down, instead of working one great beam as they usually do, works *eight*, poking out on different sides of the round tower, and each driving a pump 6 feet in diameter. I am glad to have seen it. Then by railway here.

1852

" Galvanic communication was now established with Lewisham station (thus giving power of communicating with London, Deal, &c.).—From the Report to the Board of Visitors it appears that, in the case of the Transit Circle, the azimuth of the Instrument as determined by opposite

passages of the Pole Star had varied four seconds; and in
the case of the Altazimuth, there was a discordance in the
azimuthal zeros of the Instrument, as determined from observations of stars.   In both cases it was concluded that the
discordances arose from small movements of the ground.—
Under the head of 'General Remarks' in the Report, the
following paragraph occurs: 'It will be perceived that the
number of equatoreal observations made here at present is
small: and that they are rarely directed to new comets and
similar objects which sometimes excite considerable interest.
This omission is intentional.   It is not because the instrumental means are wanting (for our Equatoreals, though not
comparable to those of either Cambridge, or of Pulkowa, are
fully equal to those usually directed to such objects), but it
is because these observations are most abundantly supplied
from other observatories, public and private, and because the
gain to those observations from our taking a part in them
would, probably, be far less than the loss to the important
class of observations which we can otherwise follow so well.
Moreover, I am unwilling to take any step which could be
interpreted as attempting to deprive the local and private
observatories of honours which they have so nobly earned.
And, finally, in this act of abstinence, I am desirous of giving
an example of adhesion to one principle which, I am confident, might be extensively followed with great advantage
to astronomy:—the principle of division of labour.'—Discoveries of small planets were now not infrequent: but the
only one of interest to me is Melpomene, for the following
reason.   On 1852 June 24 I lost my most dear, amiable,
clever daughter Elizabeth: she died at Southampton, two
days after landing from Madeira.   On that evening Mr Hind
discovered the planet; and he requested me to give a name.
I remembered Horace's 'Præcipe lugùbres cantus, Melpomene,' and Cowley's 'I called the buskin'd muse Melpomene and told her what sad story I would write,' and
suggested Melpomene, or Penthos: Melpomene was adopted.

—The first move about the Deal Time Ball was in a letter from Commander Baldock to the Admiralty, suggesting that a Time Ball, dropped by galvanic current from Greenwich, should be attached to one of the South Foreland Lighthouses. The Admiralty sent this for my Report.   I went to the place, and I suggested in reply (Nov. 15th) that a better place would be at an old signal station on the chalk downs.   The decisive change from this was made in 1853.—As the result of my examination and enquiries into the subject of sympathetic clocks, I established 8 sympathetic clocks in the Royal Observatory, one of which outside the entrance gate had a large dial with Shepherd's name as Patentee.   Exception was taken to this by the solicitor of a Mr Bain who had busied himself about galvanic clocks.   After much correspondence I agreed to remove Shepherd's name till Bain had legally established his claim.   This however was never done: and in 1853 Shepherd's name was restored.—In Nov. 1851, Denison had consented to join me in the preparation of the Westminster Clock.   In Feb. 1852 we began to have little disagreements.   However on Apr. 6th I was going to Madeira, and requested him to act with full powers from me. —I communicated to the Royal Society my Paper on the Eclipses of Agathocles, Thales, and Xerxes.—In the British Association, I had presided at the Ipswich Meeting in 1851, and according to custom I ought to attend at the 1852 Meeting (held at Belfast) to resign my office.   But I was broken in spirit by the death of my daughter, and the thing generally was beyond my willing enterprise.   I requested Sir Roderick Murchison to act generally for me: which he did, as I understood, very gracᵔfully.—In this year a proposal was made by the Government for shifting all the Meeting Rooms of the Scientific Societies to Kensington Gore, which was stoutly resisted by all, and was finally abandoned."

Of private history: "I was at Playford in January, and went thence to Chester on the enquiry about the tides of the Dee; and made excursions to Halton Castle and to

Holyhead.—From Apr. 8th to May 14th I was on the voyage to and from Madeira, and on a short visit to my wife and daughter there.—On June 23rd I went to Southampton to meet my wife and daughter just landed from Madeira: on June 24th my dear daughter Elizabeth died: she was buried at Playford on June 29th.—I was at Playford also in July and December.—From Sept. 16th to 24th I went to Cumberland, viâ Fleetwood and Peel."

1853

"On May 3rd 1853 I issued an address to the individual Members of the Board of Visitors, proposing the extension of the Lunar Reductions from 1830. From this it appears that 'Through the whole period (from 1830 to 1853), the places of the Moon, deduced from the observations, are compared with the places computed in the Nautical Almanac: that is, with Burckhardt's tables, which have been used for many years in computing the places of the Nautical Almanac.......Very lately, however, Mr Adams has shewn that Burckhardt's Parallax is erroneous in formula and is numerically incorrect, sometimes to the amount of seven seconds. In consequence of this, every reduction of the Observations of the Moon, from 1830 to the present time, is sensibly erroneous. And the error is of such a nature that it is not easy, in general, to introduce its correction by any simple process....The number of observations to the end of 1851 (after which time the parallax will be corrected in the current reductions) is about 2560. An expense approaching to £400 might be incurred in their reduction.' Subsequently I made application to the Admiralty, and the £400 was granted on Dec. 12th.—In the Report to the Visitors it is stated that with regard to the Transit Circle, changes are under contemplation in its reflection-apparatus: one of these changes relates to the material of the trough. 'Several years ago, when I was at Hamburgh, my revered friend Prof. Schumacher exhibited to

me the pacifying effect of a copper dish whose surface had
been previously amalgamated with quicksilver.......The Rev.
Charles Pritchard has lately given much attention to this
curious property of the metals, and has brought the practical
operation of amalgamation to great perfection. Still it is
not without difficulty, on account of a singular crystallization
of the amalgam.'—With regard to the Chronograph, the
Report states: 'The Barrel Apparatus for the American
method of observing transits is not yet brought into use.....I
have, however, brought it to such a state that I am beginning
to try whether the Barrel moves with sufficient uniformity to
be itself used as the Transit Clock. This, if perfectly secured, would be a very great convenience, but I am not very
sanguine on that point.'—A change had been made in the
Electrometer-apparatus: 'A wire for the collection of atmospheric electricity is now stretched from a chimney on the
north-west angle of the leads of the Octagon Room to the
Electrometer pole....There appears to be no doubt that a
greater amount of electricity is collected by this apparatus
than by that formerly in use.'—As regards the Magnetical
Observations : 'The Visitors at their last Meeting, expressed
a wish that some attempt should be made to proceed further
in the reduction or digest of the magnetical results, if any
satisfactory plan could be devised. I cannot say that I have
yet satisfied myself on the propriety of any special plan
that I have examined....I must, however, confess that, in
viewing the capricious forms of the photographic curves, my
mind is entirely bewildered, and I sometimes doubt the
possibility of extracting from them anything whatever which
can be considered trustworthy.'—Great progress had been
made with the distribution of time. 'The same Normal
Clock maintains in sympathetic movement the large clock at
the entrance gate, two other clocks in the Observatory, and a
clock at the London Bridge Terminus of the South-Eastern
Railway....It sends galvanic signals every day along all the
principal railways diverging from London. It drops the

Greenwich Ball, and the Ball on the Offices of the Electric
Telegraph Company in the Strand;...All these various effects
are produced without sensible error of time; and I cannot
but feel a satisfaction in thinking that the Royal Observatory
is thus quietly contributing to the punctuality of business
through a large portion of this busy country. I have the
satisfaction of stating to the Visitors that the Lords Commissioners of the Admiralty have decided on the erection of
a Time-Signal Ball at Deal, for the use of the shipping in the
Downs, to be dropped every day by a galvanic current from
the Royal Observatory. The construction of the apparatus
is entrusted to me. Probably there is no roadstead in the
world in which the knowledge of true time is so important.'
—The Report includes an account of the determination of
the Longitude of Cambridge Observatory by means of galvanic signals, which appear to have been perfectly successful.
—Under the head of General Remarks the following passage
appears: 'The system of combining the labour of unattached
computers with that of attached Assistants tends materially
to strengthen our powers in everything relating to computation. We find also, among the young persons who are
engaged merely to serve as computers, a most laudable
ambition to distinguish themselves as observers; and thus
we are always prepared to undertake any observations which
may be required, although necessarily by an expenditure of
strength which would usually be employed on some other
work.'—Considerable work was undertaken in preparing a
new set of maps of our buildings and grounds.—On Apr. 23rd
there was a small fire in the magnetic observatory, which did
little mischief.—In December I wrote my description of the
Transit Circle.—Lieut. Stratford, the Editor of the Nautical
Almanac, died, and there was some competition for the office.
I was willing to take it at a low rate, for the addition to my
salary: Mr Main—and I think Mr Glaisher—were desirous
of exchanging to it: Prof. Adams was anxious for it. The
Admiralty made the excellent choice of Mr Hind.—In

October Faraday and I, at Lothbury, witnessed some remarkable experiments by Mr Latimer Clark on a galvanic current carried four times to and from Manchester by subterranean wires (more than 2000 miles) shewing the retardation of visible currents (at their maximum effect) and the concentration of active power. I made investigations of the velocity of the Galvanic Current.—I was engaged on the preliminary enquiries and arrangements for the Deal Time Ball.—With respect to the Westminster Clock; an angry paper was issued by Mr Vulliamy. In October I expostulated with Denison about his conduct towards Sir Charles Barry: on November 7th I resigned.—On Feb. 11th I was elected President of the Royal Astronomical Society.—In the Royal Institution I lectured on the Ancient Eclipses.—On Dec. 15th I was elected to the Academy of Brussels.—After preliminary correspondence with Sir W. Molesworth (First Commissioner of Works, &c.) and Sir Charles Barry (Architect of the Westminster Palace), I wrote, on May 14th, to Mr Gladstone about depositing the four Parliamentary Copies of Standards, at the Royal Observatory, the Royal Mint, the Royal Society, and within a wall of Westminster Palace. Mr Gladstone assented on June 23rd.—On Mar. 26th I wrote to Mr Gladstone, proposing to take advantage of the new copper coinage for introducing the decimal system. I was always strenuous about preserving the Pound Sterling. On May 10th I attended the Committee of the House of Commons on decimal coinage: and in May and September I wrote letters to the Athenæum on decimal coinage.—I had always something on hand about Tides. A special subject now was, the cry about intercepting the tidal waters of the Tyne by the formation of the Jarrow Docks, in Jarrow Slake; which fear I considered to be ridiculous."

Of private history: "From Jan. 15th to 24th I was at Playford.—On Mar. 4th I went to Dover to try time-signals. —From June 24th to Aug. 6th I was at Little Braithwaite near Keswick, where I had hired a house, and made expedi-

tions with members of my family in all directions. On July
28th I went, with my son Wilfrid, by Workington and Mary-
port to Rose Castle, the residence of Bishop Percy (the
Bishop of Carlisle), and on to Carlisle and Newcastle, looking
at various works, mines, &c.—On Dec. 24th I went to
Playford."

1854

The chronograph Barrel-Apparatus for the American
method of transits had been practically brought into use:
" I have only to add that this apparatus is now generally
efficient. It is troublesome in use ; consuming much time in
the galvanic preparations, the preparation of the paper, and
the translation of the puncture-indications into figures. But
among the observers who use it there is but one opinion on
its astronomical merits—that, in freedom from personal
equation and in general accuracy, it is very far superior to
the observations by eye and ear."—The printing and publication of the Observations, which was always regarded by
Airy as a matter of the first importance, had fallen into
arrear: " I stated in my last Report that the printing of
the Observations for 1852 was scarcely commenced at the
time of the last meeting of the Visitors. For a long time the
printing went on so slowly that I almost despaired of ever
again seeing the Observations in a creditable state. After a
most harassing correspondence, the printers were at length
persuaded to move more actively, ... but the volume is
still very much behind its usual time of publication."—The
Deal Time-Ball has now been erected by Messrs Maudslays
and Field, and is an admirable specimen of the workmanship
of those celebrated engineers. The galvanic connection with
the Royal Observatory (through the telegraph wires of the
South Eastern Railway) is perfect. The automatic changes
of wire-communications are so arranged that, when the Ball
at Deal has dropped to its lowest point, it sends a message to

Greenwich to acquaint me, not with the time of the beginning of its fall (which cannot be in error) but with the fact that it has really fallen.   The Ball has several times been dropped experimentally with perfect success ; and some small official and subsidiary arrangements alone are wanting for bringing it into constant use."—The operations for the galvanic determination of the longitude of Brussels are described, with the following conclusion : " Thus, about 3000 effective signals were made, but only 1000 of these were admissible for the fundamental objects of the operation.   The result, I need scarcely remark, claims a degree of accuracy to which no preceding determination of longitude could ever pretend.   I apprehend that the probable error in the difference of time corresponds to not more than one or two yards upon the Earth's surface.—A careful scheme had been arranged for the determination of the longitude of Lerwick, but ' unfortunately, the demand for chronometers caused by our large naval armament has been so considerable that I cannot reckon on having at my disposal a sufficient number to carry on this operation successfully ; and I have, therefore, unwillingly deferred it to a more peaceful time.'—The covering stone of Halley's Tomb in Lee Churchyard was much shattered, and I applied to the Admiralty for funds for its complete restoration : these were granted on Feb. 3rd.—In this year, under my cognizance, £100 was added to the Hansen grant.—I had much correspondence and work in connection with the printing of Maclear's work at the Cape of Good Hope.   In June, all accounts, &c. about the Transit Circle were closed at the Admiralty, and the instrument was completely mounted at the Cape.—Dr Scoresby (who in his own way was very imperious) had attacked my methods of correcting the compass in iron ships : I replied in a letter to the Athenæum on Oct. 17th.—I made enquiries about operations for determining the longitude of Vienna, but was utterly repelled by the foreign telegraph offices. — In the Royal Astronomical Society ; I prepared the Address on presenting

the Medal to Rümker.—In Melbourne University: The first
letter received was from the Chancellor of the University
dated Jan. 26th, requesting that Sir John Herschel, Prof.
Malden, Mr Lowe (subsequently Chancellor of the Exchequer), and I would select professors. We had a great
deal of correspondence, meetings, examination of testimonials,
&c., and on August 14th we agreed on Wilson, Rowe, McCoy,
and Hearn.—On Feb. 17th I received the Prussian Order of
Merit.—I had correspondence with the Treasury on the scale
to be adopted for the Maps of the British Survey. I proposed
$\frac{1}{3000}$, and for some purposes $\frac{1}{600}$.—I printed a Paper on the
Deluge, in which I shewed (I believe to certainty) that the
Deluge of Genesis was merely a Destructive Flood of the
Nile.—Being well acquainted with the mountains of Cumberland, I had remarked that a 'man' or cairn of stones erected
by the Ordnance Surveyors on the Great Gable had covered
up a curious natural stone trough, known as one of the
remarkable singularities of the country. This year, without
giving any notice to the Ordnance Surveyors, I sent two
wallers from Borrowdale to the mountain top, to remove the
'man' about 10 feet and expose the trough. Sir Henry
James afterwards approved of my act, and refunded the
expense.—I investigated the optical condition of an eye with
conical cornea.

   "The Harton Colliery Experiment: I had long wished to
repeat the experiment which I had attempted unsuccessfully
in 1826 and 1828, of determining by pendulum-vibrations the
measure of gravity at the bottom of a mine. Residing near
Keswick this summer, and having the matter in my mind, I
availed myself of an introduction from Dr Leitch to some
gentlemen at South Shields, for inspection of the Harton
Colliery. I judged that it would answer pretty well. I find
that on Aug. 11th I wrote to Mr Anderson (lessee of the
mine), and on the same day to the Admiralty requesting
authority to employ a Greenwich Assistant, and requesting
£100 for part payment of expenses. On August 16th the

Admiralty assent.   There were many preparations to be made, both personal and instrumental.   My party consisted of Dunkin (Superintendant), Ellis, Criswick, Simmons, Pogson, and Rümker: I did not myself attend the detail of observations.   The observations began on Oct. 2nd and ended on Oct. 21st: supplementary observations were subsequently made at Greenwich for examining the coefficient of temperature-correction.   On Oct. 24th I gave a Lecture at South Shields on the whole operation.   In 'Punch' of Nov. 18th there was an excellent semi-comic account of the experiment, which as I afterwards found was written by Mr Percival Leigh."

Of private history: " On Jan. 18th I returned from Playford.   From Mar. 10th to 13th I was at Deal, and visited Sir John Herschel at Hawkhurst.—From June 28th to Aug. 7th I was staying with my family at The Grange, in Borrowdale near Keswick: and also made an expedition to Penrith, Carlisle, Newcastle, Jarrow, &c.; and descended the Harton Pit.—In September and also in October I was at South Shields on the Harton Experiments.—From Dec. 14th to 18th I was at Cambridge, and on the 26th I went to Playford."

The following letter, written in answer to a lady who had asked him to procure permission from Lord Rosse for her to observe with his telescope, is characteristic:

ROYAL OBSERVATORY, GREENWICH.
1854, September 20.

DEAR MADAM,

The state of things with regard to Lord Rosse's Telescope is this.   If a night is fine, it is wanted for his use or for the use of professional astronomers.   If it is not fine, it is of no use to anybody.   Now considering this, and considering that the appropriation of the telescope on a fine night to any body but a technical astronomer is a misapplication of an enormous capital of money and intellect which is invested in this unique instrument—it is against my conscience to ask Lord Rosse to place it at the service of any person except an experienced astronomer. No introduction, I believe, is necessary for seeing it in the day-time. The instrument stands un-enclosed in the Castle Demesne, to which strangers are admitted without question, I believe..............

Faithfully yours,

G. B. AIRY.

1855

"On May 9th it was notified to me (I think through the Hydrographer) that the Admiralty were not unwilling to increase my salary. I made application therefore; and on Jan. 21st 1856 Sir Charles Wood notified to me that the Admiralty consented to have it raised from £800 to £1000.— In the Report to the Board of Visitors it appears that 'At the instance of the Board of Trade, acting on this occasion through a Committee of the Royal Society, a model of the Transit Circle (with the improvement of perforated cube, &c. introduced in the Cape Transit Circle) has been prepared for the Great Exhibition at Paris.'—Under the head of Reduction of Astronomical Observations it is stated that 'During the whole time of which I have spoken, the galvanic-contact method has been employed for transits, with the exception of a few days, when the galvanic apparatus was out of order. From the clock errors, I have deduced the personal equations of the observers in our usual way. . . . The result is that the magnitude of the personal equations in the galvanic-touch method is not above half of that in the eye and ear method.'— With regard to the Reduction of the Magnetical Observations, 'I have not yet felt sufficiently satisfied with any proposed method of discussing the magnetic results to devote any time to their further treatment.'—'The Time-Signal Ball at Deal was brought into regular use at the beginning of the present year. In a short time, however, its action was interrupted, partly by derangement of the apparatus, and partly by the severity of the weather, which froze the sulphuric acid to the state of jelly. I sent an assistant and workman to put it in order, and since that time it has generally acted very well.—Application has been made to me from one of the important offices of Government (the Post Office) for the galvanic regulation of their clocks.— On considering the risks to which various galvanic communications are liable, and the financial necessity for occupying wires as little as possible, I perceived that it was necessary to devise constructions which should satisfy the following conditions. First, that a current sent once a day should suffice for adjusting the clock, even if it had gone ten or more seconds wrong. Secondly, that an occasional failure of the current should not stop the clock. I have arranged constructions which possess these characters, and the artist (Mr C. Shepherd) is now engaged in preparing estimates of the expense. I think it likely that this may prove to be the beginning of a very extensive system of clock regulation."— With respect to the operations for determining the longitude of Paris, it is stated that, " The whole number of days of signal transmission was eighteen, and the whole number of signals transmitted was 2530. The number of days considered available for longitude, in consequence of transits of stars having been observed at both Observatories, was twelve, and the number of signals was 1703. Very great care was taken on both sides, for the adjustments of the instruments. The resulting difference of longitude, $9^{m \cdot} 20 \cdot 63^{s \cdot}$, is probably very accurate. It is less by nearly $1^{s \cdot}$ of time than that determined in 1825 by rocket-signals, under the superintendance of Sir John Herschel and Col. Sabine. The time occupied by the passage of the galvanic current appears to be $\frac{1}{12}$th of a second."—With regard to the Pendulum Experiments in the Harton Colliery, after mentioning that personal assistance had been sought and obtained from the Observatories of Cambridge, Oxford, Durham, and Red Hill, the Report states that " The experiments appear to have been in every point successful, shewing beyond doubt that gravity is increased at the depth of 1260 feet by $\frac{1}{19000}$th part. I trust that this combination may prove a valuable precedent for future associations of the different Observatories of the kingdom, when objects requiring extensive personal organization shall present themselves."—On Oct. 18th the Astronomer Royal printed an Address to the Individual Members of the Board of Visitors on the subject of a large new Equatoreal for the Observatory. After a brief statement of the existing equipment of the Observatory in respect of equatoreal instruments, the Address continues thus: "It is known to the Visitors that I have uniformly objected to any luxury of extrameridional apparatus, which would materially divert us from a steady adherence to the meridional system which both reason and tradition have engrafted on this Observatory. But I feel that our present instruments are insufficient even for my wishes; and I cannot overlook the consideration that due provision must be made for future interests, and that we are nearer by twenty years to the time when another judgment must decide on the direction which shall be given to the force of the Observatory."—"In August I had some correspondence about the Egyptian wooden astronomical tablets with Mr Gresswell and others: they were fully examined by Mr Ellis.—In this year I was much engaged on schemes for compasses, and in June I sent my Paper on Discussions of Ships' Magnetism to the Royal Society.—On Dec. 6th the mast of the Observatory time-ball broke, and the Ball fell in the Front Court.—On Aug. 4th my valued friend Mr Sheepshanks died; and on Aug. 14th I went to London to see the Standard Bars as left by him. Afterwards, on Oct. 25th I went to Reading to collect the papers about Standards left by Mr Sheepshanks.—I made a mechanical construction for Euclid I. 47, with which I was well satisfied.—On Apr. 13th I joined a deputation to the Chancellor of the Exchequer (Sir G. Cornewall Lewis) on Decimal Coinage."

Of private history: "I was at Playford for a large part of

January.—On Mar. 26th I went to Reading, to visit Mr Sheepshanks, and afterwards to Silchester and Hereford.— On June 21st I went with my wife and two eldest sons to Edinburgh and other places in Scotland, but residing principally at Oban, where I hired a house.   Amongst other expeditions, I and my son Wilfrid went with the 'Pharos' (Northern Lights Steamer) to the Skerry Vohr Lighthouse, &c.   I also visited Newcastle, &c., and returned to Greenwich on Aug. 2nd.—From Oct. 12th to 17th I was at Cambridge.—On Dec. 24th I went to Playford."

CHAPTER VII.

At Greenwich Observatory—1856 to 1866.

1856

" In the Report to the Visitors there is an interesting account of the difficulties experienced with the Reflex Zenith Tube in consequence of the tremors of the quicksilver transmitted through the ground. Attempts were made to reduce the tremor by supporting the quicksilver trough on a stage founded at a depth of 10 feet below the surface, but it was not in the smallest degree diminished, and the Report states that 'The experience of this investigation justifies me in believing that no practicable depth of trench prevents the propagation of tremor when the soil is like that of Greenwich Hill, a gravel, in all places very hard, and in some, cemented to the consistency of rock.'—With respect to the regulation of the Post Office clocks, 'One of the galvanic clocks in the Post Office Department, Lombard Street, is already placed in connection with the Royal Observatory, and is regulated at noon every day...other clocks at the General Post Office are nearly prepared for the same regulation, and I expect that the complete system will soon be in action.'—Under the head of General Remarks a careful summary is given of the work of the Observatory, and the paragraph concludes as follows: ' Lastly there are employments which connect the scientific Observatory with the practical world ; the distribution of accurate time, the improvement of marine timekeepers, the observations and communications which tend to the advantage of Geography and Navigation, and the study, in a practical sense, of the modifications of Magnetism; a careful attention to these is likely to prove useful to the world, and conducive to the material prosperity of the Observatory: and these ought not to be banished from our system.'—In September I prepared the first specification for the building to carry the S.E. Dome.—In September, learning that Hansen's Lunar Tables were finished in manuscript, I applied to Lord Clarendon and they were conveyed to me through the Foreign Office: in October I submitted to the Admiralty the proposal for printing the Tables, and in November I learned that the Treasury had assented to the expense.—Lieut. Daynou's eclipses and occultations for longitudes of points in South Africa, observed in 1854 and 1855, were calculated here in this year.—On Feb. 16th I made my first application to Sir C. Wood (First Lord of the Admiralty) for assistance to C. Piazzi Smyth to carry out the Teneriffe Experiment: grounding it in part on the failure of attempts to see the solar prominences. He gave encouragement, and on Mar. 18th I transmitted Piazzi Smyth's Memorial to the Admiralty: on May 2nd the Admiralty authorized an expense of £500. I drew up suggestions.— The Sheepshanks Fund: After the death of my friend Richard Sheepshanks, his sister Miss Anne Sheepshanks wished to bestow some funds in connection with the University of Cambridge, Trinity College, and Astronomy, to which his name should be attached. There must have been some conversation with me, but the first letter is one from De Morgan in August. In September I had a conversation with Miss Sheepshanks, and sent her my first draft of a scheme, to which she assented. On Sept. 30th I wrote to Whewell (Master of Trinity) who was much trusted by Miss Sheepshanks: he consented to take part, and made some suggestions. There was further correspondence, but the business did not get into shape in this year.—In connection with the Correction of the Compass in Iron Ships: I discussed the observations made in the voyage of the Royal Charter. On Feb. 13th I proposed to the Admiralty a system of mounting the compasses with adjustable magnets, and it was ordered to be tried in the Trident and Transit.—In February I reported to the Admiralty that the Deal Time-Ball had been successful, and I proposed time-balls at Portsmouth, Plymouth, and Sheerness. There was much correspondence in various directions about Portsmouth and Devonport, and in March I went to Devonport and specially examined Mount Wise and the Devonport Column.—I had correspondence with Sir Howard Douglas about the sea breaking over the unfinished Dover Pier. I have an idea that this followed evidence given by me to a Harbour Commission, in which I expressed as a certainty that the sea will not be made to break by a vertical wall."

Of private history: "I returned from Playford on Jan. 18th.—From June 16th to August 5th I was, with my son Wilfrid, on an expedition to South Italy and Sicily: on our return from Sicily, we remained for three days ill at Marseilles from a touch of malaria.—On Dec. 22nd I went to Playford.—In acknowledgment of the pleasure which I had derived from excursions in the Cumberland Passes, I made a foot-bridge over a troublesome stream on the Pass of the Sty Head."

1857

"In the Report to the Visitors, when on the subject of the Altazimuth, the following paragraph occurs: 'I alluded in a preceding section to the cutting away of a very small portion of one of the rays of the three-armed pier which carries the Altazimuth. The quality of the brickwork is the best that I have ever seen, and not a single brick was disturbed beyond those actually removed. Yet the effect was to give the Altazimuth an inclination of about 23″.

This inclination evidently depends on the elasticity of the brickwork.'—With reference to the new S.E. Equatoreal the Report states that 'The support of the north or upper end of the polar axis has been received, and is planted within the walls of the building in a position convenient for raising it to its ultimate destination. It is one piece of cast-iron, and weighs nearly 5 tons.'—Small changes as previously mentioned had been noticed with regard to the Zero of Azimuth of the Transit Circle, and the Report states that 'In regard to the Azimuth of the Transit Circle, and the Azimuth of its Collimator, Mr Main has brought together the results of several years, and the following law appears to hold. There is a well-marked annual periodical change in the position of the Transit Circle, the southerly movement of the eastern pivot having its minimum value in September, and its maximum in March, the extreme range being about 14 seconds; and there is a similar change, but of smaller amount, in the position of the Collimator. I cannot conjecture any cause for these changes, except in the motion of the ground. There is also a well-marked connection between the state of level of the axis and the temperature. The eastern, pivot always rises when the temperature rises, the extreme range being about 6 seconds. I cannot offer any explanation of this.'—Under the head of Extraneous Works the Report states that 'The British Government had for some years past contributed by pecuniary grants to the preparation of Prof. Hansen's Lunar Tables. In the last winter they undertook the entire expense of printing a large impression of the Tables. The reading of the proof-sheets (a very considerable labour) has been effected entirely at the Observatory. I may take this opportunity of stating that the use of these Tables has enabled me, as I think, incontestably to fix the capture of Larissa to the date B.C. 557, May 19. This identification promises to prove valuable, not merely for its chronological utility, but also for its accurate determination of an astronomical epoch, the point eclipsed being exactly known, and the shadow having been very small.'—In April I gave a lecture to the Royal Astronomical Society on the methods available through the next 25 years for the determination of the Sun's parallax.—Dr Livingstone's observations for African longitudes were computed at the Observatory.—The Admiralty enquire of me about the feasibility of adopting Piazzi Smyth's construction for steadying telescopes on board ship: I gave a Report, of mixed character, on the whole discouraging.—I had correspondence with G. P. Bond and others about photographing the Stars and Moon.—On Feb. 17th Piazzi Smyth's books, &c. relating to the Teneriffe Experiment were sent to me: I recommended that an abridged Report should be sent to the Royal Society.—Respecting the Sheepshanks Fund: there was correspondence with Miss Sheepshanks and Whewell, but nothing got into shape this year: Miss Sheepshanks transferred to me £10,000 lying at Overend and Gurney's.—In November experiments were made for the longitude of Edinburgh, which failed totally from the bad state of the telegraph wire between Deptford and the Admiralty.—In June the first suggestion was made to me by Capt. Washington for time-signals on the Lizard Point: which in no long time I changed for the Start Point. —The Admiralty call for estimates for a time-ball at Portsmouth: on receiving them they decline further proceeding.— I was engaged in speculations and correspondence about the Atlantic Submarine Cable.—In the Royal Astronomical Society, I presented Memoirs and gave lectures on the three great chronological eclipses (Agathocles, Thales, Larissa)."— On Dec. 5th Airy wrote to the Vice-Chancellor of the University of Cambridge, objecting to the proposed changes regarding the Smith's Prizes—a subject in which he took much interest, and to which he ascribed great importance.— "On Apr. 27th I was in correspondence with G. Herbert of the Trinity House, about floating beacons.—In July I reported to the Treasury on the Swedish Calculating Engine (I think on the occasion of Mr Farr, of the Registrar-

General's Office, applying for one).—In November I had correspondence about the launch of the Great Eastern, and the main drainage of London."

Of private history: "On Jan. 14th I returned from Playford.—From June 27th to Aug. 5th I was travelling in Scotland with my wife and two eldest sons, chiefly in the West Highlands. On our return we visited Mrs Smith (my wife's mother) at Brampton.—On Dec. 26th I went to Playford."

1858

"In the Minutes of the Visitors it is noted that the new Queen's Warrant was received. The principal change was the exclusion of the Astronomer Royal and the other Observatory Officers from the Board.—In the Report to the Visitors it is stated that 'The Papers of the Board of Longitude are now finally stitched into books. They will probably form one of the most curious collections of the results of scientific enterprise, both normal and abnormal, which exists.'—It appears that the galvanic communications, external to the Observatory, had been in a bad state, the four wires to London Bridge having probably been injured by a thunderstorm in the last autumn, and the Report states that 'The state of the wires has not enabled us to drop the Ball at Deal. The feeble current which arrives there has been used for some months merely as giving a signal, by which an attendant is guided in dropping the Ball by hand.'—Regarding the new Equatoreal the Report states that 'For the new South-East Equatoreal, the object-glass was furnished by Messrs Merz and Son in the summer of last year, and I made various trials of it in a temporary tube carried by the temporary mounting which I had provided, and finally I was well satisfied with it. I cannot yet say that I have certainly divided the small star of γ Andromedæ; but, for such a test, a combination of favourable circumstances is required. From what I have

seen, I have no doubt of its proving a first-rate object-glass.'
—On March 15th was an annular eclipse of the Sun, for the
observation of which I sent parties fully equipped to Bedford,
Wellingborough, and Market Harborough. The observations
failed totally in consequence of the bad weather: I myself
went to Harrowden near Wellingborough.—Respecting the
Altazimuth, the Report states that with due caution as to the
zero of azimuth 'the results of observation are extremely
good, very nearly equal to those of the meridional instrument; perhaps I might say that three observations with the
Altazimuth are equivalent to two with the Transit Circle.'—
Respecting Meteorological Observations the Report states
that 'The observations of the maximum and minimum thermometers in the Thames, interrupted at the date of the last
Report, have been resumed, and are most regularly maintained. Regarding the Thames as the grand climatic agent
on London and its neighbourhood, I should much regret the
suppression of these observations.'—After much trouble the
longitude of Edinburgh had been determined: 'the retard of
the current is 0·04^s very nearly, and the difference of longitudes 12^m 43·05^s, subject to personal equations.'—The Report
concludes thus: 'With regard to the direction of our labours,
I trust that I shall always be supported by the Visitors in
my desire to maintain the fundamental and meridional system
of the Observatory absolutely intact. This, however, does
not impede the extension of our system in any way whatever,
provided that such means are arranged for carrying out the
extension as will render unnecessary the withdrawal of
strength from what are now the engrossing objects of the
Observatory.'—I had much correspondence on Comets, of
which Donati's great Comet was one: the tail of this Comet
passed over Arcturus on October 5th.—Respecting the Sheepshanks Fund: In September I met Whewell at Leeds, and
we settled orally the final plan of the scheme. On Oct. 27th
I saw Messrs Sharp, Miss Sheepshanks's solicitors, and drew
up a Draft of the Deed of Gift. There was much corre-

spondence, and on Nov. 20th I wrote to the Vice-Chancellor of Cambridge University. A counter-scheme was proposed by Dr Philpott, Master of St Catharine's College. By arrangement I attended the Council of the University on Dec. 3rd, and explained my views, to which the Council assented. On Dec. 9th the Senate accepted the gift of Miss Sheepshanks.— I had much correspondence throughout this year, with the Treasury, Herschel, Sabine, and the Royal Society, about the continuation of the Magnetic Establishments. The Reductions of the Magnetic Observations 1848–1857 were commenced in February of this year, under the direction of Mr Lucas, a computer who had been engaged on the Lunar Reductions.—In this year I came to a final agreement with the South Eastern Railway Company about defining the terms of our connection with them for the passage of Time Signals. I was authorized by the Admiralty to sign the 'protocol' or Memorandum of Agreement, and it was signed by the South Eastern Railway Directors.—On Aug. 28th I made my first proposal to Sir John Packington (First Lord of the Admiralty) for hourly time signals on the Start Point, and in September I went to the Start to examine localities, &c. On Dec. 23rd the Admiralty declined to sanction it.—I presented to the Royal Society a Paper about drawing a great-circle trace on a Mercator's chart.—In October I gave a Lecture on Astronomy in the Assembly Room at Bury.— On Jan. 25th I was busied with my Mathematical Tracts for republication."—In this year Airy published in the Athenæum very careful and critical remarks on the Commissioners' Draft of Statutes for Trinity College. He was always ready to take action in the interests of his old College. This Paper procured him the warmest gratitude from the Fellows of the College.

Of private history: "On Jan. 23rd I returned from Playford. From July 5th to Aug. 6th I was on an expedition in Switzerland with my two eldest sons. At Paris we visited Le Verrier, and at Geneva we visited Gautier, De La Rive, and Plantamour.   We returned by Brussels.—On Dec. 23rd I went to Playford."—In this year was erected in Playford Churchyard a granite obelisk in memory of Thomas Clarkson.   It was built by subscription amongst a few friends of Clarkson's, and the negociations and arrangements were chiefly carried out by Airy, who zealously exerted himself in the work which was intended to honour the memory of his early friend.   It gave him much trouble during the years 1856 to 1858.

Here is a letter to the Editor of the Athenæum on some other Trinity matters:

1858, *November* 22.

Dear Sir,

In the Athenæum of November 20, page 650, column 3, paragraph 4, there is an account of the erection of the statue of Barrow in Trinity College Antechapel (Cambridge) conceived in a spirit hostile to the University, and written in great ignorance of the facts.   On the latter I can give the writer some information.

The Marquis of Lansdowne, who was a Trinity man and whose son was of Trinity, intimated to the authorities of the College that he was desirous of placing in the antechapel a statue of *Milton*. This, regard being had to the customs and the college-feelings of Cambridge, was totally impossible.   The antechapel of every college is sacredly reserved for memorials of the men of that college only; and Milton was of Christ's College.   The Marquis of Lansdowne, on hearing this objection, left the choice of the person to be commemorated, to certain persons of the college, one of whom (a literary character of the highest eminence and a profound admirer of Milton) has not resided in Cambridge for many years.   Several names were carefully considered, and particularly one (not mentioned by your correspondent) of very great literary celebrity, but in whose writings there is ingrained so much of ribaldry and licentiousness that he was at length given up.   Finally the choice rested on Barrow, not as comparable to Milton, but as a person of reputation in his day and as the best who could be found under all the circumstances.

Cromwell never was mentioned; he was a member of Sidney

College: moreover it would have been very wrong to select the exponent of an extreme political party. But Cromwell has I believe many admirers in Cambridge, to which list I attach myself.

I had no part in the negociations above mentioned, but I saw the original letters, and I answer for the perfect correctness of what I have stated. But as I am not a principal, I decline to appear in public.

It is much to be desired, both for the Athenæum and for the public, that such an erroneous statement should not remain uncorrected. And I would suggest that a correction by the Editor would be just and graceful, and would tend to support the Athenæum in that high position which it has usually maintained.

I am, dear Sir,

Yours very faithfully,

G. B. AIRY.

Hepworth Dixon, Esq.

1859

" The Report to the Visitors states that ' The Lunar Reductions with amended elements (especially parallax) for correction of Observations from 1831 to 1851 are now completed. It is, I think, matter of congratulation to the Observatory and to Astronomy, that there are now exhibited the results of uninterrupted Lunar Observations extending through more than a century, made at the same place, reduced under the same superintendence and on the same general principles, and compared throughout with the same theoretical Tables.' —After reference to the great value of the Greenwich Lunar Observations to Prof. Hansen in constructing his Tables, and to the liberality of the British Government in their grants to Hansen, the Report continues thus: ' A strict comparison of Hansen's Tables with the Greenwich Observations of late years, both meridional and extra-meridional, was commenced. The same observations had, in the daily routine of the Observatory, been compared with the Nautical Almanac or Burckhardt's Tables. The result for one year only (1852) has yet reached me, but it is most remarkable. The sum of squares of residual errors with Hansen's Tables is only one-eighth part of that with Burckhardt's Tables. When it is remembered that in this is included the entire effect of errors and irregularities of observation, we shall be justified in considering Hansen's Tables as nearly perfect. So great a step, to the best of my knowledge, has never been made in numerical physical theory. I have cited this at length, not only as interesting to the Visitors from the circumstance that we have on our side contributed to this great advance, but also because an innovation, peculiar to this Observatory, has in no small degree aided in giving a decisive character to the comparison. I have never concealed my opinion that the introduction and vigorous use of the Altazimuth for observations of the Moon is the most important addition to the system of the Observatory that has been made for many years. The largest errors of Burckhardt's Tables were put in evidence almost always by the Altazimuth Observations, in portions of the Moon's Orbit which could not be touched by the meridional instruments; they amounted sometimes to nearly 40″ of arc, and they naturally became the crucial errors for distinction between Burckhardt's and Hansen's Tables. Those errors are in all cases corrected with great accuracy by Hansen's Tables.'—The Report concludes with the following paragraph: 'With the inauguration of the new Equatoreal will terminate the entire change from the old state of the Observatory. There is not now a single person employed or instrument used in the Observatory which was there in Mr Pond's time, nor a single room in the Observatory which is used as it was used then. In every step of change, however, except this last, the ancient and traditional responsibilities of the Observatory have been most carefully considered : and, in the last, the substitution of a new instrument was so absolutely necessary, and the importance of tolerating no instrument except of a high class was so obvious, that no other course was open to us. I can only trust that, while the use of the Equatoreal within legitimate limits may enlarge the utility and the reputation of the Observatory, it may never be permitted to interfere with that which has always been the staple and standard work here.'—Concerning the Sheep-shanks Fund: There was much correspondence about settling the Gift till about Feb. 21st. I took part in the first examination for the Scholarship in October of this year, and took my place with the Trinity Seniority, as one of their number on this foundation, for some general business of the Fund.— With respect to the Correction of the Compass in Iron Ships: I sent Mr Ellis to Liverpool to see some practice there in the correction of the Compass. In September I urged Mr Rundell to make a voyage in the Great Eastern (just floated) for examination of her compasses, and lent him instruments: very valuable results were obtained. Mr Archibald Smith had edited Scoresby's Voyage in the Royal Charter, with an introduction very offensive to me: I replied fully in the Athenæum of Nov. 7th.—The Sale of Gas Act: An Act of Parliament promoted by private members of the House of Commons had been passed, without the knowledge or recollection of the Government. It imposed on the Government various duties about the preparation of Standards. Suddenly, at the very expiration of the time allowed this came to the knowledge of Government. On Oct. 1st Lord Monteagle applied to me for assistance. On Oct. 15th and 22nd I wrote to Mr Hamilton, Secretary of the Treasury, and received authority to ask for the assistance of Prof. W. H. Miller.—I made an examination of Mr Ball's eyes (long-sighted and short-sighted I think).—In February I made an Analysis of the Cambridge Tripos Examination, which I communicated to some Cambridge residents." In a letter on this subject to one of his Cambridge friends Airy gives his opinion as follows: " I have looked very carefully over the Examination Papers, and think them on the whole very bad. They are utterly perverted by the insane love of Problems, and by the foolish importance given to wholly useless parts of Algebraical Geometry. For the sake of these, every Physical Subject and every useful application of pure mathematics are cut down or not mentioned." This led to much discussion at Cambridge. In this year the Smith's Prizes were awarded to the 4th and 6th Wranglers.

Of private history: "On Apr. 29th Mrs Smith (my wife's mother) died at Brampton.—From July 4th to Aug. 2nd I was in France (Auvergne and the Vivarais) with my two eldest sons. Maclear travelled with us to Paris.—On Dec. 23rd I went to Playford."—Antiquities and historical questions connected with military movements had a very great attraction for Airy. On his return from the expedition in France above-mentioned, he engaged in considerable correspondence with military authorities regarding points connected with the battle of Toulouse. And in this year also he had much correspondence with the Duke of Northumberland concerning his Map of the Roman Wall, and the military points relating to the same.

1860

" In June Mr Main accepted the office of Radcliffe Observer at Oxford (Mr Johnson having died) and resigned the First Assistancy at Greenwich: in October Mr Stone was appointed First Assistant.—At an adjourned Meeting of the Visitors on June 18th there were very heavy discussions on Hansen's merits, and about the grant to him. Papers were read from Sir J. Lubbock, Babbage, South, Whewell, and me. Finally it was recommended to the Government to grant £1000 to Hansen, which was paid to him.—In the Report to the Board of Visitors the following remark occurs: ' The apparent existence of a discordance between the results of Direct Observations and Reflection Observations (after the application of corrections for flexure, founded upon observations of the horizontal collimator wires) to an extent far greater than can be explained by any disturbance of the direction of gravity on the quicksilver by its distance from the vertical, or by the attraction of neighbouring masses, perplexes me much.'— With respect to the discordance of dips of the dipping-needles, which for years past had been a source of great trouble and puzzle, the Report states that ' The dipping-needles are still a source of anxiety. The form which their anomalies appear to take is that of a special or peculiar value of the dip given by each separate needle. With one of the 9-inch needles, the result always differs about a quarter of a degree from that of the others. I can see nothing in its mechanical construction to explain this.—Reference is made to the spontaneous currents through the wires of telegraph companies, which are frequently violent and always occur at the times of magnetic storms, and the Report continues ' It may be worth considering whether it would ever be desirable to establish in two directions at right angles to each other (for instance, along the Brighton Railway and along the North Kent Railway) wires which would photographically register in the Royal Observatory the currents that pass in these directions, exhibiting their indications by photographic curves in close juxtaposition with the registers of the magnetic elements.'—In connection with the Reduction of the Greenwich Lunar Observations from 1831 to 1851, the Report states that ' The comparison of Hansen's Lunar Tables with the Greenwich Observations, which at the last Visitation had been completed for one year only, has now been finished for the twelve years 1847 to 1858. The results for the whole period agree entirely, in their general spirit, with those for the year 1852 cited in the last Report. The greatest difference between the merits of Burckhardt's and Hansen's Tables appears in the Meridional Longitudes 1855, when the proportion of the sum of squares of errors is as 31 (Burckhardt) to 2 (Hansen). The nearest approach is in the Altazimuth Latitudes 1854, when the proportion of the sum of squares of errors is as 12 (Burckhardt) to 5 (Hansen).'—A special Address to the Members of the Board of Visitors has reference to the proposals of M. Struve for (amongst other matters) the improved determination of the longitude of Valencia, and the galvanic determination of the extreme Eastern Station of the British triangles.—On Sept. 13th I circulated amongst the Visitors my Remarks on a Paper entitled 'On the Polar Distances of the Greenwich Transit-Circle, by A. Marth,' printed in the Astronomische Nachrichten; the Paper by Mr Marth was an elaborate attack on the Greenwich methods of observation, and my Remarks were a detailed refutation of his statements. —On Oct. 20th I made enquiry of Sabine as to the advantage of keeping up magnetic observations. On Oct. 22nd he wrote, avoiding my question in some measure, but saying that our instruments must be changed for such as those at Kew (his observatory): I replied, generally declining to act on that advice.—In March and April I was in correspondence with Mr Cowper (First Commissioner of Works, &c.) about the bells of the Westminster Clock; also about the smoky chimneys of the various apartments of the Palace. On Apr. 21st I made my Report on the clock and bells, 20 foolscap pages. I employed a professional musician to examine the tones of the bells.—In November I was writing my book on Probable Errors, &c.—I was engaged on the Tides of Kurrachee and Bombay.—The first examination of Navy telescopes was made for the Admiralty.—Hoch's Paper on Aberration appeared in the Astronomische Nachrichten. This (with others) led to the construction of the water-telescope several years later.—In September I wrote in the Athenæum against a notion of Sir H. James on the effect of an upheaval of a mountain in changing the Earth's axis. In October I had drawn up a list of days for a possible evagation of the Earth's poles: but apparently nothing was done upon them.

"In this year I was a good deal occupied for the Lighthouse Commission. On Feb. 21st Admiral Hamilton (chairman) applied to me for assistance. In April I went to Chance's Factory in Birmingham on this business. In May

I made my report on the Start Lighthouse, after inspection
with the Commission.   In June, with my son Hubert, I
visited the Whitby Lighthouses, and discovered a fault of a
singular kind which most materially diminished their power.
This discovery led to a general examination of lighthouses
by the Trinity Board, to a modification of many, and to a
general improvement of system.   On June 25th I reported
on the Lights at Calais, Cap de Valde, Grisnez, South Foreland, and North Foreland.   In August I had been to the
North Foreland again, and in September to Calais and the
Cap d'Ailly.   In October I went with my son Hubert to
Aberdeen to see the Girdleness Lighthouse.   On Nov. 10th
I made a General Report.

"This was the year of the great total solar eclipse visible
in Spain.   At my representation, the Admiralty placed at
my command the large steamship 'Himalaya' to carry about
60 astronomers, British and Foreign.   Some were landed at
Santander: I with many at Bilbao.   The Eclipse was fairly
well observed : I personally did not do my part well.   The
most important were Mr De La Rue's photographic operations.   At Greenwich I had arranged a very careful series of
observations with the Great Equatoreal, which were fully
carried out."

The eclipse expedition to Spain, shortly referred to
above, was most interesting, not merely from the importance
of the results obtained (and some of the parties were very
fortunate in the weather) but from the character of the expedition.   It was a wonderful combination of the astronomers
of Europe, who were all received on board the 'Himalaya,'
and were conveyed together to the coast of Spain.   The
polyglot of languages was most remarkable, but the utmost
harmony and enthusiasm prevailed from first to last, and this
had much to do with the general success of the expedition.
Those who landed at Bilbao were received in the kindest and
most hospitable manner by Mr C. B. Vignoles, the engineer-
in-chief of the Bilbao and Tudela Railway, which was then

under construction.   This gentleman made arrangements for
the conveyance of parties to points in the interior of the
country which were judged suitable for the observation of the
eclipse, and placed all the resources of his staff at the disposal
of the expedition in the most liberal manner.   The universal
opinion was that very great difficulty would have been experienced without the active and generous assistance of Mr
Vignoles.   It is needless to say that the vote of thanks to
Mr Vignoles, proposed by the Astronomer Royal during the
return voyage, was passed by acclamation and with a very
sincere feeling of gratitude: it was to the effect that 'without
the great and liberal aid of Mr C. B. Vignoles, and the disinterested love of science evinced by him on this occasion,
the success of the "Himalaya" eclipse expedition could not
have been ensured.'   There is a graphic and interesting
account of the reception of the party at Bilbao given in the
'Life of C. B. Vignoles, F.R.S., Soldier and Civil Engineer,'
by O. J. Vignoles, M.A.

Of private history: "On May 26th my venerable friend
Arthur Biddell died.   He had been in many respects more
than a father to me: I cannot express how much I owed to
him, especially in my youth.—From June 12th to 15th I
visited the Whitby Lighthouses with my son Hubert.—From
July 6th to 28th I was in Spain, on the 'Himalaya' expedition, to observe the total eclipse: I was accompanied by my
wife, my eldest son, and my eldest daughter.—From Oct. 5th
to 18th I went with my son Hubert to Aberdeen to see the
Girdleness Lighthouse, making lateral trips to Cumberland
in going and returning.—On Dec. 21st I went to Playford."

1861

" In the Report to the Visitors there is great complaint of
want of room. 'With increase of computations, we want
more room for computers; with our greatly increased business of Chronometers and Time-Distribution, we are in want

of a nearly separate series of rooms for the Time-Department: we want rooms for book-stores; and we require rooms for the photographic operations and the computations of the Magnetic Department.'—The Report gives a curious history of Dr Bradley's Observations, which in 1776 had been transferred to the University of Oxford, and proceeds thus: 'More lately, I applied (in the first instance through Lord Wrottesley) to the Vice-Chancellor, Dr Jeune, in reference to the possibility of transferring these manuscripts to the Royal Observatory....Finally, a decree for the transfer of the manuscript observations to the Royal Observatory, without any condition, was proposed to Convocation on May 2nd, and was passed unanimously. And on May 7th my Assistant, Mr Dunkin, was sent to Oxford to receive them. And thus, after a delay of very nearly a century, the great work of justice is at length completed, and the great gap in our manuscript observations is at length filled up.'—With reference to the Transit Circle, it had been remarked that the Collimators were slightly disturbed by the proximity of the gas-flames of their illuminators, and after various experiments as to the cause of it, the Report proceeds thus: 'To my great surprise, I found that the disturbance was entirely due to the radiation of the flame upon a very small corner (about 16 square inches) of the large and massive stone on which the collimator is planted. The tin plates were subsequently shaped in such a manner as to protect the stone as well as the metal; and the disturbance has entirely ceased.'—Regarding the large S.E. Equatoreal, the Report states that 'On the character of its object-glass I am now able to speak, first, from the examination of Mr Otto Struve, made in a favourable state of atmosphere; secondly, from the examinations of my Assistants (I have not myself obtained a sight of a test-object on a night of very good definition). It appears to be of the highest order. The small star of γ Andromedæ is so far separated as to shew a broad dark space between its components. Some blue colour is shewn about the bright planets.'—It is noted in the Report that 'The Equatoreal observations of the Solar Eclipse are completely reduced; and the results are valuable. It appears from them that the error in right ascension of Burckhardt's Lunar Tables at the time of the eclipse amounted to about $38''$; while that of Hansen's (ultimately adopted by Mr Hind for the calculation of the eclipse) did not exceed $3''$.'—With regard to Chronometers it is stated that 'By use of the Chronometer Oven, to which I have formerly alluded, we have been able to give great attention to the compensation. I have reason to think that we are producing a most beneficial effect on the manufacture and adjustment of chronometers in general.'—With regard to the Cape of Good Hope Observatory and Survey, the Admiralty enquire of me when the Survey work will be completed, and I enquire of Maclear 'How is the printing of your Survey Work?' In 1862 I began to press it strongly, and in 1863 very strongly.—I introduced a method (constantly pursued since that time at the Royal Observatory) for computing interpolations without changes of sign.—I had correspondence with Herschel and Faraday, on the possible effect of the Sun's radiant heat on the sea, as explaining the curve of diurnal magnetic inequality. (That diurnal inequality was inferred from the magnetic reductions 1848—1857, which were terminated in 1860.)—Regarding the proposal of hourly time-signals on the Start Point, I consulted telegraph engineers upon the practical points, and on Dec. 21st I proposed a formal scheme, in complete detail. (The matter has been repeatedly brought before the Admiralty, but has been uniformly rejected.)—I was engaged on the question of the bad ocular vision of two or three persons.—The British Association Meeting was held at Manchester: I was President of Section A. I gave a Lecture on the Eclipse of 1860 to an enormous attendance in the Free Trade Hall." The following record of the Lecture is extracted from Dr E. J. Routh's Obituary Notice of Airy written for the Proceedings of the Royal Society. "At the meeting of the British Association at Manchester in 1861, Mr Airy delivered a Lecture on the Solar Eclipse of 1860 to an assembly of perhaps 3000 persons. The writer remembers the great Free Trade Hall crowded to excess with an immense audience whose attention and interest, notwithstanding a weak voice, he was able to retain to the very end of the lecture....The charm of Professor Airy's lectures lay in the clearness of his explanations. The subjects also of his lectures were generally those to which his attention had been turned by other causes, so that he had much that was new to tell.   His manner was slightly hesitating, and he used frequent repetitions, which perhaps were necessary from the newness of the ideas.   As the lecturer proceeded, his hearers forgot these imperfections and found their whole attention rivetted to the subject matter."

Of private history: " On Jan. 2nd there was a most remarkable crystallization of the ice on the flooded meadows at Playford: the frost was very severe.—From June 20th to Aug. 1st I was at the Grange near Keswick (where I hired a house) with my wife and most of my family.—From Nov. 5th to 14th I was on an expedition in the South of Scotland with my son Wilfrid : we walked with our knapsacks by the Roman Road across the Cheviots to Jedburgh.—On Dec. 21st I went to Playford."

1862

" The Report to the Board of Visitors states that ' A new range of wooden buildings (the Magnetic Offices) is in progress at the S.S.E. extremity of the Magnetic Ground.   It will include seven rooms.'—Also ' I took this opportunity (the relaying of the water-main) of establishing two powerful fire-plugs (one in the Front Court, and one in the Magnetic Ground); a stock of fire-hose adapted to the "Brigade-Screw" having been previously secured in the Observatory.'—'Two wires, intended for the examination of spontaneous earth-currents, have been carried from the Magnetic Observatory to the Railway Station in the town of Greenwich. From this point one wire is to be led to a point in the neighbourhood of Croydon, the other to a point in the neighbourhood of Dartford. Each wire is to be connected at its two extremities with the Earth. The angle included between the general directions of these two lines is nearly a right angle.' —'The Kew unifilar magnetometer, adapted to the determination of the horizontal part of terrestrial magnetic force in absolute measure, was mounted in the summer of 1861; and till 1862 February, occasional observations (14 in all) were taken simultaneously with the old and with the new instrument. The comparison of results shewed a steady but very small difference, not greater probably than may correspond to the omission of the inverse seventh powers of distance in the theoretical investigation; proving that the old instrument had been quite efficient for its purpose.'—Great efforts had been made to deduce a law from the Diurnal Inequalities in Declination and Horizontal Force, as shewn by the Magnetic observations; but without success: the Report states that 'The results are most amazing, for the variation in magnitude as well as in law. What cosmical change can be indicated by them is entirely beyond my power of conjecture.'—'I have alluded, in the two last Reports, to the steps necessary, on the English side, for completing the great Arc of Parallel from Valencia to the Volga. The Russian portion of the work is far advanced, and will be finished (it is understood) in the coming summer. It appeared to me therefore that the repetition of the measure of astronomical longitude between Greenwich and Valencia could be no longer delayed. Two Assistants of the Royal Observatory (Mr Dunkin and Mr Criswick) will at once proceed to Valencia, for the determination of local time and the management of galvanic signals.'—'I now ask leave to press the subject of Hourly Time Signals at the Start Point on the attention of the Board, and to submit the advantage of their addressing the Board of Admiralty upon it. The great majority of outward-bound ships pass within sight of the Start, and, if an hourly signal were exhibited, would have the means of regulating their chronometers at a most critical part of their voyage. The plan of the entire system of operations is completely arranged. The estimated expense of outfit is £2017, and the estimated annual expense is £326; both liable to some uncertainty, but sufficiently exact to shew that the outlay is inconsiderable in comparison with the advantages which might be expected from it. I know no direction of the powers of the Observatory which would tend so energetically to carry out the great object of its establishment, viz. "the finding out the so much desired Longitude at Sea."'—The attention of the Visitors is strongly drawn to the pressure on the strength of the Observatory caused by the observation of the numerous small planets, and the paragraph concludes thus: 'I shall, however, again endeavour to effect a partition of this labour with some other Observatory.'—A small fire having occurred in the Magnetic Observatory, a new building of zinc, for the operation of naphthalizing the illuminating gas, is in preparation, external to the Observatory: and thus one of the possible sources of accidental fire will be removed.—Miss Sheepshanks added, through me, £2000 to her former gift: I transferred it, I believe, to the Master and Seniors of Trinity College."—In this year Airy contributed to the Royal Society two Papers, one "On the Magnetic properties of Hot-Rolled and Cold-Rolled Malleable Iron," the other "On the Strains in the Interior of Beams." He gave evidence before the Select Committee on Weights and Measures, and also before the Public Schools Commission.

In the latter part of 1862 a difference arose between Airy and Major-General Sabine, in consequence of remarks made by the latter at a meeting of the Committee of Recommendations of the British Association. These remarks were to the effect "That it is necessary to maintain the complete system of self-registration of magnetic phænomena at the

Kew Observatory, because no sufficient system of magnetic record is maintained elsewhere in England"; implying pointedly that the system at the Royal Observatory of Greenwich was insufficient. This matter was taken up very warmly by Airy, and after a short and acrimonious correspondence with Sabine, he issued a private Address to the Visitors, enclosing copies of the correspondence with his remarks, and requesting the Board to take the matter of this attack into their careful consideration. This Address is dated November 1862, and it was followed by another dated January 1863, which contains a careful reply to the various points of General Sabine's attack, and concludes with a distinct statement that he (the Astronomer Royal) can no longer act in confidence with Sabine as a Member of the Board of Visitors.

Of private history: There were the usual short visits to Playford at the beginning and end of the year.—From June 28th to Aug. 5th he was in Scotland (chiefly in the Western Highlands) with his wife and his sons Hubert and Osmund. In the course of this journey he visited the Corryvreckan whirlpool near the island of Scarba, and the following paragraph relating to this expedition is extracted from his journal: "Landed in Black Mile Bay, island of Luing, at 10.30. Here by previous arrangement with Mr A. Brown, agent of the steam-boat company, a 4-oared boat was waiting to take us to Scarba and the Corryvreckan. We were pulled across to the island of Lunga, and rowed along its length, till we came to the first channel opening from the main sea, which the sailors called the Little Gulf. Here the sea was rushing inwards in a manner of which I had no conception. Streams were running with raving speed, sometimes in opposite directions side by side, with high broken-headed billows. Where the streams touched were sometimes great whirls (one not many yards from our boat) that looked as if they would suck anything down. Sometimes among all this were great smooth parts of the sea, still in a whirling trouble, which were surrounded by the mad currents.   We seemed entirely powerless among all these."

In the beginning of this year (1862) the Duke of Manchester, in writing to the Rev. W. Airy, had said, " I wish your brother, the Astronomer Royal, could be induced to have investigations made as to whether the aspects of the Planets have any effect on the weather." This enquiry produced the following reply :

A subject like that of the occult influences of the planets (using the word occult in no bad sense but simply as meaning not *thoroughly* traced) can be approached in two ways—either by the à priori probability of the existence of such influences, or by the à posteriori evidence of their effects.   If the two can be combined, the subject may be considered as claiming the dignity of a science. Even if the effects alone are certain, it may be considered that we have a science of inferior degree, wanting however that definiteness of law and that general plausibility which can only be given when true causes, in accordance with antecedent experience in other cases, can be suggested.

Now in regard to the à priori probability of the existence of planetary influences, I am far from saying that such a thing is impossible.   The discoveries of modern philosophy have all tended to shew that there may be many things about us, unknown even to the scientific world, but which well-followed accidents reveal with the most positive certainty.   It is known that every beam of light is accompanied by a beam of chemical agency, totally undiscoverable to the senses of light or warmth, but admitting of separation from the luminous and warm rays ; and producing photogenic effects. We know that there are disturbances of magnetism going on about us, affecting whole continents at a time, unknown to men in general, but traceable with facility and certainty, and which doubtless affect even our brains and nerves (which are indisputably subject to the influence of magnetism).

Now in the face of these things I will not undertake to say that there is any impossibility, or even any want of plausibility in the supposition that bodies external to the earth may affect us.   It may well be cited in its favour that it is certain that the sun affects our magnetism (it is doubtful whether it does so *im*mediately, or mediately by giving different degrees of warmth to different parts of the earth), and it is believed on inferior evidence that the moon also affects it. It may therefore seem not impossible or unplausible that other celestial bodies may affect perhaps others of the powers of nature about us. But there I must stop. The denial of the impossibility is no assertion of the truth or probability, and I absolutely decline to take either side—either that the influences are real, or that the influences are unreal—till I see evidence of their effects.

Such evidence it is extremely difficult to extract from ordinary facts of observation. I have alluded to the sun's daily disturbance of the magnet as one of the most certain of influences, yet if you were to observe the magnet for a single day or perhaps for several days, you might see no evidence of that influence, so completely is it involved with other disturbances whose causes and laws are totally unknown.

I believe that, in addition to the effects ascribable to Newtonian gravitation (as general motion of the earth, precession of the equinoxes, and tides), this magnetic disturbance is the only one yet established as depending on an external body. Men in general, however, do not think so. It appears to be a law of the human mind, to love to trace an effect to a cause, and to be ready to assent to any specious cause. Thus all practical men of the lower classes, even those whose pecuniary interests are concerned in it, believe firmly in the influence of the moon upon the winds and the weather. I believe that every careful examiner of recorded facts (among whom I place myself as regards the winds) has come to the conclusion that the influence of the moon is not discoverable.

I point out these two things (magnetic disturbances and weather) as tending to shew that notoriety or the assumed consent of practical men, are of no value. The unnotorious matter may be quite certain, the notorious matter may have no foundation. Everything must stand on its own evidence, as completely digested and examined.

Of such evidence the planetary influence has not a particle.

My intended short note has, in the course of writing, grown up into a discourse of very unreasonable length ; and it is possible that a large portion of it has only increased obscurity. At any rate I can add nothing, I believe, which can help to explain more fully my views on this matter.

In this year (1862, June 9th) Airy received the Honorary Degree of LL.D. in the University of Cambridge. He was nominated by the Duke of Devonshire, as appears from the following letter:

LISMORE CASTLE, IRELAND,
April 19*th*, 1862.

MY DEAR SIR,

It is proposed according to usage to confer a considerable number of Honorary Degrees on the occasion of my first visit to Cambridge as Chancellor of the University.

I hope that you will allow me to include your name in that portion of the list which I have been invited to draw up.

The ceremony is fixed for the 10th of June.

I am, my dear Sir,

Yours very truly,

DEVONSHIRE.

The Astronomer Royal.

Airy's reply was as follows:

ROYAL OBSERVATORY, GREENWICH,
LONDON, S.E.
1862, *April* 21.

MY LORD DUKE,

I am exceedingly gratified by your communication this day received, conveying a proposal which I doubt not is suggested by your Grace's recollection of transactions now many years past.

I have always been desirous of maintaining my connection with my University, and have in various ways interested myself practically in its concerns. It would give me great pleasure to have the connection strengthened in the flattering way which you propose.

I had conceived that alumni of the University were not admissible to honorary degrees; but upon this point the information possessed by your Grace, as Chancellor of the University, cannot be disputed.

I am, my Lord Duke,

Your Grace's very faithful servant,

G. B. AIRY.

His Grace
The Duke of Devonshire.

There were in all 19 Honorary Degrees of Doctor of Laws conferred on the 9th of June, including men of such eminence as Armstrong, Faraday, and Fairbairn.

1863

In this year there were several schemes for a Railway through the lower part of Greenwich Park, the most important being the scheme of the London, Chatham and Dover Railway Company. In reference to this scheme the Report to the Visitors states " I may say briefly that I believe that it would be possible to render such a railway innocuous to the Observatory; it would however be under restrictions which might be felt annoying to the authorities of the Railway, but whose relaxation would almost ensure ruin to the Observatory."—" The meridional observations of Mars in the Autumn of 1862 have been compared with those made at the Observatory of Williamstown, near Melbourne, Australia, and they give for mean solar parallax the value $8\cdot932''$, exceeding the received value by about $\frac{1}{24}$th part. (A value nearly identical with this $8\cdot93''$ has also been found by comparing the Pulkowa and Cape of Good Hope Observations.)"—" The results of the new Dip-Instrument in 1861 and 1862 appear to give a firm foundation for speculations on the state and change of the dip. As a general result,

I may state as probable that the value of dip in the middle of 1843 was about 69°1′, and in the middle of 1862 about 68°11′. The decrease of dip appears to be more rapid in the second half of this interval than in the first; the dip at beginning of 1853 being about 68°44′."—With reference to the re-determination of the longitude of Valencia, it is stated that " The concluded longitude agrees almost exactly with that determined by the transmission of chronometers in 1844; and entitles us to believe that the longitudes of Kingstown and Liverpool, steps in the chronometer conveyance, were determined with equal accuracy."—" The computations, for inferring the direction and amount of movement of the Solar System in space from the observed proper motions of 1167 stars, have been completed. The result is, that the Sun is moving towards a point, R.A. 264°, N.P.D. 65° (not very different from Sir W. Herschel's, but depending much in N.P.D. on the accuracy of Bradley's quadrant observations), and that its annual motion subtends, at the distance of a star of the first magnitude, the angle 0·4″. But the comparison, of the sum of squares of apparent proper motions uncorrected, with the sum of squares of apparent proper motions corrected for motion of Sun, shews so small an advance in the explanation of the star's apparent movements as to throw great doubt on the certainty of results; the sum of squares being diminished by only $\frac{1}{25}$th part."—" I had been writing strongly to Maclear on the delays in publishing both the geodetic work and the Star Catalogue at the Cape of Good Hope : he resolves to go on with these works. In December I am still very urgent about the geodesy."

Of private history: There was the usual short visit to Playford at the beginning and end of the year.—" From June 27th to August 10th I was travelling in the North and West of Scotland with my wife, my youngest son Osmund, and my daughter Annot."

In this year the offer of Knighthood (for the third time) was made to Airy through the Rt Hon. Sir George C. Lewis, Bart. The offer was accepted on Feb. 12th, 1863, but on the same day a second letter was written as follows:

1863, Feb. 12.

Dear Sir,

I am extremely ignorant of all matters connected with court ceremonial, and in reference to the proposed Knighthood would ask you :—

1. I trust that there is no expense of fees. To persons like myself of small fortune an honour may sometimes be somewhat dear.

2. My highest social rank is that given by my Academical Degree of D.C.L. which I hold in the Universities of Oxford and Cambridge. In regard to costume, would it be proper that I should appear in the scarlet gown of that degree? or in the ordinary Court Dress?

I am, Dear Sir,

Yours very faithfully,

G. B. AIRY.

The Right Honourable
  Sir George C. Lewis, Bart.,
  &c.   &c.   &c.

To this letter Sir G. C. Lewis replied that the fees would amount to about £30, an intimation which produced the following letter:

Royal Observatory, Greenwich, S.E.

1863, Feb. 19th.

Dear Sir,

I have to acknowledge your letter of yesterday : and I advert to that part of it in which it is stated that the Fees on Knighthood amount to about £30.

Twenty-seven years ago the same rank was offered to me by Lord John Russell and Mr Spring Rice (then Ministers of the Crown), with the express notice that no fees would be payable. I suppose that the usage (whatever it be) on which that notice was founded still subsists.

To a person whose annual income little more than suffices to meet the annual expenses of a very moderate establishment, an unsought honour may be an incumbrance. It appears, at any rate, opposed to the spirit of such an honour, that it should be loaded with Court Expenses in its very creation.

I hope that the principle stated in 1835 may serve as precedent on this occasion.

I am, dear Sir,
Your very faithful servant,
G. B. AIRY.

The Right Honourable
Sir G. C. Lewis, Bart.,
&c.   &c.   &c.

No intimation however was received that the fees would be remitted on the present occasion, and after consideration the proposed Knighthood was declined in the following letter :

ROYAL OBSERVATORY, GREENWICH, S.E.
1863, April 15.

DEAR SIR,

I have frequently reflected on the proposal made by you of the honour of Knighthood to myself. I am very grateful to you for the favourable opinion which you entertain in regard to my supposed claims to notice, and for the kindness with which you proposed publicly to express it. But on consideration I am strongly impressed with the feeling that the conditions attached by established regulation to the conferring of such an honour would be unacceptable to me, and that the honour itself would in reality, under the circumstances of my family-establishment and in my social position, be an incumbrance to me. And finally I have thought it best most respectfully, and with a full sense of the kindness of yourself and of the Queen's Government towards me, to ask that the proposal might be deferred.

There is another direction in which a step might be made, affecting my personal position in a smaller degree, but not tending to incommode me, which I would ask leave to submit to your consideration. It is, the definition of the Rank of the Astronomer Royal. The singular character of the office removes it from ordinary rules of rank, and sometimes may produce a disagreeable contest of opinions. The only offices of similar character corresponding in other conditions to that of the British Astronomer Royal are those of the Imperial Astronomers at Pulkowa (St Petersburg) and Paris. In Russia, where every rank is clearly defined by that of military grade, the Imperial Astronomer has the rank of Major-General. In France, the definition is less precise, but the present Imperial Astronomer has been created (as an attachment of rank to the office) a Senator of the Empire.

I am, dear Sir,

Your very faithful servant,

G. B. AIRY.

The Rt Hon. Sir George C. Lewis, Bart.,
&c.   &c.   &c.

Sir G. C. Lewis died before receiving this letter, and the letter was afterwards forwarded to Lord Palmerston. Some correspondence followed between Lord Palmerston and Airy on the subject of attaching a definite rank to the office of Astronomer Royal, as proposed in the above letter. But the Home Office (for various reasons set forth) stated that the suggestion could not be complied with, and the whole subject dropped.

1864

The following remarks are extracted from the Report of the Astronomer Royal to the Board of Visitors.—" In a very heavy squall which occurred in the gale of December 2 of last year, the stay of the lofty iron pillar ouside of the Park Rails, which carried our telegraph wires, gave way, and the pillar and the whole system of wires fell."—" An important alteration has been made in the Magnetic Observatory. For several years past, various plans have been under consideration for preventing large changes of temperature in the room which contains the magnetic instruments. At length I determined to excavate a subterraneous room or cellar under the original room. The work was begun in the last week in January, and in all important points it is now finished."—"In the late spring, some alarm was occasioned by the discovery that the Parliamentary Standard of the Pound Weight had become coated with an extraneous substance produced by the decomposition of the lining of the case in which it was preserved. It was decided immediately to compare it with the three Parliamentary Copies, of which that at the Observatory is one. The National Standard was found to be entirely uninjured."—"On November 16 of last year, the Transit Instrument narrowly escaped serious injury from an accident. The plate chain which carries the large western counterpoise broke. The counterpoise fell upon the pier, destroying the massive gun-metal wheels of the lifting machinery, but was prevented from falling further by the iron stay of the gas-burner flue."—"The Prismatic Spectrum-Apparatus had been completed in 1863. Achromatic object-glasses are placed on both sides of the prism, so that each pencil of light through the prism consists of parallel rays; and breadth is given to the spectrum by a cylindrical lens. The spectral lines are seen straighter than before, and generally it is believed that their definition is improved."—"For observation of the small planets, a convention has been made with M. Le Verrier. From new moon to full moon, all the small planets visible to 13^h are observed at the Royal Observatory of Greenwich. From full moon to new moon, all are observed at the Imperial Observatory of Paris. The relief gained in this way is very considerable."—"In determining the variations in the power of the horizontal-force and vertical-force magnets depending on temperature, it was found by experiment that this depended materially on whether the magnet was heated by air or by water, and 'The result of these experiments (with air) is to give a coefficient for temperature correction four or five times as great as that given by the water-heatings.'"—"With regard to the discordances of the results of observations of dip-

A. B.   17 needles, experiments had been made with needles whose breadth was in the plane passing through the axis of rotation, and it appeared that the means of extreme discordances were, for an ordinary needle 11' 45", and for a flat needle 3' 27"," and the Report continues thus : " After this I need not say that I consider it certain that the small probable errors which have been attributed to ordinary needles are a pure delusion." —The Report states that in the various operations connected with the trials and repairs of chronometers, and the system of time-signals transmitted to various time-balls and clocks, about one-fourth of the strength of the Observatory is employed, and it continues thus : " Viewing the close dependence of Nautical Astronomy upon accurate knowledge of time, there is perhaps no department of the Observatory which answers more completely to the original utilitarian intentions of the Founder of the Royal Observatory."—" With regard to the proposal of time-signals at the Start Point, it appears that communications referring to this proposal had passed between the Board of Admiralty and the Board of Trade, of which the conclusion was, that the Board of Trade possessed no funds applicable to the defraying of the expenses attending the execution of the scheme. And the Admiralty did not at present contemplate the establishment of these time-signals under their own authority."—Amongst other Papers in this year, Airy's Paper entitled " First Analysis of 177 Magnetic Storms," &c., was read before the Royal Society.

Of private history : " There was the usual visit to Playford in the beginning of the year.—From June 8th to 23rd I made an excursion with my son Hubert to the Isle of Man, and the Lake District.—From Sept. 7th to 14th I was on a trip to Cornwall with my two eldest sons, chiefly in the mining district.—In August of this year my eldest (surviving) daughter, Hilda, was married to Mr E. J. Routh, Fellow of St Peter's College, Cambridge, at Greenwich Parish Church. They afterwards resided at Cambridge."

1865

" Our telegraphic communications of every kind were again destroyed by a snow-storm and gale of wind which occurred on Jan. 28th, and which broke down nearly all the posts between the Royal Observatory and the Greenwich Railway Station.—The Report to the Visitors states that ' The only change of Buildings which I contemplate as at present required is the erection of a fire-proof Chronometer Room. The pecuniary value of Chronometers stored in the Observatory is sometimes perhaps as much as £8000.'—The South Eastern and London Chatham and Dover scheme for a railway through the Park was again brought forward. There was a meeting of Sir J. Hanmer's Committee at the Observatory on May 26th. Mr Stone was sent hastily to Dublin to make observations on Earth-disturbance by railways there. I had been before the Committee on May 25th. On Sept. 1st I approved of an amended plan. In reference to this matter the Report states that ' It is proper to remark that the shake of the Altazimuth felt in the earthquake of 1863, Oct. 5th, when no such shake was felt with instruments nearer to the ground (an experience which, as I have heard on private authority, is supported by observation of artificial tremors), gives reason to fear that, at distances from a railway which would sufficiently defend the lower instruments, the loftier instruments (as the Altazimuth and the Equatoreals) would be sensibly affected.'—Some of the Magnets had been suspended by steel wires, instead of silk, of no greater strength than was necessary for safety, and the Report states that ' Under the pressure of business, the determination of various constants of adjustment was deferred to the end of the year. The immediate results of observation, however, began to excite suspicion ; and after a time it was found that, in spite of the length of the suspending wire (about 8 feet) the torsion-coefficient was not much less than $\frac{1}{6}$. The wires were promptly dismounted, and silk skeins substituted for them. With these, the torsion-coefficient is about $\frac{1}{210}$.'—The Dip-Instrument, which had given great trouble by the irregularities of the dip-results, had been compared with two dip-instruments from Kew Observatory, which gave very good and accordant results. 'It happened that Mr Simms, by whom our instruments now in use were prepared, and who had personally witnessed our former difficulties, was present during some of these experiments. Our own instrument being placed in his hands (Nov. 10th to 19th) for another purpose, he spontaneously re-polished the apparently faultless agate-bearings. To my great astonishment, the inconsistencies of every kind have nearly or entirely vanished. On raising and lowering the needles, they return to the same readings, and the dips with the same needle appear generally consistent.' Some practical details of the polishing process by which this result had been secured are then given.—After numerous delays, the apparatus for the self-registration of Spontaneous Earth Currents was brought into a working state in the month of March. A description of the arrangement adopted is given in the Report.—'All Chronometers on trial are rated every day, by comparison with one of the clocks sympathetic with the Motor Clock. Every Chronometer, whether on trial or returned from a chronometer-maker as repaired, is tried at least once in the heat of the Chronometer-Oven, the temperature being usually limited to 90° Fahrenheit; and, guided by the results of very long experience, we have established it as a rule, that every trial in heat be continued through three weeks.'—'The only employment extraneous to the Observatory which has occupied any of my time within the last year is the giving three Lectures on the Magnetism of Iron Ships (at the request of the Lords of the Committee of Council on Education) in the Theatre of the South Kensington Museum. The preparations, however, for these Lectures, to be given in a room ill-adapted to them, occupied a great deal of my own time, and of the time of an Assistant of the Observatory.'—

' Referring to a matter in which the interests of Astronomy are deeply concerned, I think it right to report to the Visitors my late representation to the Government, to the effect that, in reference to possible observation of the Transit of Venus in 1882, it will be necessary in no long time to examine the coasts of the Great Southern Continent.' "

Of private history: " There were the usual visits to Playford at the beginning and end of the year.—From June 18th to 26th I was on a trip in Wales with my sons Hubert and Osmund.—From Sept. 6th to Oct. 2nd I was staying with most of my family at Portinscale near Keswick : we returned by Barnard Castle, Rokeby, &c."

CHAPTER VIII.

AT GREENWICH OBSERVATORY—1866 TO 1876.

1866

IN this year the cube of the Transit Circle was pierced, to permit reciprocal observations of the Collimators without raising the instrument. This involved the construction of improved Collimators, which formed the subject of a special Address to the Members of the Board of Visitors on Oct. 21st 1865.—From the Report to the Visitors it appears that " On May 23rd 1865, a thunderstorm of great violence passed very close to the Observatory. After one flash of lightning, I was convinced that the principal building was struck. Several galvanometers in the Magnetic Basement were destroyed. Lately it has been remarked that one of the old chimneys of the principal building had been dislocated and slightly twisted, at a place where it was surrounded by an iron stay-band led from the Telegraph Pole which was planted upon the leads of the Octagon Room."—" On consideration of the serious interruptions to which we have several times been exposed from the destruction of our open-air Park-wires (through snow-storms and gales), I have made an arrangement for leading the whole of our wires in underground pipes as far as the Greenwich Railway Station."—" The Committee of the House of Commons, to whom the Greenwich and Woolwich Line of the South Eastern Railway was referred, finally assented to the adoption of a line which I indicated, passing between the buildings of the Hospital Schools and

the public road to Woolwich."—"The Galvanic Chronometer attached to the S. E. Equatoreal often gave us a
great deal of trouble. At last I determined, on the proposal
of Mr Ellis, to attempt an extension of Mr R. L. Jones's
regulating principle. It is well known that Mr Jones has
with great success introduced the system of applying galvanic
currents originating in the vibrations of a normal pendulum,
not to drive the wheelwork of other clocks, but to regulate to
exact agreement the rates of their pendulums which were,
independently, nearly in agreement; each clock being driven
by weight-power as before. The same principle is now
applied to the chronometer. . . . . The construction is perfectly
successful ; the chronometer remains in coincidence with the
Transit Clock through any length of time, with a small
constant error as is required by mechanical theory."—"The
printed volume of Observations for 1864 has two Appen-
dixes ; one containing the calculations of the value of the
Moon's Semi-diameter deduced from 295 Occultations
observed at Cambridge and Greenwich from 1832 to 1860,
and shewing that the Occultation Semi-diameter is less
than the Telescopic Semi-diameter by 2″; the other containing the reduction of the Planetary Observations made at
the Royal Observatory in the years 1831—1835 ; filling up
the gap, between the Planetary Reductions 1750—1830 made
several years ago under my superintendence, and the Reductions contained in the Greenwich Volumes 1836 to the present
time : and conducted on the same general principles."—"Some
trouble had been found in regulating the temperature of the
Magnetic Basement, but it was anticipated that in future
there would be no difficulty in keeping down the annual variation within about 5° and the diurnal variation within 3°.—
Longitudes in America were determined in this year by way
of Valencia and Newfoundland: finished by Nov. 14th."

Of private history: In April he made a short visit to
Ventnor in the Isle of Wight.—From June 15th to July 23rd
he was on an expedition in Norway with his son Osmund

and his nephew Gorell Barnes.—There was probably a short stay at Playford in the winter.

In this and in the previous year (1865) the free-thinking investigations of Colenso, the Bishop of Natal, had attracted much notice, and had procured him the virulent hostility of a numerous section. His income was withheld from him, and in consequence a subscription fund was raised for his support by his admirers. Airy, who always took the liberal side in such questions, was a subscriber to the fund, and wrote the following letter to the Bishop:

ROYAL OBSERVATORY, GREENWICH, S.E.,
1865, July 24.

MY LORD,

With many thanks I have to acknowledge your kind recollection of me in sending as a presentation copy the work on Joshua, Judges, and especially on the divided authorship of Genesis; a work whose investigations, founded in great measure on severe and extensive verbal criticism, will apparently bear comparison with your Lordship's most remarkable examination of Deuteronomy. I should however not do justice to my own appreciation if I did not remark that there are other points considered which have long been matters of interest to me.

On several matters, some of them important, my present conclusions do not absolutely agree with your Lordship's. But I am not the less grateful for the amount of erudition and thought carefully directed to definite points, and above all for the noble example of unwearied research and freedom in stating its consequences, in reference to subjects which scarcely ever occupy the attention of the clergy in our country.

I am, My Lord,
Yours very faithfully,
G. B. AIRY.

The Lord Bishop of Natal.

Here also is a letter on the same subject, written to Professor Selwyn, Professor of Divinity at Cambridge:—

ROYAL OBSERVATORY, GREENWICH,
LONDON, S.E.,
1866, *May* 5.

MY DEAR SIR,

The MS. concerning Colenso duly arrived.

I note your remarks on the merits of Colenso.   I do not write to tell you that I differ from you, but to tell you why I differ.

I think that you do not make the proper distinction between a person who invents or introduces a tool, and the person who uses it.

The most resolute antigravitationist that ever lived might yet acknowledge his debt to Newton for the Method of Prime and Ultimate Ratios and the Principles of Fluxions by which Newton sought to establish gravitation.

So let it be with Colenso.   He has given me a power of tracing out truth to a certain extent which I never could have obtained without him.   And for this I am very grateful.

As to the further employment of this power, you know that he and I use it to totally different purposes.   But not the less do I say that I owe to him a new intellectual power.

I quite agree with you, that the sudden disruption of the old traditional view seems to have unhinged his mind, and to have sent him too far on the other side.   I would not give a pin for his judgment.

Nevertheless, I wish he would go over the three remaining books of the Tetrateuch.

I know something of Myers, but I should not have thought him likely to produce anything sound on such things as the Hebrew Scriptures.   I never saw his " Thoughts."

I am, my dear Sir,

Yours very truly,

G. B. AIRY.

Professor Selwyn.

The following letter has reference to Airy's proposal to introduce certain Physico-Mathematical subjects into the Senate-House Examination for B.A. Honors at Cambridge. On various occasions he sharply criticized the Papers set for the Senate-House Examination and the Smith's Prize Examination, and greatly lamented the growing importance of pure mathematics and the comparative exclusion of physical questions in those examinations. His proposal as finally submitted in the letter that follows was somewhat modified (as regards the mode of introducing the subjects) from his original draft, in deference to the opinions of Whewell, Adams, Routh, and other friends to whom he had submitted it. His proposal was favourably received by the Mathematical Board, and recommendations were made in the direction, though not to the extent, that he desired, and he subsequently submitted a Memorandum on those recommendations:

ROYAL OBSERVATORY, GREENWICH,
1866, *May* 11.

MY DEAR SIR,

You will perceive, from perusal of the enclosed paper, that I have acted on the permission which you kindly gave me, to transmit to you my proposal for extension of the mathematical education of the University in the Physical direction.

It is an unavoidable consequence of the structure of the University that studies there will have a tendency to take an unpractical form depending much on the personal tastes of special examiners. I trust that, as a person whose long separation from the daily business of the University has enabled him to see in some measure the wants of the external scientific and practical world, I may be forgiven this attempt to bring to the notice of the University my ideas on the points towards which their attention might perhaps be advantageously turned.

I am, my dear Sir,
Very faithfully yours,
G. B. AIRY.

The Rev. Dr Cartmell,
Master of Christ's College
and Vice-Chancellor.

ROYAL OBSERVATORY, GREENWICH,
1866, *May* 11.

MY DEAR MR VICE-CHANCELLOR,

About two years ago, by the kindness of the University, an opportunity was presented to me of orally stating what I conceived to be deficiencies in the educational course of the University as regards mathematical physics. Since that time, the consideration of those deficiencies, which had long been present to me, has urged itself on my attention with greater force : and finally I have entertained the idea that I might without impropriety communicate to you my opinion, in a less fugitive form than on the occasion to which I have alluded : with the request that, if you should deem such a course appropriate, you would bring it before the Board of Mathematical Studies, and perhaps ultimately make it known to the Resident Members of the Senate.

I will first give the list of subjects, which I should wish to see introduced, and to the prosecution of which the generally admirable course of the University is remarkably well adapted : and I will then, without entering into every detail, advert to the process by which I think it probable the introduction of these subjects could be effected.

In the following list, the first head is purely algebraical, and the second nearly so : but they are closely related to observational science, and to the physical subjects which follow. Some of the subjects which I exhibit on my list are partially, but in my opinion imperfectly, taught at present. I entirely omit from my list Physical Optics, Geometrical Astronomy, and Gravitational Astronomy of Points : because, to the extent to which Academical Education ought to go, I believe that there is no teaching on these sciences comparable to that in the University of Cambridge. (It is, of course, still possible that improvements may be made in the books commonly used.) It might, however, be a question, whether, as regards the time and manner of teaching them, some parts of these subjects might ultimately be associated with the other subjects included in my list.

I. *List of subjects proposed for consideration.*

(1) Partial Differential Equations to the second order, with their arbitrary functions : selected principally with reference to the physical subjects.

(2) The Theory of Probabilities as applied to the combination of Observations.

(3) Mechanics (including Hydraulic Powers) in the state which verges upon practical application, and especially including that part in which the abstract ideas of *power* and *duty* occur.

(4) Attractions. This subject is recognized in the existing course of the University : but, so far as I can infer from examination-papers, it appears to be very lightly passed over.

(5) The Figure of the Earth, and its consequences, Precession, &c. I believe that the proposal is sanctioned, of adopting some part of this theory in the ordinary course; but perhaps hardly so far as is desirable.

(6) The Tides.

(7) Waves of Water.

(8) Sound (beginning with Newton's investigation); Echoes; Pipes and Vibrating Strings; Acoustics; the Mathematical part of Music.

(9) Magnetism, terrestrial and experimental, and their connection.

(I omit for the present Mineralogy and Mathematical Electricity.)

This list of subjects appears formidable : but they are in reality easy, and would be mastered in a short time by the higher Wranglers.

II. *Mode of introducing these subjects into the University.*

After much consideration, and after learning the opinions of several persons whose judgment claims my deepest respect, I propose the gradual introduction of these subjects into the Examination for Honors at admission to the B.A. Degree, as soon as the preparation of Books and the readiness of Examiners shall enable the University to take that step. I conceive that, by a judicious pruning of the somewhat luxuriant growth of Pure Algebra, Analytical Geometry, and Mere Problems, sufficient leisure may be gained for the studies of the undergraduates, and sufficient time for the questions of the examiners. I do not contemplate that the students could advance very far into the subjects; but I know the importance of beginning them ; and, judging from the train of thoughts, of reading, and of conversation, among the Bachelors with whom I associated many years ago, I believe that there is quite a sufficient number who will be anxious to go deep into the subjects if they have once entered into them. If six Wranglers annually would take them up, my point would be gained. The part which these gentlemen might be expected, in a short time, to take in the government of the University, would enable them soon to act steadily upon the University course : the efficiency of the University instruction would be increased ; and the external character of the University would be raised.

The real difficulties, and they are not light ones, would probably be found in providing Examiners and Books. At present, both are wanting within the University. Where there is a great and well-founded objection to intrusting examinations to persons foreign to the University, and where the books have to be created with labour and with absolute outlay of money (for their sale could never be remunerative), the progress must be slow. Still progress would be certain, if the authorities of the University should think the matter deserving of their hearty encouragement.

Requesting that you and the Members of the University will accept this proposal as an indication of my deep attachment to my University,

I am,

My dear Mr Vice-Chancellor,

Your very faithful servant,

G. B. AIRY.

The Rev. Dr Cartmell,
&c. &c.
Vice-Chancellor of the University of Cambridge.

1867

" In this year it was arranged that my Treasury accounts were to be transferred to the Admiralty, making the simplification which I had so long desired.—From the Report to the Visitors it appears that a relic of the Geodetic operations commenced in 1787 for connecting the Observatories of Greenwich and Paris, in the shape of an observing cabin on the roof of the Octagon Room, was shifted and supported in such a manner that the pressure on the flat roof was entirely avoided.—With regard to the Transit Circle, the new Collimators with telescopes of seven inches aperture had been mounted. When the Transit Telescope directed vertically is interposed, the interruptions in the central cube impair the sharpness of definition, still leaving it abundantly good for general use. It had been regarded as probable that the astronomical flexure of the telescope, after cutting away small portions of the central cube, would be found sensibly changed : and this proved to be the case. The difference of flexures of the two ends has been altered more than a second of arc.—Referring to a new Portable Altazimuth which had lately been tested, the Report states as follows : 'I may mention that a study of defects in the vertical circle of a small Altazimuth formerly used by me, and an inspection of the operations in the instrument-maker's work-shop, have convinced me that the principal error to be feared in instruments of this class is ovality of the graduated limb ; this cannot be eliminated by two microscopes, and such an instrument should never be fitted with two only. Our instrument has four.'—' In Osler's Anemometer, a surface of 2 square feet is now exposed to the wind instead of one foot as formerly ; and the plate is supported by weak vertical springs instead of rods running on rollers. Its indications are much more delicate than formerly.'—' The Meteors on Nov. 14th were well observed. Eight thousand and three hundred were registered. The variations of frequency at different times were very well noted. The points of divergence were carefully determined.'—Referring to the gradual improvement in the steadiness of chronometers from 1851 to 1866, it appears that from 1851 to 1854 the 'trial number' (which is a combination of changes of weekly rate representing the fault of the chronometer) varied from 34.8^s to 52.5^s, while from 1862 to 1866 it varied from 21.2^s to 25.8^s.—The following statement will shew the usual steadiness of the Great Clock on the Westminster Palace : On 38 per cent. of days of observation, the clock's error was below 1^s. On 38 per cent. the error was between 1^s and 2^s. On 21 per cent. it was between 2^s and 3^s. On 2 per cent. between 3^s and 4^s. On 1 per cent.

between 4^s and 5^s.—The Report contains an account of the determination of the longitude of Cambridge U.S. by Dr B. A. Gould, by means of galvanic currents through the Atlantic Cable, in the spring of 1867: and advantage was taken of this opportunity for re-determining the longitude of Feagh Main near Valencia in Ireland. The longitude of Feagh Main, found by different methods is as follows: By chronometers in 1844, 41^m $23\cdot23^s$; by galvanic communication with Knight's Town in 1862, 41^m $23\cdot37^s$; by galvanic communication with Foilhommerum in 1866, 41^m $23\cdot19^s$. The collected results for longitude of Cambridge U.S. from different sources are: By moon-culminators (Walker in 1851, and Newcomb in 1862–3), 4^h 44^m $28\cdot42^s$ and 4^h 44^m $29\cdot56^s$ respectively; by Eclipses (Walker in 1851), 4^h 44^m $29\cdot64^s$; by occultations of Pleiades (Peirce 1838—1842, and 1856—1861), 4^h 44^m $29\cdot91^s$ and 4^h 44^m $30\cdot90^s$ respectively; by chronometers (W. C. Bond in 1851, and G. P. Bond in 1855), 4^h 44^m $30\cdot66^s$ and 4^h 44^m $31\cdot89^s$ respectively; by Atlantic Cable 1866, 4^h 44^m $30\cdot99^s$.—After noticing that many meteorological observatories had suddenly sprung up and had commenced printing their observations in detail, the Report continues thus : ' Whether the effect of this movement will be that millions of useless observations will be added to the millions that already exist, or whether something may be expected to result which will lead to a meteorological theory, I cannot hazard a conjecture. This only I believe, that it will be useless, at present, to attempt a process of mechanical theory; and that all that can be done must be, to connect phænomena by laws of induction. But the induction must be carried out by numerous and troublesome trials in different directions, the greater part of which would probably be failures.'—There was this year an annular eclipse; I made large preparations at the limits of the annularity; failed entirely from very bad weather."—In this year Airy contributed a Paper to the Institution of Civil Engineers ' On the use of the Suspension Bridge with stiffened roadway for

Railway and other Bridges of Great Span,' for which a Telford Medal was awarded to him by the Council of the Institution. And he communicated several Papers to the Royal Society and the Royal Astronomical Society.

Of private history: There was the usual visit to Playford in January.—In April there was a short run to Alnwick and the neighbourhood, in company with Mr and Mrs Routh.— From June 27th to July 4th he was in Wales with his two eldest sons, visiting Uriconium, &c. on his return.—From August 8th to Sept. 7th he spent a holiday in Scotland and the Lake District of Cumberland with his daughter Christabel, visiting the Langtons at Barrow House, near Keswick, and Isaac Fletcher at Tarn Bank.

In June of this year (1867) Airy was elected an Honorary Fellow of his old College of Trinity in company with Connop Thirlwall, the Bishop of St David's. They were the first Honorary Fellows elected by the College. The announcement was made in a letter from the Master of Trinity (W. H. Thompson), and Airy's reply was as follows:

> ROYAL OBSERVATORY, GREENWICH,
> LONDON, S.E.
> 1867, *June 12th.*

MY DEAR MASTER,

I am very much gratified by your kind note received this morning, conveying to me the notice that the Master and Sixteen Senior Fellows had elected me, under their new powers, as Honorary Fellow of the College.

It has always been my wish to maintain a friendly connection with my College, and I am delighted to receive this response from the College. The peculiar form in which the reference to the Statute enables them to put it renders it doubly pleasing.

As the Statute is new, I should be obliged by a copy of it. And, at any convenient time, I should be glad to know the name of the person with whom I am so honorably associated.

> I am, My dear Master,
> Very faithfully yours,
> G. B. AIRY.

Consequent on Airy's proposals in 1866 for the introduction of new physical subjects into the Senate-House Examination and his desire that the large number of questions set in Pure Mathematics, or as he termed it "Useless Algebra," should be curtailed, there was a smart and interesting correspondence between him and Prof. Cayley, who was the great exponent and advocate of Pure Mathematics at Cambridge. Both of them were men of the highest mathematical powers, but diametrically opposed in their views of the use of Mathematics. Airy regarded mathematics as simply a useful machine for the solution of practical problems and arriving at practical results. He had a great respect for Pure Mathematics and all the processes of algebra, so far as they aided him to solve his problems and to arrive at useful results; but he had a positive aversion to mathematical investigations, however skilful and elaborate, for which no immediate practical value could be claimed. Cayley on the contrary regarded mathematics as a useful exercise for the mind, apart from any immediate practical object, and he considered that the general command of mathematics gained by handling abstruse mathematical investigations (though barren in themselves) would be valuable for whatever purpose mathematics might be required: he also thought it likely that his researches and advances in the field of Pure Mathematics might facilitate the solution of physical problems and tend to the progress of the practical sciences. Their different views on this subject will be seen from the letters that follow:

ROYAL OBSERVATORY, GREENWICH,
LONDON, S.E.
1867, Nov. 8.

MY DEAR SIR,

I think it best to put in writing the purport of what I have said, or have intended to say, in reference to the Mathematical Studies in the University.

First, I will remark on the study of Partial Differential Equations. I do not know that one branch of Pure Mathematics can be con-

sidered higher than another, except in the utility of the power which
it gives. Measured thus, the Partial Differential Equations are very
useful and therefore stand very high, as far as the Second Order.
They apply, to that point, in the most important way, to the great
problems of nature concerning *time*, and *infinite division of matter*,
and *space*: and are worthy of the most careful study. Beyond that
Order they apply to nothing. It was for the purpose of limiting the
study to the Second Order, and at the same time working it carefully, philosophically, and practically, up to that point, that I drew
up my little work.

On the general question of Mathematical Studies, I will first give
my leading ideas on what I may call the moral part. I think that a
heavy responsibility rests on the persons who influence most strongly
the course of education in the University, to direct that course in
the way in which it will be most useful to the students—in the two
ways, of disciplining their powers and habits, and of giving them
scientific knowledge of the highest and most accurate order (applying
to the phenomena of nature) such as will be useful to them through
life. I do not think that the mere personal taste of a teacher is
sufficient justification for a special course, unless it has been adopted
under a consideration of that responsibility. Now I can say for
myself that I have, for some years, inspected the examination
papers, and have considered the bearing of the course which they
imply upon the education of the student, and am firmly convinced
that as regards men below the very few first—say below the ten
first—there is a prodigious loss of time without any permanent good
whatever. For the great majority of men, such subjects as abstract
Analytical Geometry perish at once. With men like Adams and
Stokes they remain, and are advantageous; but probably there is
not a single man (beside them) of their respective years who
remembers a bit, or who if he remembers them has the leisure
and other opportunities of applying them.

I believe on the other hand that a careful selection of physical
subjects would enable the University to communicate to its students
a vast amount of information; of accurate kind and requiring the
most logical treatment; but so bearing upon the natural phenomena
which are constantly before us that it would be felt by every student
to possess a real value, that (from that circumstance) it would dwell
in his mind, and that it would enable him to correct a great amount
of flimsy education in the country, and, so far, to raise the national
character.

The consideration of the education of the reasoning habits suggests ideas far from favourable to the existing course. I am old enough to remember the time of mere geometrical processes, and I do not hesitate to say that for the cultivation of accurate mental discipline they were far superior to the operations in vogue at the present day. There is no subject in the world more favourable to logical habit than the Differential Calculus in all its branches *if logically worked in its elements:* and I think that its applications to various physical subjects, compelling from time to time an attention to the elementary grounds of the Calculus, would be far more advantageous to that logical habit than the simple applications to Pure Equations and Pure Algebraical Geometry now occupying so much attention.

I am, my dear Sir,

Yours very truly,

G. B. AIRY.

Professor Cayley.

DEAR SIR,

I have been intending to answer your letter of the 8th November. So far as it is (if at all) personal to myself, I would remark that the statutory duty of the Sadlerian Professor is that he shall explain and teach the principles of Pure Mathematics and apply himself to the advancement of the Science.

As to Partial Differential Equations, they are "high" as being an inverse problem, and perhaps the most difficult inverse problem that has been dealt with. In regard to the limitation of them to the second order, whatever other reasons exist for it, there is also the reason that the theory to this order is as yet so incomplete that there is no inducement to go beyond it ; there could hardly be a more valuable step than anything which would give a notion of the form of the general integral of a Partial Differential Equation of the second order.

I cannot but differ from you *in toto* as to the educational value of Analytical Geometry, or I would rather say of Modern Geometry generally. It appears to me that in the Physical Sciences depending on Partial Differential Equations, there is scarcely anything that a student can do for himself:—he finds the integral of the ordinary equation for Sound—if he wishes to go a step further and integrate the non-linear equation $\left(\dfrac{dy}{dx}\right)^2 \dfrac{d^2y}{dt^2} = a^2 \dfrac{d^2y}{dx^2}$, he is simply unable to do so; and so in other cases there is nothing that he can add to what he finds in his books. Whereas Geometry (of course to an intelligent student) is a real inductive and deductive science of inexhaustible extent, in which he can experiment for himself—the very tracing of a curve from its equation (and still more the consideration of the cases belonging to different values of the parameters) is the construction of a theory to bind together the facts—and the selection of a curve or surface proper for the verification of any general theorem is the selection of an experiment in proof or disproof of a theory.

I do not quite understand your reference to Stokes and Adams, as types of the men who alone retain their abstract Analytical Geometry. If a man when he takes his degree drops mathematics, he drops geometry—but if not I think for the above reasons that he is more likely to go on with it than with almost any other subject—and any mathematical journal will shew that a very great amount of attention is in fact given to geometry. And the subject is in a very high degree a progressive one; quite as much as to Physics, one may apply to it the lines, Yet I doubt not thro' the ages one increasing purpose runs, and the thoughts of men are widened with the progress of the suns.

I remain, dear Sir,

Yours very sincerely,

A. CAYLEY.

CAMBRIDGE,
   6 *Dec.*, 1867.

ROYAL OBSERVATORY, GREENWICH,
LONDON, S.E.
1867, December 9.

MY DEAR SIR,

I have received with much pleasure your letter of December 6. In this University discussion, I have acted only in public, and have not made private communication to any person whatever till required to do so by private letter addressed to me. Your few words in Queens' Hall seemed to expect a little reply.

Now as to the Modern Geometry. With your praises of this science—as to the room for extension in induction and deduction, &c. ; and with your facts—as to the amount of space which it occupies in Mathematical Journals; I entirely agree. And if men, after leaving Cambridge, were designed to shut themselves up in a cavern, they could have nothing better for their subjective amusement. They might have other things as good; enormous complication and probably beautiful investigation might be found in varying the game of billiards with novel islands on a newly shaped billiard table. But the persons who devote themselves to these subjects do thereby separate themselves from the world. They make no step towards natural science or utilitarian science, the two subjects which the world specially desires. The world could go on as well without these separatists.

Now if these persons lived only for themselves, no other person would have any title to question or remark on their devotion to this barren subject. But a Cambridge Examiner is not in that position. The University is a national body, for education of young men : and the power of a Cambridge Examiner is omnipotent in directing the education of the young men ; and his responsibility to the cause of education is very distinct and very strong. And the question for him to consider is—in the sense in which mathematical education is desired by the best authorities in the nation, is the course taken by this national institution satisfactory to the nation?

I express my belief that it is *not* satisfactory. I believe that many of the best men of the nation consider that a great deal of time is lost on subjects which they esteem as puerile, and that much of that time might be employed on noble and useful science.

You may remember that the Commissions which have visited Cambridge originated in a Memorial addressed to the Government by men of respected scientific character : Sabine was one, and I may take him as the representative. He is a man of extensive knowledge of the application of mathematics as it has been employed for many years in the science of the world ; but he has no profundity of science. He, as I believe, desired to find persons who could enter accurately into mathematical science, and naturally looked to the Great Mathematical University ; but he must have been much disappointed. So much time is swallowed up by the forced study of the Pure Mathematics that it is not easy to find anybody who can really enter on these subjects in which men of science want assist-

ance.   And so Sabine thought that the Government ought to interfere, probably without any clear idea of what they could do.

I am, my dear Sir,
Yours very truly,
G. B. AIRY.

Professor Cayley.

DEAR SIR,

   I have to thank you for your last letter.   I do not think
everything should be subordinated to the educational element : my
idea of a University is that of a place for the cultivation of all
science.   Therefore among other sciences Pure Mathematics; including whatever is interesting as part of this science.   I am bound
therefore to admit that your proposed extension of the problem of
billiards, *if it* were found susceptible of interesting mathematical
developments, would be a fit subject of study.   But in this case I do
not think the problem could fairly be objected to as puerile—a more
legitimate objection would I conceive be its extreme speciality.   But
this is not an objection that can be brought against Modern Geometry as a whole : in regard to any particular parts of it which may
appear open to such an objection, the question is whether they are
or are not, for their own sakes, or their bearing upon other parts of
the science to which they belong, worthy of being entered upon and
pursued.

   But admitting (as I do not) that Pure Mathematics are only to
be studied with a view to Natural and Physical Science, the question
still arises how are they best to be studied in that view.   I assume
and admit that as to a large part of Modern Geometry and of the
Theory of Numbers, there is no present probability that these will
find any physical applications.   But among the remaining parts of
Pure Mathematics we have the theory of Elliptic Functions and of
the Jacobian and Abelian Functions, and the theory of Differential
Equations, including of course Partial Differential Equations.   Now
taking for instance the problem of three bodies—unless this is to be
gone on with by the mere improvement in detail of the present
approximate methods—it is at least conceivable that the future treatment of it will be in the direction of the problem of two fixed centres,
by means of elliptic functions, &c. ; and that the discovery will be
made not by searching for it directly with the mathematical resources

now at our command, but by "prospecting" for it in the field of
these functions. Even improvements in the existing methods are
more likely to arise from a study of differential equations in general
than from a special one of the equations of the particular problem :
the materials for such improvements which exist in the writings of
Hamilton, Jacobi, Bertrand, and Bour, have certainly so arisen.
And the like remarks would apply to the physical problems which
depend on Partial Differential Equations.

I think that the course of mathematical study at the University
is likely to be a better one if regulated with a view to the cultivation
of Science, as if for its own sake, rather than directly upon considerations of what is educationally best (I mean that the best educational
course will be so obtained), and that we have thus a justification for
a thorough study of Pure Mathematics. In my own limited experience of examinations, the fault which I find with the men is a want
of analytical power, and that whatever else may have been in defect
Pure Mathematics has certainly not been in excess.

I remain, dear Sir,

Yours sincerely,

A. CAYLEY.

CAMBRIDGE,
  10th Dec., 1867.

1867, December 17.

MY DEAR SIR,

Since receiving your letter of 9th I positively have not
had time to express the single remark which I proposed to make
on it.

You state your idea that the educational element ought not to be
the predominating element in the University. "I do not think that
every thing should be subordinated to the educational element." I
cannot conceal my surprise at this sentiment. Assuredly the founders
of the Colleges intended them for education (so far as they apply to
persons in statu pupillari), the statutes of the University and the
Colleges are framed for education, and fathers send their sons to the
University for education. If I had not had your words before me, I
should have said that it is impossible to doubt this.

It is much to be desired that Professors and others who exercise
no control by force should take every method, not only of promoting

science in themselves, but also of placing the promoted science before students : and it is much to be desired that students who have passed the compulsory curriculum should be encouraged to proceed into the novelties which will be most agreeable to them. But this is a totally different thing from using the Compulsory Force of Examination to drive students in paths traced only by the taste of the examiner. For them, I conceive the obligation to the nation and the duty to follow the national sense on education (as far as it can be gathered from its best representatives) to be undoubted ; and to be, in the intensity of the obligation and duty, most serious.

I am, my dear Sir,

Yours very truly,

G. B. AIRY.

Professor Cayley.

1868

" In the South-East Dome, the alteration proposed last year for rendering the building fire-proof had been completely carried out. The middle room, which was to be appropriated to Chronometers, was being fitted up accordingly.—From the Report it appears that 'our subterranean telegraph wires were all broken by one blow, from an accident in the Metropolitan Drainage Works on Croom's Hill, but were speedily repaired.'—In my office as Chairman of successive Commissions on Standards, I had collected a number of Standards, some of great historical value (as Ramsden's and Roy's Standards of Length, Kater's Scale-beam for weighing great weights, and others), &c. These have been transferred to the newly-created Standards Department of the Board of Trade."
—In the Report is given a detailed account of the system of preserving and arranging the manuscripts and correspondence of the Observatory, which was always regarded by Airy as a matter of the first importance.—From a careful discussion of the results of observation Mr Stone had concluded that the refractions ought to be diminished. ' Relying on this, we

have now computed our mean refractions by diminishing those of Bessel's Fundamenta in the proportion of 1 to 0·99797.'—The Magnetometer-Indications for the period 1858—1863 had been reduced and discussed, with remarkable results. It is inferred that magnetic disturbances, both solar and lunar, are produced mediately by the Earth, and that the Earth in periods of several years undergoes changes which fit it and unfit it for exercising a powerful mediate action.—The Earth-current records had been reduced, and the magnetic effect which the currents would produce had been computed. The result was, that the agreement between the magnetic effects so computed and the magnetic disturbances really recorded by the magnetometers was such as to leave no doubt on the general validity of the explanation of the great storm-disturbances of the magnets as consequences of the galvanic currents through the earth.—Referring to the difficulty experienced in making the meteorological observations practically available the Report states thus: 'The want of Meteorology, at the present time, is principally in suggestive theory.'—In this year Airy communicated to the Royal Astronomical Society a Paper 'On the Preparatory Arrangements for the Observation of the Transits of Venus 1874 and 1882': this subject was now well in hand.—The First Report of the Commissioners (of whom he was Chairman) appointed to enquire into the condition of the Exchequer Standards was printed: this business took up much time.—He was in this year much engaged on the Coinage Commission.

Of private history: There was the usual winter visit to Playford, and a short visit to Cambridge in June.—From about Aug. 1st to Sept. 3rd he was travelling in Switzerland with his youngest son and his two youngest daughters. In the course of this journey they visited Zermatt. There had been much rain, the rivers were greatly flooded, and much mischief was done to the roads. During the journey from Visp to Zermatt, near St Nicholas, in a steep part of the gorge, a large stone rolled from the cliffs and knocked their baggage horse over the lower precipice, a fall of several hundred feet. The packages were all burst, and many things were lost, but a good deal was recovered by men suspended by ropes.

In this year also Airy was busy with the subject of University Examination, which in previous years had occupied so much of his attention, as will be seen from the following letters:

Royal Observatory, Greenwich,
London, S.E.
1868, *March* 12.

My dear Master,

I have had the pleasure of corresponding with you on matters of University Examination so frequently that I at once turn to you as the proper person to whom I may address any remarks on that important subject.

Circumstances have enabled me lately to obtain private information of a most accurate kind on the late Mathematical Tripos: and among other things, I have received a statement of every individual question answered or partly answered by five honour-men. I have collected the numbers of these in a small table which I enclose.

I am struck with the *almost* nugatory character of the five days' honour examination as applied to Senior Optimes, and I do not doubt that it is *totally* nugatory as applied to Junior Optimes. It appears to me that, for all that depends on these days, the rank of the Optimes is mere matter of chance.

In the examinations of the Civil Service, the whole number of marks is published, and also the number of marks gained by each candidate. I have none of their papers at hand, but my impression is that the lowest candidates make about 1 in 3; and the fair candidates about 2 in 3, instead of 1 in 10 or 1 in 13 as our good Senior Optimes.

I am, my dear Master,
Very truly yours,
G. B. AIRY.

The Rev. Dr Cookson,
   Master of St Peter's College,
      &c. &c.

The Table referred to in the above letter is as follows :

Number of Questions, and numbers of Answers to Questions as given by several Wranglers and Senior Optimes, in the Examination of Mathematical Tripos for Honours, 1868, January 13, 14, 15, 16, 17.

Number of Questions and Riders in the Printed Papers.

	Questions.	Riders.	Aggregate.
In the 10 Papers of the 5 days	123	101	224

NUMBER OF QUESTIONS AND RIDERS ANSWERED.

	Questions.	Riders.	Aggregate.	
By a Wrangler, between the 1st and 7th	$69\frac{1}{2}$	$25\frac{1}{2}$	95	1 in 2·36
By a Wrangler, between the 12th and 22nd.............................	$48\frac{1}{2}$	$12\frac{1}{2}$	61	1 in 3·68
By a Wrangler, between the 22nd and 32nd	36	$12\frac{1}{2}$	$48\frac{1}{2}$	1 in 4·62
By a Sen. Opt. between the 1st and 10th....................................	$17\frac{1}{2}$	5	$22\frac{1}{2}$	1 in 9·95
By a Sen. Opt. between the 10th and 20th.................................	$14\frac{1}{2}$	2	$16\frac{1}{2}$	1 in 13·60

G. B. AIRY.

1868, *March* 12.

ST PETER'S COLLEGE LODGE, CAMBRIDGE,
March 13*th*, 1868.

MY DEAR SIR,

I am much obliged by your letter and enclosed paper.

Any thing done in the last five days by a Junior Optime only shews (generally) that he has been employing some of his time *mischievously*, for he must have been working at subjects which he is quite unable to master or cramming them by heart on the chance of meeting with a stray question which he may answer.

The chief part of the Senior Optimes are in something of the same situation.

I think that the proposed addition of a day to the first part of the Examination, in which "easy questions in physical subjects" may be set, is, on this account, a great improvement.

Our new Scheme comes on for discussion on Friday next, March 20, at 2 p.m. in the Arts School. It is much opposed by

private tutors, examiners and others, and may possibly be thrown out
in the Senate this year, though I hope that with a little patience it
may be carried, in an unmutilated form, eventually.

The enclosed Report on the Smith's Prize Examination will be
discussed at the same time.

I will consider what is best to be done on the subject to which
your note refers, without delay.   With many thanks,

I am,

Very faithfully yours,
H. W. COOKSON.

The Astronomer Royal.

In this year certain Members of the Senate of the University of Cambridge petitioned Parliament against the abolition of religious declarations required of persons admitted to
Fellowships or proceeding to the degree of M.A.   The document was sent to Airy for his signature, and his reply was as
follows :

ROYAL OBSERVATORY, GREENWICH,
LONDON, S.E.
1868, <i>March</i> 18.</div>

MY DEAR SIR,

Though I sympathize to a great extent ·with the
prayer of the petition to Parliament which you sent to me yesterday,
and assent to most of the reasons, I do not attach my signature to
it, for the following considerations :

1.   I understand, from the introductory clause, and from the
unqualified character of the phrase "any such measures" in the
second clause, that the petition objects to granting the M.A. degree
without religious declaration.   I do not see any adequate necessity
for this objection, and I cannot join in it.

2.   It appears to me that the Colleges were intended for two
collateral objects :—instruction by part of the Fellows, on a religious basis; and support of certain Fellows for scientific purposes, without the same ostentatious connection with religion.   I
like this spirit.well, and should be glad to maintain it.

3.   I therefore think (as I have publicly stated before) that the
Master of the College ought to be in holy orders; and so ought

those of the Fellows who may be expected to be usually resident and to take continuous part in the instruction. But there are many who, upon taking a fellowship, at once lay aside all thoughts of this : and I think that such persons ought not to be trammelled with declarations.

4. My modification of existing regulations, if it once got into shape, would I dare say be but a small fraction of that proposed by the "measures in contemplation." Still I do not like to join in unqualified resistance to interference in the affairs of the Established Colleges, with that generality of opposition to interference which the petition seems to intimate.

I agree with articles 3, 4, and 5 ; and I am pleased with the graceful allusion in article 4 to the assistance which has been rendered by the Colleges, and by none perhaps so honourably as Trinity, to the parishes connected with it. And I could much wish that the spirit of 3 and 5 could be carried out, with some concession to my ideas in *my* paragraph 3, above.

I am, my dear Sir,

Yours very truly,

G. B. AIRY.

Rev. Dr Lightfoot.

1869

From the Report to the Board of Visitors it appears that application had been made for an extension of the grounds of the Observatory to a distance of 100 feet south of the Magnetic Ground, and that a Warrant for the annexation of this space was signed on 1868, Dec. 8. The new Depôt for the Printed Productions of the Observatory had been transferred to its position in the new ground, and the foundations for the Great Shed were completed.—"The courses of our wires for the registration of spontaneous terrestrial galvanic currents have been entirely changed. The lines to Croydon and Deptford are abandoned ; and for these are substituted, a line from Angerstein Wharf to Lady Well Station, and a line from North Kent Junction to Morden

College Tunnel.   At each of these points the communication with Earth is made by a copper plate 2 feet square. The straight line connecting the extreme points of the first station intersects that connecting the two points of the second station, nearly at right angles, and at little distance from the Observatory.—The question of dependence of the measurable amount of sidereal aberration upon the thickness of glass or other transparent material in the telescope (a question which involves, theoretically, one of the most delicate points in the Undulatory Theory of Light) has lately been agitated on the Continent with much earnestness.   I have calculated the curvatures of the lenses of crown and flint glass (the flint being exterior) for correcting spherical and chromatic aberration in a telescope whose tube is filled with water, and have instructed Mr Simms to proceed with the preparation of an instrument carrying such a telescope. I have not finally decided whether to rely on Zenith-distances of γ Draconis or on right-ascensions of Polaris.   In any form the experiment will probably be troublesome.—The transit of Mercury on 1868, Nov. 4th, was observed by six observers.   The atmospheric conditions were favourable; and the singular appearances usually presented in a planetary transit were well seen.—Mr Stone has attached to the South-East Equatoreal a thermo-multiplier, with the view of examining whether heat radiating from the principal stars can be made sensible in our instruments.   The results hitherto obtained are encouraging, but they shew clearly that it is vain to attempt this enquiry except in the most superb weather; and there has not been a night deserving that epithet for some months past.—The preparations for observing the Transits of Venus were now begun in earnest. I had come to the conclusion, that after every reliance was placed on foreign and colonial observatories, it would be necessary for the British Government to undertake the equipment of five or six temporary stations.   On Feb. 15th I sent a pamphlet on the subject to Mr Childers (First

Lord of Admiralty), and in April I wrote to the Secretary, asking authority for the purchase of instruments. On June 22nd authority is given to me for the instruments: the Treasury assent to £10,500. On August 9th I had purchased 3 equatoreals.—I have given a short course of Lectures in the University of Cambridge on the subject of Magnetism, with the view of introducing that important physical science into the studies of the University. The want of books available to Students, and the novelty of the subject, made the preparation more laborious than the duration of the lectures would seem to imply."—In this year there was much work on the Standards Commission, chiefly regarding the suggested abolition of Troy Weight, and several Papers on the subject were prepared by Airy.— He also wrote a long and careful description of the Great Equatoreal at Greenwich.

Of private history: There was the usual visit to Playford in the winter. Mrs Airy was now becoming feebler, and did not now leave Greenwich: since April of this year her letters were written in pencil, and with difficulty, but she still made great efforts to keep up the accustomed correspondence.—In April Airy went to Cambridge to deliver his lectures on magnetism to the undergraduates: the following passage occurs in one of his letters at this time: "I have a mighty attendance (there were 147 names on my board yesterday), and, though the room is large with plenty of benches, I have been obliged to bring in some chairs. The men are exceedingly attentive, and when I look up I am quite struck to see the number of faces staring into mine. I go at 12, and find men at the room copying from my big papers: I lecture from 1 to 2, and stop till after 3, and through the last hour some men are talking to me and others are copying from the papers; and I usually leave some men still at work. The men applaud and shew their respect very gracefully. There are present some two or three persons who attended my former lectures, and they say that I lecture exactly as I did formerly. One of my attendants is a man that they say cannot, from years and infirmity and habit, be induced to go anywhere else: Dr Archdall, the Master of Emmanuel. I find that some of my old lecturing habits come again on me. I drink a great deal of cold water, and am very glad to go to bed early."—From June 10th—30th he was travelling in Scotland, and staying at Barrow House near Keswick (the residence of Mr Langton), with his son Hubert.—Subsequently, from Aug. 17th to 31st, he was again in the Lake District, with his daughter Christabel, and was joined there by his son Hubert on the 24th. The first part of the time was spent at Tarn Bank, near Carlisle, the residence of Mr Isaac Fletcher, M.P. From thence he made several expeditions, especially to Barrow in Furness and Seascale, where he witnessed with great interest the Bessemer process of making steel. From Barrow House he made continual excursions among the Cumberland mountains, which he knew so well.

1870

"In this year Mr Stone, the First Assistant, was appointed to the Cape of Good Hope Observatory, and resigned his post of First Assistant. Mr Christie was appointed in his place.—From the Report to the Visitors it appears that 'A few months since we were annoyed by a failure in the illumination of the field of view of the Transit Circle. The reflector was cleaned, but in vain; at last it was discovered that one of the lenses (the convex lens) of the combination which forms the object-glass of a Reversed Telescope in the interior of the Transit-axis, and through which all illuminating light must pass, had become so corroded as to be almost opaque.'— The South-East Equatoreal has been partly occupied with the thermo-multiplier employed by Mr Stone for the measure of heat radiating from the principal

stars. Mr Stone's results for the radiation from Arcturus and
a Lyræ appear to be incontrovertible, and to give bases for
distinct numerical estimation of the radiant heat of these
stars.—In my last Report I alluded to a proposed systematic
reduction of the meteorological observations during the whole
time of their efficient self-registration. Having received from
the Admiralty the funds necessary for immediate operations,
I have commenced with the photographic registers of the
thermometers, dry-bulb and wet-bulb, from 1848 to 1868.—
Our chronometer-room contains at present 219 chronometers,
including 37 chronometers which have been placed here by
chronometer-makers as competing for the honorary reputation
and the pecuniary advantages to be derived from success in
the half-year's trial to which they are subjected. I take this
opportunity of stating that I have uniformly advocated the
policy of offering good prices for the chronometers of great
excellence, and that I have given much attention to the
decision on their merits; and I am convinced that this
system has greatly contributed to the remarkably steady
improvement in the performance of chronometers. In the
trial which terminated in August 1869, the best chronometers
(taking as usual the average of the first six) were superior in
merit to those of any preceding year.—With the funds placed
at my disposal for the Transit of Venus 1874 I purchased
three 6-inch equatoreals, and have ordered two: I have also
ordered altazimuths (with accurate vertical circles only), and
clocks sufficient, as I expect, to equip five stations. For
methods of observation, I rely generally on the simple eye-
observation, possibly relieved of some of its uncertainty by the
use of my colour-correcting eyepiece. But active discussion
has taken place on the feasibility of using photographic and
spectroscopic methods; and it will not be easy for some time
to announce that the plan of observations is settled.—There
can be no doubt, I imagine, that the first and necessary duty
of the Royal Observatory is to maintain its place well as an
Observing Establishment; and that this must be secured, at

whatever sacrifice, if necessary, of other pursuits. Still the question has not unfrequently presented itself to me, whether the duties to which I allude have not, by force of circumstances, become too exclusive; and whether the cause of Science might not gain if, as in the Imperial Observatory of Paris for instance, the higher branches of mathematical physics should not take their place by the side of Observatory routine. I have often felt the desire practically to refresh my acquaintance with what were once favourite subjects: Lunar Theory and Physical Optics. But I do not at present clearly see how I can enter upon them with that degree of freedom of thought which is necessary for success in abstruse investigations."

Of private history: There was a longer visit than usual to Playford, lasting till Jan. 27th.—In April he made a short excursion (of less than a week) with his son Hubert to Monmouth, &c.—From June 14th to July 2nd he was staying at Barrow House, near Keswick, with his son Hubert: during this time he was much troubled with a painful skin-irritation of his leg and back, which lasted in some degree for a long time afterwards.—From Sept. 25th to Oct. 6th he made an excursion with his daughter Christabel to Scarborough, Whitby, &c., and again spent a few days at Barrow House.

1871

" In April 1870 the Assistants had applied for an increase of salary, a request which I had urged strongly upon the Admiralty. On Jan. 27 of this year the Admiralty answered that, on account of Mr Childers's illness, the consideration must be deferred to next year! The Assistants wrote bitterly to me: and with my sanction they wrote to the First Lord. On Jan. 31st I requested an interview with Mr Baxter (secretary of the Admiralty), and saw him on Feb. 3rd, when I obtained his consent to an addition of

£530. There was still a difficulty with the Treasury, but on June 27th the liberal scale was allowed.—Experiments made by Mr Stone shew clearly that a local elevation, like that of the Royal Observatory on the hill of Greenwich Park, has no tendency to diminish the effect of railway tremors.— The correction for level error in the Transit Circle having become inconveniently large, a sheet of very thin paper, $\frac{1}{270}$ inch in thickness, was placed under the eastern Y, which was raised from its bed for the purpose. The mean annual value of the level-error appears to be now sensibly zero.—As the siege and war operations in Paris seriously interfered with the observations of small planets made at the Paris Observatory, observations of them were continued at Greenwich throughout each entire lunation during the investment of the city.—The new Water-Telescope has been got into working order, and performs most satisfactorily. Observations of γ Draconis have been made with it, when the star passed between 20^{h} and 17^{h}, with some observations for adjustment at a still more advanced time. As the astronomical latitude of the place of observation is not known, the bearing of these observations on the question of aberration cannot be certainly pronounced until the autumn observations shall have been made; but supposing the geodetic latitude to be accordant with the astronomical latitude, the result for aberration appears to be sensibly the same as with ordinary telescopes.—Several years since, I prepared a barometer, by which the barometric fluctuations were enlarged, for the information of the public; its indications are exhibited on the wall, near to the entrance gate of the Observatory. A card is now also exhibited, in a glass case near the public barometer, giving the highest and lowest readings of the thermometer in the preceding twenty-four hours.—Those who have given attention to the history of Terrestrial Magnetism are aware that Halley's Magnetic Chart is very frequently cited; but I could not learn that any person, at least in modern times, had seen it. At last I discovered a copy in the library of the British Museum, and have been allowed to take copies by photolithography. These are appended to the Magnetical and Meteorological Volume for 1869.—The trials and certificates of hand-telescopes for the use of the Royal Navy have lately been so frequent that they almost become a regular part of the work of the Observatory. I may state here that by availing myself of a theory of eyepieces which I published long since in the Cambridge Transactions, I have been able to effect a considerable improvement in the telescopes furnished to the Admiralty.— The occurrence of the Total Eclipse of the Sun in December last has brought much labour upon the Observatory. As regards the assistants and computers, the actual observation on a complicated plan with the Great Equatoreal (a plan for which few equatoreals are sufficiently steady, but which when properly carried out gives a most complete solution of the geometrical problem) has required, in observation and in computation, a large expenditure of time.—My preparations for the Transit of Venus have respect only to eye-observation of contact of limbs. With all the liabilities and defects to which it is subject, this method possesses the inestimable advantage of placing no reliance on instrumental scales. I hope that the error of observation may not exceed four seconds of time, corresponding to about $0\cdot13''$ of arc. I shall be very glad to see, in a detailed form, a plan for making the proper measures by heliometric or photographic apparatus; and should take great interest in combining these with the eye-observations, if my selected stations can be made available. But my present impression is one of doubt on the certainty of equality of parts in the scale employed. An error depending on this cause could not be diminished by any repetition of observations."—After referring to the desirability of vigorously prosecuting the Meteorological Reductions (already begun) and of discussing the Magnetic Observations, the Report concludes thus: "There is another consideration which very often presents itself to my mind; the waste of labour in the repetition of observations at different observatories. . . . . I think that this consideration ought not to be put out of sight in planning the courses of different Observatories."—In this year De Launay's Lunar Theory was published. This valuable work was of great service to Airy in the preparation of the Numerical Lunar Theory, which he subsequently undertook.—In the latter part of this year Airy was elected President of the Royal Society, and held the office during 1872 and 1873. At this time he was much pressed with work, and could ill afford to take up additional duties, as the following quotation from a letter to one of his friends shews: "The election to the Presidency of R. S. is flattering, and has brought to me the friendly remembrances of many persons; but in its material and laborious connections, I could well have dispensed with it, and should have done so but for the respectful way in which it was pressed on me."

Of private history: There was the usual winter visit to Playford.—In April he made a short trip to Cornwall with his daughter Annot.—In June he was appointed a Companion of the Bath, and was presented at Court on his appointment. —Mrs Airy was staying with her daughter, Mrs Routh, at Hunstanton, during June, her state of health being somewhat improved. — From August 1st to 28th he was chiefly in Cumberland, at Barrow House, and at Grange, Borrowdale, where his son Osmund was staying for a holiday.

1872

"From the Report to the Board of Visitors it appears that 'The Normal Sidereal Clock for giving sidereal time by galvanic communication to the Astronomical Observatory was established in the Magnetic Basement in 1871, June; that locality being adapted for it on account of the uniformity of temperature, the daily changes of temperature rarely exceeding 1° Fahrenheit. Its escapement is one which I suggested many years ago in the Cambridge Transactions; a detached escapement, very closely analogous to the ordinary chronometer escapement, the pendulum receiving an impulse only at alternate vibrations....The steadiness of rate is very far superior to any that we have previously attained.'—The aspect of railway enterprise is at present favourable to the Park and to the Observatory. The South-Eastern Railway Company has made an arrangement with the Metropolitan Board of Works for shifting the course of the great Southern Outfall Sewer. This enables the Company to trace a new line for the railway, passing on the north side of London Street, at such a distance from the Observatory as to remove all cause of alarm. I understand that the Bill, which was unopposed, has passed the Committee of the House of Commons. I trust that the contest, which has lasted thirty-seven years, is now terminated.—The observations of γ Draconis with the Water-Telescope, made in the autumn of 1871, and the spring of 1872, are reduced, the latter only in their first steps....Using the values of the level scales as determined by Mr Simms (which I have no reason to believe to be inaccurate) the spring and autumn observations of 1871 absolutely negative the idea of any effect being produced on the constant of aberration by the amount of refracting medium traversed by the light.—The great Aurora of 1872 Feb. 4 was well observed. On this occasion the term Borealis would have been a misnomer, for the phenomenon began in the South and was most conspicuous in the South. Three times in the evening it exhibited that umbrella-like appearance which has been called (perhaps inaccurately) a corona. I have very carefully compared its momentary phenomena with the corresponding movements of the magnetometers. In some of the most critical times, the comparison fails on account of the violent movements and consequent faint traces of the magnetometers. I have not been able to connect the phases of aurora and those of magnetic disturbance very distinctly.—The Report contains a detailed account of the heavy preparations for the observation of the Transit of Venus 1874, including the portable buildings for the instruments, the instruments themselves (being a transit-instrument, an altazimuth, and an equatoreal, for each station), and first class and second-class clocks, all sufficient for the equipment of 5 stations, and continues thus: I was made aware of the assent of the Government to the wish of the Board of Visitors, as expressed at their last meeting, that provision should be made for the application of photography to the observation of the Transit of Venus. It is unnecessary for me to remark that our hope of success is founded entirely on our confidence in Mr De La Rue. Under his direction, Mr Dallmeyer has advanced far in the preparation of five photoheliographs.... The subject is recognized by many astronomers as not wholly free from difficulties, but it is generally believed that these difficulties may be overcome, and Mr De La Rue is giving careful attention to the most important of them.—I take this opportunity of reporting to the Board that the Observatory was honoured by a visit of His Majesty the Emperor of Brazil, who minutely examined every part."—After referring to various subjects which in his opinion might be usefully pursued systematically at the Observatory, the Report proceeds thus: "'The character of the Observatory would be somewhat changed by this innovation, but not, as I imagine, in a direction to which any objection can be màde. It would become, pro tanto, a physical observatory; and possibly in time its operations might be extended still further in a physical direction.'—The consideration of possible changes in the future of the Observatory leads me to the recollection of actual changes in the past. In my Annual Reports to the Visitors I have endeavoured to chronicle these; but still there will be many circumstances which at present are known only to myself, but which ought not to be beyond the reach of history. I have therefore lately employed some time in drawing up a series of skeleton annals of the Observatory (which unavoid-

ably partakes in some measure of the form of biography),
and have carried it through the critical period, 1836—1851.
If I should command sufficient leisure to bring it down to
1861, I think that I might then very well stop." (The skeleton
annals here referred to are undoubtedly the manuscript notes
which form the basis of the present biography.   Ed.)—"On
Feb. 23rd in this year I first (privately) formed the notion of
preparing a numerical Lunar Theory by substituting Delaunay's numbers in the proper Equations and seeing what
would come of it."

Of private history : There was the usual visit to Playford
—in this year later than usual—from Feb. 4th to Mar. 4th.
The letters written during this visit are, as usual, full of
freshness and delight at finding himself in his favourite
country village.—On June 5th he went to Barrow House,
near Keswick, to be present at the marriage of his second
son Hubert to Miss S. C. Langton, daughter of Z. Langton
Esq., of Barrow House.—After the wedding he made a trip
through the Trossachs district of Scotland with his daughter
Annot, and returned to Greenwich on June 17th.

On the 26th June 1872 Airy was appointed a Knight
Commander of the Most Honourable Order of the Bath : he
was knighted by the Queen at Osborne on the 30th of July.
In the course of his official career he had three times been
offered Knighthood, and had each time declined it : but it
seemed now as if his scruples on the subject were removed,
and it is probable that he felt gratified by the public recognition of his services. Of course the occasion produced
many letters of congratulation from his friends : to one
of these he replied as follows : " The real charm of these
public compliments seems to be, that they excite the
sympathies and elicit the kind expressions of private friends
or of official superiors as well as subordinates. In every
way I have derived pleasure from these." From the Assistants of the Royal Observatory he received a hearty letter
of congratulation containing the following paragraph. "Our

position has naturally given us peculiar opportunities for perceiving the high and broad purposes which have characterized your many and great undertakings, and of witnessing the untiring zeal and self-denial with which they have been pursued."

On the 18th of March 1872 Airy was nominated a Foreign Associate of the Institut de France, to fill the place vacant by the death of Sir John Herschel. The following letter of acknowledgment shews how much he was gratified by this high scientific honour :

ROYAL OBSERVATORY, GREENWICH,
1872, *March* 23.

À Messieurs
    Messieurs ELIE DE BEAUMONT,
        et J. B. DUMAS,
    Secrétaires perpetuels de l'Académie
        des Sciences, Institut de France.

GENTLEMEN,

        I am honoured with your letter of March 18, communicating to me my nomination by the Academy of Sciences to the place rendered vacant in the class of Foreign Associates of the Academy by the decease of Sir John Herschel, and enclosing Copy of the Decree of the President of the French Republic approving the Election.

    It is almost unnecessary for me to attempt to express to you the pride and gratification with which I receive this announcement. By universal consent, the title of *Associé Etranger de l'Académie des Sciences* is recognised as the highest distinction to which any man of science can aspire ; and I can scarcely imagine that, unless by the flattering interpretation of my friends in the Academy, I am entitled to bear it. But in any case, I am delighted to feel that the bands of friendship are drawn closer between myself and the distinguished body whom, partly by personal intercourse, partly by correspondence, and in every instance by reputation, I have known so long.

I beg that you will convey to the Academy my long-felt esteem
for that body in its scientific capacity, and my deep recognition of
its friendship to me and of the honor which it has conferred on me
in the late election.

I have the honor to be

Gentlemen,

Your very faithful servant,

G. B. AIRY.

On the 20th November 1872 Airy was nominated a
Grand Cross in the Imperial Order of the Rose of Brazil:
the insignia of the Order were accompanied by an autograph
letter from the Emperor of Brazil, of which the following is a

MONSIEUR,

Vous êtes un des doyens de la science, et le Président de
l'illustre Société, qui a eu la bienveillance d'inscrire mon nom parmi
ceux de ses associés. La manière, dont vous m'avez fait les hon-
neurs de votre Observatoire m'a imposé aussi l'agréable devoir
d'indiquer votre nom à l'empereur de Brésil pour un témoignage de
haute estime, dont je suis fort heureux de vous faire part personelle-
ment, en vous envoyant les décorations que vous garderez, au moins,
comme un souvenir de ma visite à Greenwich.

J'espère que vous m'informerez, quand il vous sera aisé, des
travaux de votre observatoire, et surtout de ce que l'on aura fait
pour l'observation du passage de Vénus et la détermination exacte
de la passage.

J'ai reçu déjà les *Proceedings de la Royal Society* lesquels m'inté-
ressent vivement.

Je voudrais vous écrire dans votre langue, mais, comme je n'en
ai pas l'habitude, j'ai craigné de ne pas vous exprimer tout-à-fait les
sentiments de

Votre affectionné,

D. PEDRO D'ALCANTARA.

RIO,
22 *Octobre*, 1872.

Airy's reply was as follows:

ROYAL OBSERVATORY, GREENWICH,
1872, *November* 26.

SIRE,

I am honoured with your Imperial Majesty's autograph letter of October 22 informing me that, on considering the attention which the Royal Society of London had been able to offer to your Majesty, as well as the explanation of the various parts of the establishment of this Observatory which I had the honor and the high gratification to communicate, You had been pleased to place my name in the Imperial Order of the Rose, and to present to me the Decorations of Grand Cross of that Order.

With pride I receive this proof of Your Majesty's recollection of your visit to the scientific institutions of Great Britain.

The Diploma of the appointment to the Order of the Rose, under the Imperial Sign Manual, together with the Decorations of the Order, have been transmitted to me by his Excellency Don Pereira de Andrada, Your Majesty's Representative at the British Court.

Your Majesty has been pleased to advert to the approaching Transit of Venus, on the preparations for which you found me engaged. It is unfortunate that the Transit of 1874 will not be visible at Rio de Janeiro. For that of 1882, Rio will be a favourable position, and we reckon on the observations to be made there. Your Majesty may be assured that I shall loyally bear in mind your desire to be informed of any remarkable enterprise of this Observatory, or of any principal step in the preparations for the Transit of Venus and of its results.

I have the honor to be

Sire,

Your Imperial Majesty's very faithful servant,

G. B. AIRY.

To His Majesty
The Emperor of Brazil.

———

Airy's old friend, Adam Sedgwick, was now very aged and infirm, but his spirit was still vigorous, and he was warm-hearted as ever. The following letter from him (probably the last of their long correspondence) was written in this year, and appears characteristic:

TRINITY COLLEGE, CAMBRIDGE,
May 10, 1872.

MY DEAR AIRY,

I have received your card of invitation for the 1st of June, and with great joy should I count upon that day if I thought that I should be able to accept your invitation : but alas I have no hope of the kind, for that humiliating malady which now has fastened upon me for a full year and a half has not let go its hold, nor is it likely to do so. A man who is journeying in the 88th year of his pilgrimage is not likely to throw off such a chronic malady. Indeed were I well enough to come I am deaf as a post and half blind, and if I were with you I should only be able to play dummy. Several years have passed away since I was last at your Visitation and I had great joy in seeing Mrs Airy and some lady friends at the Observatory, but I could not then attend the dinner. At that Meeting were many faces that I knew, but strangely altered by the rude handling of old Time, and there were many new faces which I had never seen before at a Royal Society Meeting; but worse than all, all the old faces were away. In vain I looked round for Wollaston, Davy, Davies Gilbert, Barrow, Troughton, &c. &c. ; and the merry companion Admiral Smyth was also away, so that my last visit had its sorrowful side. But why should I bother you with these old man's mopings.

I send an old man's blessing and an old man's love to all the members of your family ; especially to Mrs Airy, the oldest and dearest of my lady friends.

I remain, my dear Airy,

Your true-hearted old friend,

his
ADAM X SEDGWICK.
mark

P.S. Shall I ever again gaze with wonder and delight from the great window of your Observatory,

The body of the above letter is in the handwriting of an amanuensis, but the signature and Postscript are in Sedgwick's handwriting. (Ed.)

1873

"Chronographic registration having been established at the Paris Observatory, Mr Hilgard, principal officer of the American Coast Survey, has made use of it for determining the longitude of Harvard from Greenwich, through Paris, Brest, and St Pierre. For this purpose Mr Hilgard's Transit Instrument was planted in the Magnetic Court. I understand that the result does not sensibly differ from that obtained by Mr Gould, through Valentia and Newfoundland.—It was known to the scientific world that several of the original thermometers, constructed by Mr Sheepshanks (in the course of his preparation of the National Standard of Length) by independent calibration of the bores, and independent determination of the freezing and boiling points on arbitrary graduations, were still preserved at the Royal Observatory. It was lately stated to me by M. Tresca, the principal officer of the International Metrical Commission, that, in the late unhappy war in Paris, the French original thermometers were destroyed ; and M. Tresca requested that, if possible, some of the original thermometers made by Mr Sheepshanks might be appropriated to the use of the International Commission. I have therefore transferred to M. Tresca the three thermometers A. 6, S. 1, S. 2, with the documentary information relating to them, which was found in Mr Sheepshanks's papers ; retaining six thermometers of the same class in the Royal Observatory.—The Sidereal Standard Clock continues to give great satisfaction. I am considering (with the aid of Mr Buckney, of the firm of E. Dent and Co.) an arrangement for barometric correction, founded on the principle of action on the pendulum by means of a magnet which can be raised or lowered by the agency of a large barometer.—The Altazimuth has received some important alterations. An examination of the results of observations had made me dissatisfied

with the bearings of the horizontal pivots in their Y's.   Mr
Simms, at my request, changed the bearings in Y's for bearing
in segments of circles, a construction which has worked
admirably well in the pivots of the Transit Circle." (And in
various other respects the instrument appears to have received
a thorough overhauling. Ed.)—"With the consent of the Royal
Society and of the Kew Committee, the Kew Heliograph has
been planted in the new dome looking over the South Ground.
It is not yet finally adjusted.—Some magnetic observations
in the Britannia and Conway tubular bridges were made last
autumn.   For this purpose I detached an Assistant (Mr Carpenter), who was aided by Capt. Tupman, R.M.A.; in other
respects the enterprise was private and at private expense.—
The rates of the first six chronometers (in the annual trials)
are published, in a form which appears most likely to lead to
examination of the causes that influence their merits or
demerits.   This report is extensively distributed to British
and Foreign horologists and instrument-makers.   All these
artists appear to entertain the conviction that the careful
comparisons made at this Observatory, and the orderly form
of their publication, have contributed powerfully to the improvement of chronometers.—Very lately, application has
been made to me, through the Board of Trade, for plans and
other information regarding time-signal-balls, to assist in
guiding the authorities of the German Empire in the establishment of time signals at various ports of that State.   In
other foreign countries the system is extending, and is referred
to Greenwich as its origin.—The arrangements and preparations for the observation of the Transit of Venus occupied
much attention.   With regard to the photoheliographs it is
proposed to make trial of a plan proposed by M. Janssen, for
numerous photographs of Venus when very near to the Sun's
limb.   On Apr. 26th the engaging of photographic teachers
was sanctioned.   Observers were selected and engaged.   A
working model of the Transit was prepared, and the use of
De La Rue's Scale was practised.   There was some hostile

criticism of the stations selected for the observation of the
Transit, which necessitated a formal reply.—Reference is
made to the increase of facilities for making magnetical and
meteorological observations.   The inevitable result of it is,
that observations are produced in numbers so great that
complete reduction becomes almost impossible.   The labour
of reduction is very great, and it is concluded that, of the
enormous number of meteorological observations now made
at numerous observatories, very few can ever possess the
smallest utility.—Referring to my Numerical Lunar Theory :
on June 30th, 1873, a theory was formed, nearly but not perfectly complete.   Numerical development of powers of $a \div r$
and $r \div a$.   Factors of corrections to Delaunay first attempted,
but entirely in numerical form."—In March of this year Airy
was consulted by Mr W. H. Barlow, C.E., and Mr Thomas
Bouch (the Engineer of the Tay Bridge, which was blown
down in 1879, and of a proposed scheme for a Forth Bridge
in 1873) on the subject of the wind pressure, &c., that should
be allowed for in the construction of the bridge.   Airy's
report on this question is dated 1873, Apr. 9th: it was subsequently much referred to at the Official Enquiry into the
causes of the failure of the Tay Bridge.—At the end of this
year Airy resigned the Presidency of the Royal Society.   In
his Address to the Society on Dec. 1st he stated his reasons
in full, as follows : " the severity of official duties, which seem
to increase, while vigour to discharge them does not increase ;
and the distance of my residence....Another cause is a difficulty of hearing, which unfits me for effective action as Chairman of Council."

Of private history: There was the usual visit to Playford in January: also a short visit in May: and a third visit
at Christmas.—There was a short run in June, of about a
week, to Coniston, with one of his daughters.—And there was
a trip to Weymouth, &c., for about 10 days, with one of his
daughters, in the beginning of August.—On his return from
the last-mentioned trip, Airy found a letter from the Secretary

of the Swedish Legation, enclosing the Warrant under the Royal Sign Manual of His Majesty (Oscar), the King of Sweden and Norway, by which he was nominated as a First Class Commander of the Order of the North Star, and accompanying the Decorations of that Order.

1874

"In this year Mr Glaisher resigned his appointment: I placed his Department (Magnetical and Meteorological) under Mr Ellis.—A balance of peculiar construction has been made by Mr Oertling, from my instructions, and fixed near the public barometer at the Entrance Gate. This instrument enables the public to test any ordinary pound weight, shewing on a scale the number of grains by which it is too heavy or too light.—Fresh counterpoises have been attached to the Great Equatoreal to balance the additional weight of the new Spectroscope, which was finally received from Mr Browning's hands on May 2nd of the present year. The Spectroscope is specifically adapted to sweeping round the Sun's limb, with a view to mapping out the prominences, and is also available for work on Stars and Nebulæ, the dispersive power being very readily varied. An induction-coil, capable of giving a six-inch spark, has been made for this instrument by Mr Browning.—Some new classes of reductions of the meteorological observations from 1848 to 1868 have been undertaken and completed in the past year. The general state of this work is as follows: The diurnal changes of the dry-bulb thermometer, as depending on the month, on the temperature waves, on the barometric waves, on the overcast and cloudless states of the sky, and on the direction of the wind, have been computed and examined for the whole period; and the exhibition of the results is ready for press. The similar reductions for the wet-bulb thermometer are rapidly approaching completion.—Regarding the preparations for the Transit of Venus

Expeditions. Originally five stations were selected and fully equipped with equatoreals, transits, altazimuths, photoheliographs, and clocks; but I have since thought it desirable to supplement these by two branch stations in the Sandwich Islands and one in Kerguelen's Island; and the additional instruments thus required have been borrowed from various sources, so that there is now an abundant supply of instrumental means....There will thus be available for observation of the Transit of Venus 23 telescopes, nine of which will be provided with double-image-micrometers; and five photoheliographs; and for determination of local time, and latitude and longitude, there will be nine transits and six altazimuths. ...All the observers have undergone a course of training in photography; first, under a professional photographer, Mr Reynolds, and subsequently under Capt. Abney, R.E., whose new dry-plate process is to be adopted at all the British Stations....A Janssen slide, capable of taking 50 photographs of Venus and the neighbouring part of the Sun's limb at intervals of one second, has been made by Mr Dallmeyer for each of the five photoheliographs."—Attached to the Report to the Visitors is a copy of the Instructions to Observers engaged in the Transit of Venus Expeditions, prepared with great care and in remarkable detail.—" In the past spring I published in the Monthly Notices of the Royal Astronomical Society a statement of the fundamental points in a new treatment of the Lunar Theory, by which, availing myself of all that has been done in the best algebraical investigations of that theory, I trust to be able by numerical operations only to give greater accuracy to final results. Considerable progress has been made in the extensive numerical developments, the work being done, at my private expense, entirely by a junior computer; and I hope, at any rate, to put it in such a state that there will be no liability to its entire loss. When this was reported to the Board of Visitors, it was resolved on the motion of Prof. Stokes, that this work, as a public expense, ought to be borne by the Government; and this was forwarded to the Admiralty. On June 24th I wrote to the Secretary of the Admiralty, asking for £100 for the present year, which after the usual enquiries and explanations was sanctioned on Aug. 29th."

Of private history: There were short visits to Playford in January, June, and October, but only for a few days in each case.—In March there was a run of two or three days to Newnham (on the Severn) to see the Bore on the Severn, and to Malvern.—In July he went to Newcastle to observe with Mr Newall's great telescope, but the weather was unfavourable: he then went on to Barrow House near Keswick, and spent a few days there, with excursions among the mountains.—On Aug. 13th he went with his daughter Christabel to the Isle of Arran, and then by Glasgow to the Trosachs, where he made several excursions to verify the localities mentioned in the "Lady of the Lake."—While in Scotland he heard of the death of his brother, the Rev. William Airy, and travelled to Keysoe in Bedfordshire to attend the funeral; and returned to Greenwich on Aug. 24th.

1875

"In October of this year I wrote to the Admiralty that I had grounds for asking for an increase of my salary: because the pension which had been settled on my wife, and which I had practically recognized as part of my salary, had been terminated by her death; so that my salary now stood lower by £200 than that of the Director of Studies of the Royal Naval College. The Admiralty reply favourably, and on Nov. 27th the Treasury raise my salary to £1,200.—For the service of the Clock Movement of the Great Equatoreal, a water-cistern has been established in the highest part of the Ball-Turret, the necessity for which arose from the following circumstance: The Water Clock was supplied by a small pipe, about 80 feet in length, connected with the

3-inch Observatory main (which passes through the Park), at a distance of about 250 feet from any other branch pipe. In spite of this distance I have seen that, on stopping the water-tap in the Battery-Basement under the North-East Turret, the pressure in the gauge of the Water Clock has been instantly increased by more than 40 lbs. per square inch. The consequent derangement of the Water Clock in its now incessant daily use became intolerable. Since the independent supply was provided, its performance has been most satisfactory.—With the Spectroscope the solar prominences have been mapped on 28 days only; but the weather of the past winter was exceptionally unfavourable for this class of observation. After mapping the prominences, as seen on the C line, the other lines, especially F and b, have been regularly examined, whenever practicable. Great care has been taken in determining the position, angle, and heights of the prominences in all cases. The spectrum of Coggia's Comet was examined at every available opportunity last July, and compared directly with that of carbon dioxide, the bands of the two spectra being sensibly coincident. Fifty-four measures of the displacement of lines in the spectra of 10 stars, as compared with the corresponding lines in the spectra of terrestrial elements (chiefly hydrogen), have been made, but some of these appear to be affected by a constant error depending on faulty adjustment of the Spectroscope.—Photographs of the Sun have been taken with the Kew Photoheliograph on 186 days; and of these 377 have been selected for preservation. The Moon, Jupiter, Saturn, and several stars (including the Pleiades and some double stars) have been photographed with the Great Equatoreal, with fairly satisfactory results, though further practice is required in this class of work.— I would mention a supplemental mechanism which I have myself introduced into some chronometers. I have long remarked that, in ordinary good chronometers, the freedom from irregularities depending on mechanical causes is most

remarkable; but that, after all the efforts of the most judicious makers, there is in nearly every case a perceptible
defect of thermal compensation. There is great difficulty
in correcting the residual fault, not only because an incon-
ceivably small movement of the weights on the balance-curve
is required, but also because it endangers the equilibrium
of the balance. The mechanism adopted to remedy the
defect is described in a Paper in the Horological Journal of
July 1875 by Mr W. Ellis, and has received the approval of
some able chronometer-makers.—With respect to the Transit
of Venus Expeditions : The parties from Egypt and Rodriguez are returned. I am in continual expectation of the
arrival of the other parties. I believe the eye-observations
and the ordinary photographs to be quite successful; I
doubt the advantage of the Janssen; one of the double-
image-micrometers seems to have failed; and the Zenith-
telescope gives some trouble. At three stations at Rodriguez, and three at Kerguelen, the observations appear to
have been most successful. At the Sandwich Islands, two
of the stations appear to have been perfectly successful
(except that I fear that the Janssen has failed), and a rich
series of lunar observations for longitude is obtained. At
New Zealand, I grieve to say, the observations were totally
lost, entirely in consequence of bad weather. There has been
little annoyance from the dreaded 'black drop.' Greater
inconvenience and doubt have been caused by the unexpected luminous ring round Venus.—With regard to the
progress of my proposed New Lunar Theory: Three computers are now steadily employed on the work. It will be
remembered that the detail and mass of this work are
purely numerical; every numerical coefficient being accompanied with a symbolical correction whose value will sometimes depend on the time, but in every case is ultimately
to be obtained in a numerical form. Of these coefficients,
extracted (for convenience) from Delaunay's results, there
are 100 for parallax, 182 for longitude, 142 for latitude;

the arguments being preserved in the usual form."—After reviewing the changes that had taken place at the Observatory during the past forty years, the Report to the Board of Visitors concludes thus: " I much desire to see the system of time-signals extended, by clocks or daily signals, to various parts of our great cities and our dockyards, and above all by hourly signals on the Start Point, which I believe would be the greatest of all benefits to nautical chronometry. Should any extension of our scientific work ever be contemplated, I would remark that the Observatory is not the place for new physical investigations. It is well adapted for following out any which, originating with private investigators, have been reduced to laws susceptible of verification by daily observation. The National Observatory will, I trust, always remain on the site where it was first planted, and which early acquired the name of ' Flamsteed Hill.' There are some inconveniences in the position, arising principally from the limited extent of the hill, but they are, in my opinion, very far overbalanced by its advantages."—In a letter on the subject of the Smith's Prizes Examination at Cambridge, which was always a matter of the greatest interest to him, Airy renewed his objections to the preponderance in the Papers of a class of Pure Mathematics, which he considered was never likely under any circumstances to give the slightest assistance to Physics. And, as before, these remarks called forth a rejoinder from Prof. Cayley, who was responsible for many of the questions of the class referred to.—In this year Airy completed his " Notes on the Earlier Hebrew Scriptures," which were shortly afterwards published as a book by Messrs Longmans, Green, & Co. In his letter to the publishers introducing the subject, he says, " For many years past I have at times put together a few sentences explanatory as I conceive of the geographical and historical circumstances connected with the principal events recorded in the Hebrew Scriptures. The view which I take is free, but I trust not irreverent. They terminate with a brief review of Colenso's great work. The collection now amounts to a small book." From the references already given in previous years to his Papers and correspondence on the geography of Exodus, his correspondence with Colenso, &c. &c., it will be seen that he took a great interest in the early history of the Israelites.—On August 10th, 1875, Airy celebrated the Bicentenary of the Royal Observatory by a dinner in the Octagon Room, which was attended by the Presidents of the Royal Society and the R. Astr. Society, and by a large number of Scientific gentlemen interested in Astronomy.— In February he was revising his Treatise on " Probabilities."

Of private history: up to Jan. 16th Airy was at Playford as usual.—For about a week in April he was in the Isle of Man with his daughter Christabel.—In June there was a short trip to Salisbury, Blandford, and Wimborne.—On August 12th he started with his daughter Annot for a holiday in Cumberland, but on the next day he was recalled by a telegram with the intelligence that a change for the worse had come over his wife's health. Lady Airy died on August 13th, 1875. For the last five years of her life she had been very helpless from the effects of a paralytic stroke—a very sad ending to a bright and happy life—and had been continually nursed throughout this time by her two unmarried daughters with the greatest self-denial and devotion. Her husband had been unremitting in his care and attention. Nothing was wanting that the most thoughtful kindness could supply. And in all his trips and excursions his constant and kind letters shewed how anxious he was that she should participate in all his interests and amusements. From the nature of the case it could hardly be said that her death was unexpected, and he received the shock with the manly steadiness which belonged to him. Lady Airy was buried in Playford churchyard.—From Sept. 22nd to Oct. 4 he made a short expedition to Wales (Capel Curig, &c.).—On Dec. 15th he attended the Commemoration at Trinity College, Cambridge.—On Dec. 22nd he went as usual to Playford.

In this year Airy received the high honour of the Freedom of the City of London, in the following communication:

STONE, Mayor.—A Common Council holden in the Chamber of the Guildhall of the City of London, on Thursday the 29th day of April 1875.

Resolved Unanimously

That the Freedom of this City in a Gold Box of the value of One hundred guineas be presented to Sir George Biddell Airy, K.C.B., D.C.L., LL.D. &c., Astronomer Royal, as a recognition of his indefatigable labours in Astronomy, and of his eminent services in the advancement of practical science, whereby he has so materially benefited the cause of Commerce and Civilization.

MONCKTON.

This Resolution was forwarded with a letter from Benjamin Scott, the Chamberlain.   Airy's reply was as follows:

ROYAL OBSERVATORY, GREENWICH, S.E.
1875, *May* 1.

DEAR SIR,

I have the honour to acknowledge your letter of April 30, accompanied with Copy of the Resolution of the Common Council of the City of London passed at their Meeting of April 29, under signature of the Town Clerk, That the Freedom of the City of London in a valuable Box be presented to me, in recognition of works stated in the Resolution.   And I am requested by you to inform you whether it is my intention to accept the compliment proposed by the Corporation.

In reply, I beg you to convey to the Right Honorable the Lord Mayor and the Corporation that I accept with the greatest pride and pleasure the honour which they propose to offer to me.   The Freedom of our Great City, conferred by the spontaneous act of its Municipal Governors, is in my estimation the highest honour which it is possible to receive; and its presentation at this time is peculiarly grateful to me.

                    I have the honour to be,

                              Sir,

                                   Your very obedient servant,

                                        G. B. AIRY.

Benjamin Scott, Esq.,
        &c.   &c.   &c.
Chamberlain of the Corporation of the
        City of London.

As it was technically necessary that a Freeman of the City of London should belong to one or other of the City Companies, the Worshipful Company of Spectacle Makers through their clerk (with very great appropriateness) enquired whether it would be agreeable that that Company should have the privilege of conferring their Honorary Freedom on him, and added: "In soliciting your acquiescence to the proposal I am directed to call attention to the fact that this Guild is permitted to claim all manufacturers of Mathematical and Astronomical Instruments within the City of London, which is now pleaded as an apology for the wish that one so distinguished as yourself in the use of such Instruments should be enrolled as a Member of this Craft." In his reply, accepting the Freedom of the Company, Airy wrote thus: "I shall much value the association with a body whose ostensible title bears so close a relation to the official engagements which have long occupied me. I have had extensive experience both in arranging and in using optical and mathematical instruments, and feel that my own pursuits are closely connected with the original employments of the Company." The Freedom of the Company was duly presented, and the occasion was celebrated by a banquet at the Albion Tavern on Tuesday, July 6th.

The Freedom of the City of London was conferred at

a Court of Common Council held at the Guildhall on Thursday the 4th of November. In presenting the gold box containing the Freedom, the Chamberlain, in an eloquent speech, first referred to the fact that this was the first occasion on which the Freedom had been conferred on a person whose name was associated with the sciences other than those of war and statecraft. He then referred to the solid character of his work, in that, while others had turned their attention to the more attractive fields of exploration, the discovery of new worlds or of novel celestial phænomena, he had incessantly devoted himself to the less interesting, less obtrusive, but more valuable walks of practical astronomy. And he instanced as the special grounds of the honour conferred, the compilation of nautical tables of extraordinary accuracy, the improvement of chronometers, the correction of the compasses of iron ships, the restoration of the standards of length and weight, and the Transit of Venus Expeditions. In his reply Airy stated that he regarded the honour just conferred upon him as the greatest and proudest ever received by him. He referred to the fact that the same honour had been previously conferred on the valued friend of his youth, Thomas Clarkson, and said that the circumstance of his succeeding such a man was to himself a great honour and pleasure. He alluded to his having received a small exhibition from one of the London Companies, when he was a poor undergraduate at Cambridge, and acknowledged the great assistance that it had been to him. With regard to his occupation, he said that he had followed it in a great measure because of its practical use, and thought it fortunate that from the first he was connected with an institution in which utility was combined with science. The occasion of this presentation was celebrated by a Banquet at the Mansion House on Saturday July 3rd, 1875, to Sir George Airy (Astronomer Royal) and the Representatives of Learned Societies.

There is no doubt that Airy was extremely gratified

by the honour that he had received. It was to him the crowning honour of his life, and coming last of all it threw all his other honours into the shade. To his independent and liberal spirit there was something peculiarly touching in the unsolicited approbation and act of so powerful and disinterested a body as the Corporation of the City of London.

CHAPTER IX.

1876

"AT the door from the Front Court to the staircase of the Octagon Room (the original entrance to the Observatory as erected by Sir Christopher Wren), a small porch-shelter has been often desired. I proposed to fix there a fan-roof of quadrantal form, covering the upper flat stone of the external steps.—On a critical examination of the micrometer-screws of the Transit Circle it was found that the corrections, which range from $-1\cdot38''$ to $+0\cdot76''$, indicate considerable wear in the screws; and it was found that as much as one-hundreth part of an inch had been worn away from some of the threads. The old screws were consequently discarded, and new ones were made by Mr Simms.—The adjustment of the Spectroscope has occupied a great deal of attention. There was astigmatism of the prisms; and false light reflected from the base of the prisms, causing loss both of light and of definition. The latter defect was corrected by altering the angles, and then astigmatism was corrected by a cylindrical lens near the slit. The definition in both planes was then found to be perfect.—The number of small planets has now become so great, and the interest of establishing the elements of all their orbits so small,—while at the same time the light of all those lately discovered is very faint, and the difficulty and doubt of observation greatly increased,—that I have begun to think seriously of limiting future observations to a small number of these objects.—All observations with the Spectroscope have been completely reduced ; the measures of lines in the spectra of elements being converted into corresponding wave-lengths, and the observations of displacement of lines in the spectra of stars being reduced so as to exhibit the concluded motion in miles per second, after applying a correction for the earth's motion. Sixteen measures of the F ·line in the spectrum of the Moon as compared with hydrogen give a displacement corresponding to a motion of less than two miles a second, which seems to shew that the method of comparison now adopted is free from systematic error ; and this is supported by the manner in which motions of approach and recession are distributed among the stars examined on each night of observation. The results recently obtained appear to be on the whole as consistent as can be expected in such delicate observations, and they support in a remarkable manner the conclusions of Dr Huggins, with regard to the motions of those stars which he examined.— Photographs of the sun have been taken with the photo-heliograph on 182 days. On one of the photographs, which was accidentally exposed while the drop slit was being drawn up, there appears to be a faint image of a cloud-like prominence close to the sun's limb, though the exposure probably only amounted to a fraction of a second. A prominence of unusual brilliancy was seen with the Spectroscope about the same time and in the same position with reference to the Sun's limb. All groups of Sun-spots and faculæ have been numbered, and the dates of their first and last appearances entered up to the present time. Areas of spots have been measured, and the measures have been reduced to millionths of the Sun's visible hemisphere.—The examination of the readings of the deep-sunk thermometers from 1846 to 1873 has exhibited some laws which had been sufficiently established before this time, and some which were less known.

Among the former were the successive retardations of seasons in successive descents, amounting to about four months at the depth of 25 feet; and the successive diminutions of the annual range of temperature. Among the latter is the character of the changes from year to year, which the great length of this series of observations brings well to light. It is found that from year to year the mean temperature of the surface for the year, varying by three or four degrees of Fahrenheit, follows in its changes the mean temperature of the atmosphere for the year, and that the changes of annual temperature are propagated downwards, retarded in phase and diminishing in amount of change, in the same manner (though probably not following the same law) as the season changes. The inference from this is, that changes of temperature come entirely from the exterior and in no discoverable degree from the interior; an inference which may be important in regard both to solar action and to geology.—Referring to the Transit of Venus observations: In the astronomical part of the reductions, there has been great labour and difficulty in the determination of local sidereal times; some books of observations required extensive transcription; some instrumental errors are still uncertain; the latter determinations have perplexed us so much that we are inclined to believe that, in spite of the great facilities of reduction given by the transit instrument, it would be better to rely on the altazimuth for time-determinations....... In the photographic part, I have confined my attention entirely to measures of the distance between the centres of the Sun and Planet, a troublesome and complex operation.—Referring to the progress of the Numerical Lunar Theory: With a repetition of grant from the Treasury, I have usually maintained four junior computers on this work. The progress, though considerable, has not been so great as I had hoped, by reason of the excessive personal pressure upon me during the whole year.—I wrote a letter of congratulation to Le Verrier on the completion of his great work of Planetary Tables.—On May 13th the Queen was at South Kensington, and I attended to explain the astronomical instruments, and shewed Her Majesty one of the Transit of Venus photographs."

Of private history: He returned from his Playford visit on the 18th of January.—In April there was a two-day trip to Colchester.—From June 13th to July 12th he was travelling in the North of Scotland and the Orkneys with his daughters, staying for a short time with Mr Webster, M.P., at Aberdeen, and with Mr Newall at Newcastle.—In September there was a week's run to Birkenhead and Keswick.—In November a week's run to Playford.—From the 13th to 15th of December he was at Cambridge, and on the 28th he went to Playford for the usual winter stay there.

1877

" In April of this year I was much engaged on the subject of Mr Gill's expedition to Ascension to observe for the determination of the parallax of Mars at the approaching opposition of that planet.—A large Direct-vision Spectroscope has been quite recently made by Mr Hilger under Mr Christie's direction on a new plan, in which either great dispersion or great purity of spectrum is obtained by the use of ' Half-prisms,' according as the incident pencil falls first on the perpendicular or on the oblique face. In this Spectroscope either one or two half prisms can be used at pleasure, according to the dispersion required, and there is facility for increasing the train to three or four half-prisms, though the dispersion with two only is nearly double of that given by the large ten-prism Spectroscope. The definition in this form of Spectroscope appears to be very fine.—At the end of May 1876, spectroscopic determinations of the Sun's rotation were made by observations of the relative displacement of the Fraunhofer lines at the east and west limbs respectively. The results are in close agreement with the value of the rotation found from observations of Sun-spots. A similar determination has also been made in the case of Jupiter, with equally satisfactory results.—An Electrometer on Sir William Thomson's plan, for continuous photographic registration of atmospheric electricity has been received from Mr White of Glasgow. It was mounted in December.—The computation of the photographic records of the barometer from 1854 to 1873 has so far advanced that we can assert positively that there is no trace of lunar tide in the atmosphere; but that there is a strongly marked semi-diurnal solar tide, accompanied with a smaller diurnal tide. We are at present engaged in comparing the barometric measures with the directions of the wind.—Regarding the distribution of the printed observations : There is no extensive wish for separate magnetic observations, but general magnetic results are in great demand, especially for mining operations, and to meet this a map of magnetic declination is furnished in the newspaper called the 'Colliery Guardian.'—As regards the operations for the Transit of Venus : The computing staff has by degrees been reduced to two junior computers within the Observatory; and one or two computers external to the Observatory, who are employed on large groups of systematic calculations. The principal part of the calculations remaining at the date of the last Report was that applying to the determination of the geographical longitudes of fundamental stations. At the moment of my writing, the last of these (the longitude of Observatory Bay, Kerguelen) is not absolutely finished :...... The method of determining the geographical longitude of the principal station in each group by vertical transits of the Moon has been found very successful at Honolulu and Rodriguez. For stations in high south latitude, horizontal transits are preferable.—As regards the Numerical Lunar Theory : With the view of preserving, against the ordinary chances of destruction or abandonment, a work which is already one of considerable magnitude, I have prepared and have printed as Appendix to the Greenwich Observations (with additional copies as for a separate work) the ordinary Equations of Lunar Disturbance, the novel theory of Symbolical Variations, and the Numerical Developments of the quantities on the first side of the Equations.—At various times from February to May I was engaged on the reduction of Malta Tides, and on a Paper concerning the same.—In July I was awarded the Albert Medal for my Compass corrections, and received the same from the Prince of Wales.—In February, Campbell's instrument for the registration of sunshine was introduced: it was mounted in July."

Of private history: "I was at Playford until Jan. 19th, in close correspondence as usual with Mr Christie at the Observatory, and attending to my Numerical Lunar Theory.—From Mar. 29th to Apr. 2nd I went on a short trip to Hereford, Worcester, &c. — From June 8th to 20th I was at Playford.—From Aug. 13th to Sept. 8th Airy was on an expedition in Ireland, chiefly in the North and West, with his daughters. When at Dublin he visited Grubb's instrument factory. On the return journey he stayed for some time in the Lake District of Cumberland, and took soundings in the neighbourhood of the place of the 'floating island' in Derwentwater."

Airy took the greatest interest in antiquarian matters, whether military or ecclesiastical, and his feelings on such matters is well illustrated by the following letter:

ROYAL OBSERVATORY, GREENWICH, S.E.
1877, *February* 27.

DEAR SIR,

I venture to ask if you can assist me in the following matter.

In the Parish Church of Playford, near Ipswich, Suffolk, was a splendid brass tombstone to Sir Thomas Felbrigg. By an act of folly and barbarism, almost unequalled in the history of the world, the Incumbent and Curate nearly destroyed the brass inscription surrounding the image of the Knight.

This tombstone is figured in Gough's Sepulchral Antiquities, which, I presume, is to be found in the British Museum.

And I take the liberty to ask if you would kindly look at the engraving, and give me any suggestion as to the way in which some copies of it could be made, in a fairly durable form.   I am connected with the parish of Playford, and am anxious to preserve for it this memorial of a family of high rank formerly resident there.

I am, dear Sir,

Very faithfully yours,

G. B. AIRY.

T. Winter Jones, Esq.

To this request Mr Winter Jones immediately acceded, and the engraving was duly photographed, and copies were circulated with a historical notice of Sir George (not Sir Thomas) Felbrigg and a history of the Monument. Sir George Felbrigg was Esquire-at-Arms to Edward III., and Lord of the Manor of Playford : he died in 1400, and was buried in the North wall of Playford Church.

1878

The Report to the Board of Visitors has this paragraph : " I continue to remark the approaching necessity for Library extension.   Without having absolutely decided on a site, I may suggest that I should wish to erect a brick building, about 50 feet by 20, consisting of two very low stories (or rather of one story with a gallery running round its walls), so low that books can be moved by hand without necessity for a ladder.—In the month of December, 1877, the azimuthal error of the Transit Circle had increased to 10″.   A skilful workman, instructed by Mr Simms, easily reduced the error to about 2″·5 (which would leave its mean error nearly 0), the western Y being moved to the north so far as to reduce the reading of the transit micrometer, when pointed to the south, from $35^r\cdot500$ to $35^r\cdot000$.   The level error was not sensibly affected.—The Sidereal Standard Clock preserves a rate approaching to perfection, so long as it is left without disturbance of the galvanic-contact springs (touched by its pendulum), which transmit signals at every second of time to sympathetic clocks and the chronograph. A readjustment of these springs usually disturbs the rate.—To facilitate the observations of stars, a new working catalogue has been prepared, in which are included all stars down to the third magnitude, stars down to the fifth magnitude which have not been observed in the last two catalogues, and a list of 258 stars of about the sixth magnitude of which the places are required for the United States Coast Survey. The whole number of stars in our new working list is about 2500. It may be here mentioned that an extensive series of observations was made, during the autumn, of about 70 stars, at the request of Mr Gill, for comparison with Mars, Ariadne, and Melpomene.—On Apr. 10th last, a very heavy fall of rain took place. Between Apr. 10d. 5h. and Apr. 11d. 2h., 2·824 inch. was recorded, and 75 per cent. of this, or 2·12 inch., fell in the eight hours between 13½ h. and 21½ h.; and on May 7, 1 inch of rain fell in 50 minutes, of which ½ inch fell in 15 minutes.—The supplementary compensation continues to be applied with success to Government chronometers which offer facilities for its introduction, and a marked improvement in the performance of chronometers returned after repair by the makers appears to have resulted from the increased attention now given to the compensation. Of the 29 competitive chronometers, 25 have the supplementary compensation."—With regard to the reduction of the observations of the Transit of Venus: After reference to the difficulties arising from the errors and the interpretation of the language used by some of the observers, the Report continues thus : "Finally a Report was made to the Government on July 5th, giving as the mean result for Mean Solar Parallax 8"·76; the results from ingress and from egress, however, differing to the extent of 0"·11....After further examination and consideration, the result for parallax has been increased to 8″·82 or 8″·83. The results from photography have disappointed me much. The failure has arisen, perhaps sometimes from irregularity of limb, or from atmospheric distortion, but more frequently from faintness and from want of clear definition. Many photographs, which to the eye appeared good, lost all strength and sharpness when placed under the measuring microscope. A final result 8″·17 was obtained from Mr Burton's measures, and 8″·08 from Capt. Tupman's.—With regard to the Numerical Lunar Theory: A cursory collection of the terms relating to the Areas (in the Ecliptic) led me to suppose that there might be some error in the computations of the Annual Equation and related terms. A most jealous re-examination has however detected nothing, and has confirmed my belief in the general accuracy of the numerical computations. I dare not yet venture to assume an error in Delaunay's theory; but I remember that the Annual Equation gave great trouble to the late Sir John Lubbock, and that he more than once changed his conclusions as to its true value.—In February I was engaged on the drawings and preparations for my intended Lecture at Cockermouth on the probable condition of the interior of the Earth. The Lecture was delivered in April.—At different times in the autumn I was engaged on diagrams to illustrate the passage of rays through eye-pieces and double-image micrometers.—The miscellaneous scientific correspondence, which was always going on, was in this year unusually varied and heavy."

Of private history: He was at Playford till Jan. 26th.— In April he went to Cockermouth to deliver his Lecture above-mentioned: the journey was by Birmingham, where he stayed for two days (probably with his son Osmund, who resided there), to Tarn Bank (the residence of Isaac Fletcher, M.P.): the lecture was delivered on the 22nd: he made excursions to Thirlmere and Barrow, and to Edward I.'s Monument, and returned to Greenwich on the 27th.—From June 17th to 28th he was at Playford.—From Aug. 19th to

Sept. 17th he was travelling in Scotland, visiting the Tay Bridge, the Loch Katrine Waterworks, &c., and spent the last fortnight of his trip at Portinscale, near Keswick. On Dec. 23rd he went to Playford.

1879

"The manuscripts of every kind, which are accumulated in the ordinary transactions of the Observatory, are preserved with the same care and arranged on the same system as heretofore. The total number of bound volumes exceeds 4000. Besides these there is the great mass of Transit of Venus reductions and manuscripts, which when bound may be expected to form about 200 volumes.—With regard to the numerous group of Minor Planets, the Berlin authorities have most kindly given attention to my representation, and we have now a most admirable and comprehensive Ephemeris. But the extreme faintness of the majority of these bodies places them practically beyond the reach of our meridian instrument, and the difficulty of observation is in many cases further increased by the large errors of the predicted places.— After a fine autumn, the weather in the past winter and spring has been remarkably bad. More than an entire lunation was lost with the Transit Circle, no observation of the Moon on the meridian having been possible between January 8 and March 1, a period of more than seven weeks. Neither Sun nor stars were visible for eleven days, during which period the clock-times were carried on entirely by the preceding rate of the clock. The accumulated error at the end of this time did not exceed $0^s.3$.—Some difficulty was at first experienced with the Thomson Electrometer, which was traced to want of insulation. This has been mastered by the use of glass supporters, which carry some sulphuric acid. The instrument is now in excellent order, and the photographic registers have been perfectly satisfactory since 1879,

February, when the new insulators were applied.—From the annual curves of diurnal inequality, deduced from the Magnetic Reductions, most important inferences may be drawn, as to the connection between magnetic phenomena and sunspots. These annual curves shew a well-marked change in close correspondence with the number of sun-spots. About the epoch of maximum of sun-spots they are large and nearly circular, having the same character as the curves for the summer months; whilst about the time of sun-spot minimum they are small and lemniscate-shaped, with a striking resemblance to the curves for the winter months. The connection between changes of terrestrial magnetism and sun-spots is shewn in a still more striking manner by a comparison which Mr Ellis has made between the monthly means of the diurnal range of declination and horizontal force, and Dr R. Wolf's 'relative numbers' for frequency of sun-spots.—The records of sunshine with Campbell's Registering Sun-dial are preserved in a form easily accessible for reference, and the results are communicated weekly to the Agricultural Gazette.—Prof. Oppolzer's results for the determination of the longitudes of Vienna and Berlin, made in 1877, have now been made public. They shew a remarkable agreement of the Chronometric determination formerly made with the Telegraphic. It may be of interest to recall the fact that a similar agreement was found between the Chronometric and Telegraphic determinations of the longitude of Valentia.—For observing the Transit of Venus of 1882, the general impression appears to be that it will be best to confine our observations to simple telescopic observations or micrometer observations at Ingress and Egress, if possible at places whose longitudes are known. For the first phenomenon (accelerated ingress) the choice of stations is not good; but for the other phenomena (retarded ingress, accelerated egress, retarded egress) there appears to be no difficulty.—With regard to the Numerical Lunar Theory: Respecting the discordance of Annual Equation, I suspend my judgment. I have now discussed the theory

completely; and in going into details of secular changes, I
am at this time engaged on that which is the foundation of
all, namely, the change of excentricity of the Solar Orbit,
and its result in producing Lunar Acceleration. An important error in the theoretical formulæ for Variations of Radius
Vector, Longitude, and Latitude, was discovered; some calculations depending on them are cancelled."—Referring to the
magnitude of the printed volume of "Greenwich Observations," and the practicability of reducing the extent of it,
the Report states thus: "The tendency of external scientific
movement is to give great attention to the phenomena of the
Solar disc (in which this Observatory ought undoubtedly to
bear its part). And I personally am most unwilling to recede
from the existing course of magnetical and meteorological
observations....The general tendency of these considerations
is to increase the annual expenses of the Observatory. And
so it has been, almost continuously, for the last 42 years.
The annual ordinary expenses are now between $2\frac{1}{2}$ and 3
times as great as in my first years at the Royal Observatory.
—Mr Gill was appointed to the Cape Observatory, and I
wrote out instructions for him in March: there was subsequently much correspondence respecting the equipment and
repairs of the Cape Observatory."—In the Monthly Notices of
the Royal Astronomical Society for January an article had
appeared headed "Notes on the late Admiral Smyth's Cycle
of Celestial Objects, Vol. II." by Mr Herbert Sadler. In this
article Mr Sadler had criticized the work of Admiral Smyth
in a manner which Airy regarded as imputing bad faith to
Admiral Smyth. He at once took up the defence of his old
friend very warmly, and proposed certain Drafts of Resolutions to the Council of the Society. These Resolutions were
moved, but were amended or negatived, and Airy immediately
resigned his office of Vice-President. There was considerable
negociation on the subject, and discussion with Lord Lindsay,
and on May 9th Airy's Resolutions were accepted by the
Council.—In October Airy inspected the "Faraday" tele-

graph ship, then lying in the river near Messrs Siemens' works, and broke his finger by a fall on board the vessel.—In this year Airy wrote and circulated a letter to the Members of the Senate of the University of Cambridge, on the subject of the Papers set in the Smith's Prizes Examination. In this letter, as on former occasions, he objected much to the large number of questions in " purely idle algebra, arbitrary combinations of symbols, applicable to no further purpose." And in particular he singled out for comment the following question, which was one of those set, " Using the term circle as extending to the case where the radius is a pure imaginary, it is required to construct the common chord of two given circles." This drew forth as usual a rejoinder from Prof. Cayley, who wrote enclosing a solution of his problem, but not at all to Airy's satisfaction, who replied as follows : " I am not so deeply plunged in the mists of impossibles as to appreciate fully your explanation in this instance, or to think that it is a good criterion for University candidates."

Of private history : On Jan. 21st he returned from Playford.—On March 22nd he attended the funeral of his sister at Little Welnetham near Bury St Edmunds : Miss Elizabeth Airy had lived with him at the Observatory from shortly after his appointment.—For about a week at the end of April he was visiting Matlock, Edensor, and Buxton.—From June 14th to July 18th he was staying at Portinscale near Keswick.—He was at Playford for two or three days in October, and went there again on Dec. 23rd for his usual winter holiday.

The following letter, relating to the life of Thomas Clarkson, was written to Dr Merivale, Dean of Ely, after reading the account in the " Times " of October 10th of the unveiling of a statue of Clarkson near Ware :

ROYAL OBSERVATORY, GREENWICH,
LONDON, S.E.
1879, *October* 11.

DEAR SIR,

Pardon my intrusion on you, in reference to a transaction
which has greatly interested me—the honour paid by you to the
memory of Thomas Clarkson. With very great pleasure I have heard
of this step: and I have also been much satisfied with the remarks
on it in the "Times." I well remember, in Clarkson's "History of
the Abolition," which I read some 60 years ago, the account of the
circumstance, now commemorated by you, which determined the
action of his whole subsequent life.

It is not improbable that, among those who still remember
Clarkson, my acquaintance with him began at the earliest time of
all. I knew him, intimately, from the beginning of 1815 to his
death. The family which he represented must have occupied a very
good position in society. I have heard that he sold two good
estates to defray the expenses which he incurred in his personal
labours for Abolition : and his brother was Governor of Sierra Leone
(I know not at what time appointed). Thomas Clarkson was at St
John's College; and, as I gather from circumstances which I have
heard him mention, must have been a rather gay man. He kept a
horse, and at one time kept two. He took Orders in the Church;
and on one occasion, in the course of his Abolition struggle, he
preached in a church. But he afterwards resolutely laid aside all
pretensions to the title of Minister of the Church, and never would
accept any title except as layman. He was, however, a very earnest
reader of theology during my acquaintance with him, and appeared
to be well acquainted with the Early Fathers.

The precise words in which was announced the subject for Prize
Essay in the University were "Anne liceat invitos in servitutem
trahere."

After the first great victory on the slave trade question, he established himself in a house on the bank of Ullswater. I have not
identified the place : from a view which he once shewed me I supposed it to be near the bottom of the lake : but from an account of
the storm of wind which he encountered when walking with a lady
over a pass, it seemed to be in or near Patterdale. When the
remains of a mountaineer, who perished in Helvellyn (as described

in Scott's well-known poem), were discovered by a shepherd, it was to Mr Clarkson that the intelligence was first brought.

He then lived at Bury St Edmunds.   Mrs Clarkson was a lady of Bury.   But I cannot assign conjecturally any dates to his removals or his marriage.   His only son took his B.A. degree, I think, about 1817.

I think it was in 1814 that he began his occupation of Playford Hall—a moated mansion near Ipswich, formerly of great importance—where he lived as Gentleman Farmer, managing a farm leased from the Marquis of Bristol, and occupying a good position among the gentry of the county.   A relative of mine, with whom I was most intimately acquainted, lived in the same parish (where in defiance of school rules I spent nearly half my time, to my great advantage as I believe, and where I still retain a cottage for occasional residence), and I enjoyed much of Mr Clarkson's notice.   It was by his strong advice that I was sent to Cambridge, and that Trinity College was selected : he rode with me to Rev. Mr Rogers of Sproughton for introductory examination ; he introduced me to Rev. C. Musgrave (subsequently of Halifax), accidentally doing duty at Grundisburgh, who then introduced me to Sedgwick, Peacock, and T. Musgrave (subsequently of York).   In 1825, when I spent the summer at Keswick, he introduced me to Southey and Wordsworth.

Mr Clarkson lived about thirty years at Playford Hall, and died there, and lies interred with his wife, son, and grandson, in Playford churchyard.   I joined several friends in erecting a granite obelisk to his memory in the same churchyard.   His family is extinct : but a daughter of his brother is living, first married to T. Clarkson's son, and now Mrs Dickinson, of the Rectory, Wolferton.

I am, my dear Sir,

Very faithfully yours,

G.   B.   AIRY.

The Very Reverend,
The Dean of Ely.

1880

"The Admiralty, on final consideration of the estimates, decided not to proceed with the erection of a new Library

near the Magnetic Observatory in the present year. In the
mean time the space has been cleared for the erection of a
building 50 by 20 feet.—I have removed the Electrometer
Mast (a source of some expense and some danger), the perfect success of Sir William Thomson's Electrometer rendering
all further apparatus for the same purpose unnecessary.—
Many years ago a double-image micrometer, in which the
images were formed by the double refraction of a sphere of
quartz, was prepared by Mr Dollond for Capt. Smyth, R.N.
Adopting the same principle on a larger scale, I have had
constructed by Mr Hilger a micrometer with double refraction of a sphere of Iceland spar. Marks have been prepared
for examination of the scale, but I have not yet had opportunity of trying it.—The spectroscopic determination of Star-
motions has been steadily pursued. The stars are taken
from a working list of 150 stars, which may eventually be
extended to include all stars down to the fourth magnitude,
and it is expected that in the course of time the motions of
about 300 stars may be spectroscopically determined.—A
new pressure-plate with springs has been applied by Mr
Browning to Osler's Anemometer, and it is proposed to make
such modification as will give a scale extending to 50 lbs.
pressure on the square foot. Other parts of the instrument
have also been renewed.—As regards the reduction of the
magnetical results since 1863 : In the study of the forms of
the individual curves; their relations to the hour, the month,
the year ; their connection with solar or meteorological facts;
the conjectural physico-mechanical causes by which they are
produced ; there is much to occupy the mind. I regret that,
though in contemplation of these curves I have remarked
some singular (but imperfect) laws, I have not been able to
pursue them.—The mean temperature of the year 1879 was
46·1°, being 3·3° below the average of the preceding 38 years.
The highest temperature was 80·6° on July 30, and the lowest
13·7° on Dec. 7. The mean temperature was below the
average in every month of the year ; the months of greatest

deviation being January and December, respectively 6·8° and 7·6° below the average ; the months of April, May, July, and November were each between 4° and 5° below the average. The number of hours of bright sunshine, recorded with Campbell's Sunshine Instrument, during 1879, was only 983.—In the summer of 1879 Commander Green, U.S.N., came over to this country for the purpose of determining telegraphically the longitude of Lisbon, as part of a chain of longitudes extending from South America to Greenwich. A successful interchange of signals was made with Commander Green between Greenwich and Porthcurno on four nights, 1879, June 25 to 29. The results communicated by Commander Green shew that the longitude of Lisbon Observatory, as adopted in the Nautical Almanac, requires the large correction of + 8·54ˢ.—With regard to the coming Transit of Venus in 1882 : From the facility with which the requirements for geographical position are satisfied, and from the rapid and accurate communication of time now given by electric telegraph, the observation of this Transit will be comparatively easy and inexpensive. I have attached greater importance than I did formerly to the elevation of the Sun.......I remark that it is highly desirable that steps be taken now for determining by telegraph the longitude of some point of Australia. I have stated as the general opinion that it will be useless to repeat photographic observations.—In April Mr Barlow called, in reference to the Enquiry on the Tay Bridge Disaster. (The Bridge had been blown down on Dec. 28th, 1879.) I prepared a memorandum on the subject for the Tay Bridge Commission, and gave evidence in a Committee Room of the House of Lords on Apr. 29th." (Much of the Astronomer Royal's evidence on this occasion had reference to the opinions which he had expressed concerning the windpressure which might be expected on the projected Forth Bridge, in 1873.)—In May Airy was consulted by the Postmaster-General in the matter of a dispute which had arisen between the Post Office and the Telephone Companies, which latter were alleged to have infringed the monopoly of the Post Office in commercial telegraphs : Airy made a declaration on the subject.—In July Mr Bakhuyzen came to England to determine the longitude of Leyden, on which he was engaged till Sept. 9th, and carried on his observations at the Observatory.—In July Airy was much engaged in perusing the records of Mr Gill's work at the Cape of Good Hope.

Of private history: On Jan. 24th he returned from Playford.—From June 14th to July 4th he was again at Playford. —From September 21st to October 20th he was staying at Portinscale near Keswick.—On Dec. 23rd he went again to Playford for his winter holiday.

Respecting the agitation at Cambridge for granting University degrees to women, the following extract from a letter addressed to a young lady who had forwarded a Memorial on the subject for his consideration, and dated Nov. 10th, 1880, contains Airy's views on this matter.

"I have not signed the Memorial which you sent for my consideration: and I will endeavour to tell you why. I entirely approve of education of young women to a higher pitch than they do commonly reach. I think that they can successfully advance so far as to be able clearly to understand—with gratification to themselves and with advantage to those whose education they will superintend— much of the results of the highest class of science which have been obtained by men whose lives are in great measure devoted to it. But I do not think that their nature or their employments will permit of their mastering the *severe* steps of beginning (and indeed all through) and the *complicated* steps at the end. And I think it well that this their success should be well known—as it is sure to be —among their relatives, their friends, their visitors, and all in whom they are likely to take interest. Their connection with such a place as Girton College is I think sufficient to lead to this. But I desire above all that all this be done in entire subservience to what I regard as *infinitely* more valuable than any amount of knowledge, namely the delicacy of woman's character. And here, I think, our views totally separate. I do not imagine that the University Degree would really imply, as regards education, anything more than is known to all persons (socially concerned in the happiness of the young woman) from the less public testimonial of the able men who have the means of knowing their merits.   And thus it appears to me that the admission to University Degree would simply mean a more extended publication of their names.   I dread this."

1881

" The new line of underground telegraph wires has been completed by the officers of the General Post Office.   The new route is down Croom's Hill in Greenwich, and the result of this change, at least as regards the earth-current wires, and probably as regards the other wires, has not been satisfactory. It was soon found that the indications of the earth-current wires were disturbed by a continual series of petty fluctuations which almost completely masked the proper features of earth currents.......If this fault cannot be removed, I should propose to return to our original system of independent wires (formerly to Croydon and Dartford).—The new Azimuth-mark (for the Altazimuth), upon the parapet of the Naval College, is found to be perfectly satisfactory as regards both steadiness and visibility.   The observations of a low star for zero of azimuth have been omitted since the beginning of 1881 ; the mark, in combination with a high star, appearing to give all that is necessary for this purpose.—All the instruments have suffered from the congealing of the oil during the severe weather of the past winter, and very thorough cleaning of all the moving parts has been necessary.—The Solar Eclipse of 1880, Dec. 31, was well observed.   The first contact was observed by four observers and the last contact by two. The computations for the observations have been exceptionally heavy, from the circumstance that the Sun was very low ($86°14'$ Z. D. at the last observation) and that it has therefore been necessary to compute the refraction with great accuracy, involving the calculation of the zenith distance for every observation. And besides this, eighty-six separate computations of the tabular R. A. and N. P. D. of cusps have been required. — Amongst other interesting spectroscopic observations of the Sun, a remarkable spectrum of a sun-spot shewing 17 strong black lines or bands, each as broad as b_1, in the solar spectrum, was observed on 1880, Nov. 27 and 29. These bands to which there is nothing corresponding in the Solar Spectrum (except some very faint lines) have also been subsequently remarked in the spectrum of several spots.— The Police Ship 'Royalist' (which was injured by a collision in 1879 and had been laid up in dock) has not been again moored in the river, and the series of observations of the temperature of the Thames is thus terminated.—Part of the month of January 1881 was, as regards cold, especially severe. The mean temperature of the period January 12 to 26 (15 days) was only 24·2°, or 14·7° below the average; the temperature fell below 20° on 10 days, and rose above the freezing point only on 3 days. The highest temperature in this period was 35·3°, the lowest 12·7°. On January 17th (while staying at Playford) my son Hubert and I noticed an almost imperceptible movement in the upper clouds from the South-East. On that night began the terrible easterly gale, accompanied with much snow, which lasted to the night of the 18th. The limiting pressure of 50 lbs. on the square foot of Osler's Anemometer was twice exceeded during this storm.—With respect to the Diurnal Inequalities of Magnetic Horizontal Force : Assuming it to be certain that they originate from the Sun's power, not immediately, but mediately through his action on the Earth, it appears to me (as I suggested long ago) that they are the effects of the attraction of the red end or north end of the needle by the heated portions of our globe, especially by the heated sea, whose effect appears to predominate greatly over that of the land. I do not say that everything is thus made perfectly clear, but I think that the leading phenomena may be thus explained. And this is almost necessarily the way of beginning a science.—In the first few years after the strict and systematic examination of competitive chronometers, beginning with 1856, the accuracy of chronometers was greatly increased. For many years past it has been nearly stationary. I interpret this as shewing that the effects of bad workmanship are almost eliminated, and that future improvement must be sought in change of some points of construction. —Referring to the Transit of Venus in 1874, the printing of all sections of the Observations, with specimens of the printed forms employed, and remarks on the photographic operations, is very nearly completed. An Introduction is begun in manuscript. I am in correspondence with the Commission which is entrusted with the arrangements for observation of the Transit of 1882.—The Numerical Lunar Theory has been much interrupted by the pressure of the Transit of Venus work and other business."—In his Report to the Board of Visitors (his 46th and last), Airy remarks that it would be a fitting opportunity for the expression of his views on the general objects of the Observatory, and on the duties which they impose on all who are actively concerned in its conduct. And this he proceeds to do in very considerable detail.—On May 5th he wrote to Lord Northbrook (First Lord of the Admiralty) and to Mr Gladstone to resign his post of Astronomer Royal. From time to time he was engaged on the subject of a house for his future residence, and finally took a lease of the White House at the top of Croom's Hill, just outside one of the gates of Greenwich Park. On the 15th of August he formally resigned his office to Mr W. H. M. Christie, who had been appointed to succeed him as Astronomer Royal, and removed to the White House on the next day, August 16th.

His holiday movements in the portion of the year up to August 16th consisted in his winter visit to Playford, from which he returned on Jan. 24th: and a subsequent visit to Playford from June 7th to 18th.

The following correspondence relating to Airy's retirement from office testifies in a remarkable manner to the estimation in which his services were held, and to the good feeling which subsisted between him and his official superiors.

10, DOWNING STREET, WHITEHALL,
June 6, 1881.

DEAR SIR GEORGE AIRY,

I cannot receive the announcement of your resignation, which you have just conveyed to me, without expressing my strong sense of the distinction you have conferred upon the office of Astronomer Royal, and of the difficulty of supplying your place with a person of equal eminence. Let me add the expression of my best wishes for the full enjoyment of your retirement from responsibility.

I remain, dear Sir George Airy,

Faithfully yours,

W. E. GLADSTONE.

ADMIRALTY,
June 10*th*, 1881.

SIR,

I am commanded by my Lords Commissioners of the Admiralty to acknowledge the receipt of your letter of the 4th instant, intimating your desire to retire on the 15th August next from the office of Astronomer Royal.

2. In reply I am to acquaint you that your wishes in this matter have been communicated to the Prime Minister, and that the further necessary official intimation will in due course be made to the Treasury.

3. At the same time I am instructed by their Lordships to convey to you the expression of their high appreciation of the remarkably able and gifted manner, combined with unwearied diligence and devotion to the Public Service (especially as regards the Department of the State over which they preside), in which you have performed the duties of Astronomer Royal throughout the long period of forty-five years.

4. I am further to add that their Lordships cannot allow the present opportunity to pass without giving expression to their sense

of the loss which the Public Service must sustain by your retirement, and to the hope that you may long enjoy the rest to which you are so justly entitled.

I am, Sir,

Your obedient Servant,

ROBERT HALL.

Sir G. B. Airy, K.C.B.
&c.,   &c.,
Royal Observatory, Greenwich.

ADMIRALTY,

28th June, 1881.

SIR,

My Lords Commissioners of the Admiralty have much pleasure in transmitting copy of a resolution passed by the Board of Visitors of the Royal Observatory on the 4th June last, bearing testimony to the valuable services you have rendered to Astronomy, to Navigation, and the allied Sciences throughout the long period during which you have presided over the Royal Observatory.

I am, Sir,

Your obedient Servant,

ROBERT HALL.

Sir George Biddell Airy, K.C.B.
&c.,   &c.,   &c.,
Royal Observatory, Greenwich.

" The Astronomer Royal (Sir George B. Airy) having announced his intention of shortly retiring from his position at the Royal Observatory, the following resolution proposed by Professor J. C. Adams, and seconded by Professor G. G. Stokes, was then unanimously adopted and ordered to be recorded in the Minutes of the Proceedings.

" The Board having heard from the Astronomer Royal that he proposes to terminate his connection with the Observatory on the 15th of August next, desire to record in the most emphatic manner their sense of the eminent services which he has rendered to Astronomy, to Navigation and the allied Sciences, throughout the long period of 45 years during which he has presided over the Royal Observatory.

"They consider that during that time he has not only maintained but has greatly extended the ancient reputation of the Institution, and they believe that the Astronomical and other work which has been carried on in it under his direction will form an enduring monument of his Scientific insight and his powers of organization.

"Among his many services to Science, the following are a few which they desire especially to commemorate:

(a) "The complete re-organization of the Equipment of the Observatory.

(b) "The designing of instruments of exceptional stability and delicacy suitable for the increased accuracy of observation demanded by the advance of Astronomy.

(c) "The extension of the means of making observations of the Moon in such portions of her orbit as are not accessible to the Transit Circle.

(d) "The investigation of the effect of the iron of ships upon compasses and the correction of the errors thence arising.

(e) "The Establishment at the Observatory and elsewhere of a System of Time Signals since extensively developed by the Government.

"The Board feel it their duty to add that Sir George Airy has at all times devoted himself in the most unsparing manner to the business of the Observatory, and has watched over its interests with an assiduity inspired by the strongest personal attachment to the Institution. He has availed himself zealously of every scientific discovery and invention which was in his judgment capable of adaptation to the work of the Observatory; and the long series of his annual reports to the Board of Visitors furnish abundant evidence, if such were needed, of the soundness of his judgment in the appreciation of suggested changes, and of his readiness to introduce improvements when the proper time arrived.   While maintaining the most remarkable punctuality in the reduction and publication of the observations made under his own superintendance, he had reduced, collected, and thus rendered available for use by astronomers, the Lunar and Planetary Observations of his predecessors.   Nor can it be forgotten that, notwithstanding his absorbing occupations, his advice and assistance have always been at the disposal of Astronomers for any work of importance.

"To refer in detail to his labours in departments of Science not directly connected with the Royal Observatory may seem to lie beyond the province of the Board.   But it cannot be improper to state that its members are not unacquainted with the high estimation in which his contributions to the Theory of Tides, to the undulatory theory of Light, and to various abstract branches of Mathematics are held by men of Science throughout the world.

"In conclusion the Board would express their earnest hope, that in his retirement Sir George Airy may enjoy health and strength and that leisure for which he has often expressed a desire to enable him not only to complete the numerical Lunar Theory on which he has been engaged for some years past, but also to advance Astronomical Science in other directions."

ADMIRALTY,
27th October, 1881.

SIR,

I am commanded by my Lords Commissioners of the Admiralty to transmit to you, herewith, a copy of a Treasury Minute, awarding you a Special Pension of £1100 a year, in consideration of your long and brilliant services as Astronomer Royal.

I am, Sir,
Your obedient Servant,
ROBERT HALL.

Sir G. B. Airy, K.C.B., F.R.S., &c., &c.
The White House, Croom's Hill, Greenwich.

Copy of Treasury Minute, dated 10th October, 1881 :

My Lords have before them a statement of the services of Sir George Biddell Airy, K.C.B., F.R.S., who has resigned the appointment of Astronomer Royal on the ground of age.

Sir George Airy has held his office since the year 1835, and has also, during that period, undertaken various laborious works, demanding scientific qualifications of the highest order, and not always such as could strictly be said to be included among the duties of his office.

The salary of Sir G. Airy as Astronomer Royal is £1200 a year, in addition to which he enjoys an official residence rent free, and, under ordinary circumstances he would be entitled to a pension equal to two-thirds of his salary and emoluments.

My Lords, however, in order to mark their strong sense of the distinction which, during a long and brilliant career Sir George Airy has conferred upon his office, and of the great services which, in connection with, as well as in the discharge of, his duties, he has rendered to the Crown and the Public, decide to deal with his case under the ixth Section of the Superannuation Act, 1859, which empowers them to grant a special pension for special services.

Accordingly my Lords are pleased to award to Sir George Biddell Airy, K.C.B., F.R.S., a special Retired Allowance of £1100 per annum.

THE WHITE HOUSE,
CROOM'S HILL, GREENWICH,
1881, *October* 29.

SIR,

I have the honour to acknowledge your letter of October 27, transmitting to me, by instruction of The Lords Commissioners of Admiralty, copy of a Treasury Minute dated 1881 October 10, in which the Lords Commissioners of Her Majesty's Treasury are pleased to award to me an annual retired allowance of £1100 per annum.

Acknowledging the very liberal award of the Lords Commissioners of Treasury, and the honourable and acceptable terms in which it is announced, I take leave at the same time to offer to

Their Lordships of the Admiralty my recognition of Their Lordships' kindness and courtesy in thus handing to me copy of the Treasury Minute.

> I have the honour to be, Sir,
>
> Your very obedient Servant,
>
> G. B. AIRY.

The Secretary of the Admiralty.

From the Assistants of the Royal Observatory, with whom he was in daily communication, whose faithful and laborious services he had so often thankfully recognized in his Annual Reports to the Board of Visitors, and to whom so much of the credit and success of the Observatory was due, he received the following address :

> ROYAL OBSERVATORY, GREENWICH,
> 1881, *August* 11.

DEAR SIR,

We cannot allow the official relation which has so long existed between yourself and us to terminate without expressing to you our sense of the admirable manner in which you have, in our opinion, upheld the dignity of the office of Astronomer Royal during the many years that you have occupied that important post.

Your long continued and varied scientific work has received such universal recognition from astronomers in all lands, that it is unnecessary for us to do more than assure you how heartily we join in their appreciation of your labours. We may however add that our position has given us opportunities of seeing that which others cannot equally well know, the untiring energy and great industry which have been therein displayed throughout a long and laborious career, an energy which leads you in retirement, and at fourscore years of age, to contemplate further scientific work.

We would ask you to carry with you into private life the best wishes of each one of us for your future happiness, and that of your family, expressing the hope that the days of retirement may not be

few, and assuring you that your name will long live in our remembrance.

> We are, dear Sir,
>> Yours very faithfully,
>>> W. H. M. CHRISTIE, EDWIN DUNKIN, WILLIAM
>>> ELLIS, GEORGE STRICKLAND CRISWICK, W.
>>> C. NASH, A. M. W. DOWNING, EDWARD W.
>>> MAUNDER, W. G. THACKERAY, THOMAS LEWIS.

Sir G. B. Airy, K.C.B., &c., &c.,
Astronomer Royal.

> ROYAL OBSERVATORY, GREENWICH,
> 1881, *August* 13.

MY DEAR MR CHRISTIE,

and Gentlemen of the Royal Observatory,

With very great pleasure I have received
your letter of August 11. I thank you much for your recognition
of the general success of the Observatory, and of a portion of its
conduct which—as you remark—can scarcely be known except to
those who are every day engaged in it: but I thank you still more
for the kind tone of your letter, which seems to shew that the terms
on which we have met are such as leaves, after so many years'
intercourse, no shadow of complaint on any side.

Reciprocating your wishes for a happy life, and in your case a
progressive and successful one,

> I am,
>> My dear Mr Christie and Gentlemen,
>>> Yours faithfully,
>>>> G. B. AIRY.

Throughout his tenure of office Airy had cultivated and
maintained the most friendly relations with foreign astronomers, to the great advantage of the Observatory. Probably

all of them, at one time or another, had visited Greenwich, and to most of them he was well known. On his retirement from office he received an illuminated Address from his old friend Otto Struve and the staff of the Pulkowa Observatory, an illuminated Address from the Vorstand of the Astronomische Gesellschaft at Berlin signed by Dr Auwers and the Secretaries, a complimentary letter from the Academy of Sciences at Amsterdam, and friendly letters of sympathy from Dr Gould, Prof. Newcombe, Dr Listing, and from many other scientific friends and societies. His replies to the Russian and German Addresses were as follows:

ROYAL OBSERVATORY, GREENWICH,
1881, *August* 5.

MY DEAR SIR,

      I received, with feelings which I will not attempt to describe, the Address of yourself and the Astronomers of Pulkowa generally, on the occasion of my retirement from the office of Astronomer Royal. I can scarcely credit myself with possessing all the varied claims to your scientific regard which you detail. I must be permitted to attribute many of them to the long and warm friendship which has subsisted so long between the Directors of the Pulkowa Observatory and myself, and which has influenced the feelings of the whole body of Astronomers attached to that Institution. On one point, however, I willingly accept your favourable expressions—I have not been sparing of my personal labour—and to this I must attribute partial success on some of the subjects to which you allude.

      In glancing over the marginal list of scientific pursuits, I remark with pleasure the reference to *Optics*. I still recur with delight to the Undulatory Theory, once the branch of science on which I was best known to the world, and which by calculations, writings, and lectures, I supported against the Laplacian School. But the close of your remarks touches me much more—the association of the name of W. Struve and my own. I respected deeply the whole character of your Father, and I believe that he had confidence in me. From our first meeting in 1830 (on a Commission for improvement of the Nautical Almanac) I never ceased to regard him as superior to others. I may be permitted to add that the delivery of his authority to the hands of his son has not weakened the connection of myself with the Observatory of Poulkova.

Acknowledging gratefully your kindness, and that of all the Astronomers of the Observatory of Poulkova, and requesting you to convey to them this expression,

I am, my dear Sir,

Yours most truly,

G. B. AIRY.

To M. Otto von Struve,
  Director of the Observatory of Poulkova
  and the Astronomers of that Observatory.

ROYAL OBSERVATORY, GREENWICH,
1881, *August* 3.

MY DEAR SIR,

      With very great pleasure I received the Address of the Astronomische Gesellschaft on occasion of my intended resignation of the Office of Astronomer Royal: dated July 27, and signed by yourself as President and Messrs Schoenfeld and Winnecke as Secretaries of the Astronomische Gesellschaft. I thank you much for the delicacy of your arrangement for the transmission of this document by the hands of our friend Dr Huggins. And I think you will be gratified to learn that it arrived at a moment when I was surrounded by my whole family assembled at my *jour-de-fête*, and that it added greatly to the happiness of the party.

      I may perhaps permit myself to accept your kind recognition of my devotion of time and thought to the interests of my Science and my Office. It is full reward to me that they are so recognized. As to the success or utility of these efforts, without presuming, myself, to form an opinion, I acknowledge that the connection made by the Astronomische Gesellschaft, between my name and the advance of modern astronomy, is most flattering, and will always be remembered by me with pride.

      It is true, as is suggested in your Address, that one motive for my resignation of Office was the desire to find myself more free for the prosecution of further astronomical investigations.   Should my health remain unbroken, I hope to enter shortly upon this undertaking.

Again acknowledging the kindness of yourself and the Vorstand of the Astronomische Gesellschaft, and offering my best wishes for the continued success of that honourable institution,

I am, my dear Sir,

Yours very truly,

G. B. AIRY.

*To Dr Auwers
and the Vorstand of the
Astronomische Gesellschaft.*

CHAPTER X.

At the White House, Greenwich. From his Resignation of Office on August 15th, 1881, to his Death on January 2nd, 1892.

History of his Life after his Resignation of Office.

On the 16th of August 1881 Airy left the Observatory which had been his residence for nearly 46 years, and removed to the White House. Whatever his feelings may have been at the severing of his old associations he carefully kept them to himself, and entered upon his new life with the cheerful composure and steadiness of temper which he possessed in a remarkable degree. He was now more than 80 years old, and the cares of office had begun to weigh heavily upon him: the long-continued drag of the Transit of Venus work had wearied him, and he was anxious to carry on and if possible complete his Numerical Lunar Theory, the great work which for some years had occupied much of his time and attention. His mental powers were still vigorous, and his energy but little impaired: his strong constitution, his regular habits of life, the systematic relief which he obtained by short holiday expeditions whenever he found himself worn with work, and his keen interest in history, poetry, classics, antiquities, engineering, and other subjects not immediately connected with his profession, had combined to produce this result. And in leaving office, he had no idea of leaving off work; his resignation of office merely meant for him a change of work. It is needless to say that his interest in the welfare and progress of the Observatory was as keen as ever; his advice was always at the service of his successor, and his appointment as Visitor a year or two after his resignation gave him an official position with regard to the Observatory which he much valued. The White House, which was to be his home for the rest of his life, is just outside one of the upper gates of the Park, and about a quarter of a mile from the Observatory. Here he resided with his two unmarried daughters. The house suited him well and he was very comfortable there: he preferred to live in the neighbourhood with which he was so familiar and in which he was so well known, rather than to remove to a distance. His daily habits of life were but little altered: he worked steadily as formerly, took his daily walk on Blackheath, made frequent visits to Playford, and occasional expeditions to the Cumberland Lakes and elsewhere.

The work to which he chiefly devoted himself in his retirement was the completion of his Numerical Lunar Theory. This was a vast work, involving the subtlest considerations of principle, very long and elaborate mathematical investigations of a high order, and an enormous amount of arithmetical computation. The issue of it was unfortunate: he concluded that there was an error in some of the early work, which vitiated the results obtained: and although the whole process was published, and was left in such a state that it would be a comparatively simple task for a future astronomer to correct and complete it, yet it was not permitted to the original author of it to do this. To avoid the necessity of frequent reference to this work in the history of Airy's remaining years, it will be convenient to summarize it here. It was commenced in 1872: "On Feb. 23rd in this year I first (privately) formed the notion of preparing a Numerical Lunar Theory by substituting Delaunay's numbers in the proper Equations and seeing what would come of it." From this time forward

till his power to continue it absolutely failed, he pursued
the subject with his usual tenacity of purpose. During his
tenure of office every available opportunity was seized for
making progress with his Lunar Theory, and in every Report
to the Visitors a careful statement was inserted of the state
in which it then stood. And, after his resignation of office,
it formed the bulk of his occupation. In 1873 the Theory
was formed, and by 1874 it was so far advanced that he
published in the Monthly Notices of the Royal Astronomical
Society a statement of the fundamental points of the Theory.
In 1875, the Theory having advanced to a stage where extensive arithmetical computation was required, he obtained
a small grant from the Government in aid of the expense
of the work, and other grants were made in subsequent
years. By 1878 the calculations were so far advanced that
an opinion could be formed as to the probable accuracy of
the Theory, and the following remark is made: "A cursory
collation of the terms relating to the Areas (in the Ecliptic)
led me to suppose that there might be some error in the
computations of the Annual Equation and related terms;"
but no error could be discovered and the work proceeded.
The complex character of the Theory, and the extreme care
required in the mathematical processes, are well illustrated
by the following statement, which occurs in the Report of
1879, "An important error in the theoretical formulæ for
Variations of Radius Vector, Longitude, and Latitude, was
discovered; some calculations depending on them are cancelled." In 1880 and 1881 the work was continued, but was
"sadly interrupted by the pressure of the Transit of Venus
work and other business." After his resignation of the Office
of Astronomer Royal he had no further public assistance,
and did much of the computations himself, but a sum of
£100 was contributed by Mr De La Rue in furtherance of
the work, and this sum was spent on computers. In his
retirement the work made good progress, and on Dec. 31st,
1882, he made the following note: "I finished and put in

general order the final tables of Equations of Variations. This is a definite point in the Lunar Theory.......I hope shortly to take up severely the numerical operations of the Lunar Theory from the very beginning." The work was continued steadily through 1883, and on Mar. 24th, 1884, he made application through the Board of Visitors to the Admiralty to print the work: after the usual enquiries as to the expense this was acceded to, and copy was sent to the printers as soon as it was ready. The first printed proofs were received on Feb. 5th, 1885, and the whole book was printed by the end of 1886. From the frequent references in his journal to errors discovered and corrected during the progress of these calculations, it would seem likely that his powers were not what they had been, and that there was a probability that some important errors might escape correction. He was far too honest to blind himself to this possibility, and in the Preface to his Numerical Lunar Theory he says thus: " I have explained above that the principle of operations was, to arrange the fundamental mechanical equations in a form suited for the investigations of Lunar Theory; to substitute in the terms of these equations the numerical values furnished by Delaunay's great work; and to examine whether the equations are thereby satisfied. With painful alarm, I find that they are not satisfied; and that the discordance, or failure of satisfying the equations, is large. The critical trial depends on the great mass of computations in Section II. These have been made in duplicate, with all the care for accuracy that anxiety could supply. Still I cannot but fear that the error which is the source of discordance must be on my part. I cannot conjecture whether I may be able to examine sufficiently into this matter." He resolutely took in hand the revision of his work, and continued it till October 1888. But it is clear from the entries in his journal that his powers were now unequal to the task, and although from time to time he suspected that he had discovered errors, yet it does not appear that he determined anything with certainty. He never doubted that there were important errors in the work, and later on he left the following private note on the subject:

NUMERICAL LUNAR THEORY.

1890, *Sept.* 29.

I had made considerable advance (under official difficulties) in calculations on my favourite Numerical Lunar Theory, when I discovered that, under the heavy pressure of unusual matters (two Transits of Venus and some eclipses) I had committed a grievous error in the first stage of giving numerical value to my Theory.

My spirit in the work was broken, and I have never heartily proceeded with it since.

G. B. AIRY.

Probably the error referred to here is the suspected error mentioned above in his Report of 1878, as to which he subsequently became more certain.

Whatever may be the imperfections of the Numerical Lunar Theory, it is a wonderful work to have been turned out by a man 85 years old. In its idea and inception it embodies the experience of a long life actively spent in practical science. And it may be that it will yet fulfil the objects of its author, and that some younger astronomer may take it up, correct its errors (wherever they may be), and fit it for practical use. And then the labour bestowed upon it will not have been in vain.

Subject always to the absorbing occupations of the Lunar Theory he amused himself with reading his favourite subjects of History and Antiquities. His movements during the remainder of the year 1881 were as follows: In September he paid a two days' visit to Lady Herschel at Hawkhurst. From Oct. 4th to 17th he was at the Cumberland Lakes and engaged in expeditions in the neighbourhood. From Nov. 5th to 8th he was at Cambridge, inspecting Prof. Stuart's workshops, and other scientific institutions. On Dec. 13th he went to Playford.—Amongst miscellaneous matters: in November he wrote to Mr Rothery on the loss of the 'Teuton' at some length, with suggestions for the safer construction of such vessels.—In October he was asked for suggestions regarding the establishment of a "Standard Time" applicable to the railway traffic in the United States: he replied as follows:

1881, *Oct.* 31.

Sir,

I have to acknowledge your letter of October 17, introducing to my notice the difficulty which appears to be arising in America regarding a "Standard Time," for extensive use throughout N. America "applicable to railway traffic only." The subject, as including considerations of convenience in all the matters to which it applies, is one of difficulties probably insuperable. The certainty, however, that objections may be raised to every scheme, renders me less timid in offering my own remarks; which are much at your service.

I first comment upon your expression of "Standard Time....... applicable to railway traffic only." But do you mean this as affecting the transactions between one railway and another railway, or as affecting each railway and the local interests (temporal and others) of the towns which it touches? The difference is so great that I should be disposed to adopt it as marking very strongly the difference to be made between the practices of railways among themselves and the practices of railways towards the public; and will base a system on that difference.

As regards the practices of railways among themselves: if the various railways of America are joined and inosculated as they are in England, it appears to me indispensable that they have one common standard *among themselves*: say Washington Observatory time. But this is only needed for the office-transactions between the railways; it may be kept perfectly private; never communicated to the public at all. And I should recommend this as the first step.

There will then be no difficulty in deducing, from these private Washington times, the accurate local times at those stations (whose longitude is supposed to be fairly well known, as a sailor with a sextant can determine one in a few hours) which the railway authorities may deem worthy of that honour; generally the termini of railways. Thus we shall have a series of bases of local time, of authoritative character, through the country.

Of such bases *we* have two, Greenwich and Dublin: and they are separated by a sea-voyage. In the U.S. of America there must be a greater number, and probably not so well separated. Still it is indispensable to adopt such a system of local centers.

No people in this world can be induced to use a reckoning which does not depend clearly upon the sun. In all civilized countries it depends (approximately) on the sun's meridian passage. Even the sailor on mid-ocean refers to that phænomenon. And the solar passage, with reasonable allowance, 20m. or 30m. one way or another, must be recognized in all time-arrangements as giving the fundamental time. The only practical way of doing this is, to adopt for a whole region the fundamental time of a center of that region.

And to this fundamental time, the local time of the railway, as now entering into all the concerns of life, must be adapted. A solicitor has an appointment to meet a client by railway; a physician to a consultation. How is this to be kept if the railway uses one time and every other act of life another?

There is one chain of circumstances which is almost peculiar—that of the line from New York to San Francisco. Here I would have two clocks at every station: those on the north side all shewing San Francisco time, and those on the south all shewing New York time. Every traveller's watch would then be available to the end of his journey.

A system, fundamentally such as I have sketched, would give little trouble, and may I think be adopted with advantage.

> I am, Sir,
> Your faithful servant,
> G. B. AIRY.

Mr Edward Barrington.

1882

He returned from Playford on Jan. 17: his other movements during the year were as follows: from Apr. 27th to May 11th he was at Playford; and again from August 1st to 24th. From Oct. 9th to Nov. 1st he was travelling with his two unmarried daughters in the Lake District of Cumberland : the journey was by Furness and Coniston to Portinscale near Keswick; on Oct. 13th he fell and sprained his ankle, and his excursions for the rest of the time were mainly conducted by driving. Shortly after his return, on Nov. 11th, while walking alone on Blackheath, he was seized with a violent attack of illness, and lay helpless for some time before he was found and brought home : he seems however to have recovered to a great extent in the course of a day or two, and continued his Lunar Theory and other work as before. On June 22nd he made the following sad note, " This morning, died after a most painful illness my much-loved daughter-in-law, Anna Airy, daughter of Professor Listing of Göttingen, wife of my eldest son Wilfrid." In February he wrote out his reminiscences of the village of Playford during his boyhood.

In June he was much disturbed in mind on hearing of some important alterations made by the Astronomer Royal in the Collimators of the Transit Circle, and some correspondence ensued on the subject.—During the year he had much correspondence on the subject of the subsidences on Blackheath.

The following letter was written in reply to a gentleman who had asked whether it could be ascertained by calculation how long it is since the Glacial Period existed :

1882, July 4.

Sir,

I should have much pleasure in fully answering your questions of July 3 if I were able to do so : but the subject really is very obscure.

(1) Though it is recognized that the glacial period (or periods) is late, I do not think that any one has ventured to fix upon a rude number of years since elapsed.

(2) We have no reason to think that the mean distance of the earth from the sun has sensibly altered. There have been changes in the eccentricity of the orbit (making the earth's distance from the sun less in one month and greater in the opposite month), but I do not perceive that this would explain glaciers.

(3) I consider it to be certain that the whole surface of the earth, at a very distant period, was very hot, that it has cooled gradually, and (theoretically and imperceptibly) is cooling still. The glaciers must be later than these hot times, and later than our last consolidated strata : but this is nearly all that I can say.

I am, Sir, Your obedient Servant,

G. B. AIRY.

James Alston, Esq.

1883

From May 2nd to 29th he was at Playford. From July 10th to 20th he was travelling in South Wales with his daughters.—From Oct. 10th to Nov. 10th he was at Playford. —Between Nov. 20th of this year and Jan. 4th of the year 1884, he sat several times to Mr John Collier for his portrait : the picture was exhibited in the Academy of 1884; it is a most successful and excellent likeness.

Throughout the year he was very busy with the Numerical Lunar Theory.—In March he was officially asked to accept the office of Visitor of the Royal Observatory, which he accepted, and in this capacity attended at the Annual Visitation on June 2nd, and addressed a Memorandum to the Visitors on the progress of his Lunar Theory.—On March 12th he published in several newspapers a state-ment in opposition to the proposed Braithwaite and Buttermere Railway, which he considered would be injurious to the Lake District, in which he took so deep an interest.— In May he communicated to " The Observatory " a statement of his objections to a Theory advanced by Mr Stone (then President of the Royal Astronomical Society) to account for the recognized inequality in the Mean Motion of the Moon. This Theory, on a subject to which Airy had given his incessant attention for so many years, would naturally receive his careful attention and criticism, and it attracted much general notice at the time.—In December he wrote to the

Secretary of the Royal Astronomical Society his opinion as to the award of the Medal of the Society. In this letter he stated the principles which guided him as follows: " I have always maintained that the award of the Medal ought to be guided mainly by the originality of communications : that one advance in a new direction ought in our decision to outweigh any mass of work in a routine already established : and that, in any case, scientific utility as distinguished from mere elegance is indispensable."—In July Lieut. Pinheiro of the Brazilian Navy called with an autograph letter of introduction from the Emperor of Brazil. The Lieutenant desired to make himself acquainted with the English system of Lighthouses and Meteorology, and Airy took much trouble in providing him with introductions through which he received every facility for the thorough accomplishment of his object.—On Oct. 8th he forwarded to Prof. Cayley proofs of Euclid's Propositions I. 47 and III. 35 with the following remarks : " I place on the other side the propositions which may be substituted (with knowledge of Euclid's VI. book) for the two celebrated propositions of the geometrical books. They leave on my mind no doubt whatever that they were invented as proofs by ratios, and that they were then violently expanded into cumbrous geometrical proofs."—On June 28th he declined to sign a memorial asking for the interment of Mr Spottiswoode in Westminster Abbey, stating as his reason, " I take it, that interment possessing such a public character is a public recognition of benefits, political, literary, or philosophical, whose effects will be great and durable. Now I doubt whether it can be stated that Mr Spottiswoode had conferred such benefits on Society." But he adds at length his cordial recognition of Mr Spottiswoode's scientific services. —Throughout his life Airy was a regular attendant at church, and took much interest in the conduct of the Church services. In October of this year he wrote a long letter to the Vicar of Greenwich on various points, in which occurs the following paragraph : " But there is one matter in the present form of the Church Service, on which my feeling is very strong, namely the (so-called, I believe) Choral Service, in the Confession, the Prayer, and the Creed. I have long listened with veneration to our noble Liturgy, and I have always been struck with the deep personally religious feeling which pervades it, especially those parts of it which are for 'The People.' And an earnest Priest, earnestly pressing these parts by his vocal example on the notice of the People, can scarcely fail to excite a corresponding earnestness in them. All this is totally lost in the choral system. For a venerable persuasion there is substituted a rude irreverential confusion of voices; for an earnest acceptance of the form offered by the Priest there is substituted—in my feeling at least—a weary waiting for the end of an unmeaning form." He also objected much to singing the responses to the Commandments.

1884

From Apr. 29th to May 30th he was at Playford, concluding his Journal there with the note "So ends a pleasant Vacation."—On June 11th he went to Cambridge and attended the Trinity College Commemoration Service, and dined in Hall.—From Aug. 14th to Sept. 11th he was at Playford.—On Sept. 26th he made an expedition to Guildford and Farnham.—During this year he was closely engaged on the Numerical Lunar Theory, and for relaxation was reading theology and sundry books of the Old Testament.

On June 7th he attended at the Visitation of the Royal Observatory.—In a letter written in April to Lt.-Col. Marindin, R.A., on the subject of wind pressure there occurs the following remark: "When the heavy gusts come on, the wind is blowing in directions changing rapidly, but limited in extent. My conclusion is that in arches of small extent (as in the Tay Bridge) every thing must be calculated for full pressure; but in arches of large extent (as in the Forth Bridge) every thing may be calculated for small pressure. And for a suspension bridge the pressure is far less dangerous than for a stiff arch."—In January he had some correspondence with Professor Tyndall on the Theory of the "White Rainbow," and stated that he thoroughly agreed with Dr Young's explanation of. this phænomenon.—The following is extracted from a letter on May 1st to his old friend Otto Struve: "I received from you about 3 or 4 weeks past a sign of your friendly remembrance, a copy of your paper on the Annual Parallax of Aldebaran. It pleased me much. Especially I was delighted with your noble retention of the one equation whose result differed so sensibly from ‚that of the other equations. It is quite possible, even probable, that the mean result is improved by it. I have known such instances. The first, which attracted much attention, was Capt. Kater's attempt to establish a scale of longitude in England by reciprocal observations of azimuth between Beachy Head and Dunnose. The result was evidently erroneous. But Colonel Colby, on examination of the original papers, found that some observations had been omitted, as suspicious; and that when these were included the mean agreed well with the scale of observation inferred from other methods."—In a letter to the Rev. R. C. M. Rouse, acknowledging the receipt of a geometrical book, there occurs the following paragraph: "I do not value Euclid's Elements as a super-excellent book of instruction—though some important points are better presented in it than in any other book of geometrical instruction that I have seen. But I value it as a book of strong and distinct reasoning, and of orderly succession of reasonings. I do not think that there is any book in the world which presents so distinctly the 'because......therefore.......' And this is invaluable for the mental education of youth."—In May he was in correspondence with Professor Balfour Stewart regarding a projected movement in Terrestrial Magnetism to be submitted to the British

Association. Airy cordially approved of this movement, and supported it to the best of his ability, stating that in his opinion what was mainly wanted was the collation of existing records.—In January and February he was much pressed by Prof. Pritchard of Oxford to give his opinion as to the incorrectness of statements made by Dr Kinns in his Lectures on the Scientific Accuracy of the Bible. Airy refused absolutely to take part in the controversy, but he could not escape from the correspondence which the matter involved: and this led up to other points connected with the early history of the Israelites, a subject in which he took much interest.

1885

From May 4th to June 3rd he was at Playford.—From July 2nd to 22nd he was in the Lake District. The journey was by Windermere to Kentmere, where he made enquiries concerning the Airy family, as it had been concluded with much probability from investigations made by his nephew, the Rev. Basil R. Airy, that the family was settled there at a very early date. Some persons of the name of Airy were still living there. He then went on by Coniston and Gras- mere to Portinscale, and spent the rest of his time in expe- ditions amongst the hills and visits to friends.—On July 28th he went to Woodbridge in Suffolk and distributed the prizes to the boys of the Grammar School there.—From Oct. 9th to Nov. 12th he was again at Playford.—Throughout the year he was busily engaged on the Numerical Lunar Theory, and found but little time for miscellaneous reading.

Of printed papers by Airy in this year the most im- portant was one on the "Results deduced from the Mea- sures of Terrestrial Magnetic Force in the Horizontal Plane," &c. This was a long Paper, communicated to the Royal Society, and published in the Phil. Trans., and was the last Scientific Paper of any importance (except the Volume of the Numerical Lunar Theory) in the long list of "Papers by G. B. Airy." The preparation of this Paper took much time.—Of miscellaneous matters: In May a Committee of the Royal Society had been appointed to advise the India Office as to the publication of Col. J. Herschel's pendulum observations in India; and Airy was asked to assist the Committee with his advice. He gave very careful and anxious consideration to the subject, and it occupied much time.—In the early part of the year he was asked by Sir William Thomson to assist him with an affidavit in a lawsuit concerning an alleged infringement of one of his Patents for the improvement of the Compass. Airy declined to make an affidavit or to take sides in the dispute, but he wrote a letter from which the following is extracted: "I cannot have the least difficulty in expressing my opinion that you have made a great advance in the application of my method of correcting the compass in iron ships, by your introduction of the use of short needles for the compass-cards. In my original investigations, when the whole subject was in darkness, I could only use existing means for experiment, namely the long-needle compasses then existing. But when I applied mechanical theory to explanation of the results, I felt grievously the deficiency of a theory and the construction which it suggested (necessarily founded on assumption that the proportion of the needle-length to the other elements of measure is small) when the length of the needles was really so great. I should possibly have used some construction like yours, but the Government had not then a single iron vessel, and did not seem disposed to urge the enquiry. You, under happier auspices, have successfully carried it out, and, I fully believe, with much advantage to the science."—He wrote a Paper for the Athenæum and had various correspondence on the subject of the Badbury Rings in Dorsetshire, which he (and others) considered as identical with the "Mons Badonicus" of Gildas, the site of an ancient British battle.—In February he was in correspondence with the Astronomer Royal on Uniform Time Reckoning, and on considerations relating to it.—On June 6th he attended the Annual Visitation of the Observatory, and brought before the Board his investigations of the Diurnal Magnetic Inequalities, and the revises of his Lunar Theory.

1886

From June 8th to July 17th he was at Playford.—And again at Playford from Oct. 5th to Nov. 8th.—On March 27th he had an attack of gout in his right foot, which continued through April and into May, causing him much inconvenience.—He was busy with the Numerical Lunar Theory up to Sept. 25th, when he was reading the last proof-sheet received from the printers: during this period his powers were evidently failing, and there are frequent references to errors discovered and corrected, and to uncertainties connected with points of the Theory. But his great work on the Numerical Lunar Theory was printed in this year: and there can be no doubt that he experienced a great feeling of relief when this was accomplished.—He was in correspondence with Prof. Adams as to the effect of his reduction of the Coefficient of Lunar Acceleration on the calculation of the ancient historical eclipses.—He compiled a Paper "On the establishment of the Roman dominion in England," which was printed in 1887.—He wrote a notice concerning events in the life of Mr John Jackson of Rosthwaite near Keswick, a well-known guide and much-respected authority on matters relating to the Lake District.—He also wrote a short account of the connection of the history of Mdlle de Quéroualle with that of the Royal Observatory at Greenwich.—On June 4th he attended at the Annual Visitation of the Observatory.

1887

On May 9th to 11th he made a short visit to Eastbourne and the neighbourhood.—From June 8th to July 13th he was at Playford.—From Aug. 29th to Sept. 5th he was travelling in Dorsetshire and Wiltshire: he went first to Weymouth, a very favourite centre for excursions with him, and afterwards visited Bridport and Lyme Regis: then by Dorchester to Blandford, and visited the Hod Hill, Badbury Rings, &c.: at Wimborne he was much interested in the architecture of the church: lastly he visited Salisbury, Old Sarum, Stonehenge, &c., and returned to Greenwich.—From Oct. 11th to Nov. 12th he was at Playford.—During this year he partly occupied himself with arranging his papers and drawings, and with miscellaneous reading. But he could not withdraw his thoughts from his Lunar Theory, and he still continued to struggle with the difficulties of the subject, and was constantly scheming improvements. His private accounts also now gave him much trouble. Throughout his life he had been accustomed to keep his accounts by double entry in very perfect order. But he now began to make mistakes and to grow confused, and this distressed him greatly. It never seemed to occur to him to abandon his elaborate system of accounts, and to content himself with simple entries of receipts and expenses. This would have been utterly opposed to his sense of order, which was now more than ever the ruling principle of his mind. And so he struggled with his accounts as he did with his Lunar Theory till his powers absolutely failed. In his Journal for this year there are various entries of mental attacks of short duration and other ailments ascribable to his advanced age.

The last printed "Papers by G. B. Airy" belong to this year. One was the Paper before referred to "On the establishment of the Roman dominion in England": another was on the solution of a certain Equation: and there were early reminiscences of the Cambridge Tripos, &c.—In February he attended a little to a new edition of his Ipswich Lectures, but soon handed it over to Mr H. H. Turner of the Royal Observatory.—On May 23rd he was drawing up suggestions for the arrangement of the Seckford School, &c., at Wood-bridge.—On June 4th he attended the Visitation of the Royal Observatory, when a resolution was passed in favour of complete photography of the star-sky.

1888

From the 14th to 16th of May he made a short expedition to Bournemouth, and stopped on the way home to visit Winchester Cathedral.—From June 27th to Aug. 3rd he was at Playford; and again from Oct. 13th to Nov. 10th.—During the first half of the year he continued his examination of his Lunar Theory, but gradually dropped it. There are several references in his Journal to his feelings of pain and weakness, both mental and bodily: at the end of March he had an attack of gout in the fingers of his right hand. During the latter part of the year he was troubled with his private accounts, as before.—He does not appear to have been engaged on any miscellaneous matters calling for special notice in this year. But he kept up his astronomical correspondence— with Lockyer on the meteorite system of planetary formation; with Pritchard on the work of the Oxford University Observatory; with Adams on his Numerical Lunar Theory, &c., and with others.—On June 2nd he attended the Visitation of the Royal Observatory.—He amused himself occasionally with reading his favourite subjects of history and antiquities, and with looking over some of his early investigations of scientific questions.

1889

On June 5th he made a one-day's excursion to Colchester. —From July 2nd to 27th he was in the Cumberland Lake District, chiefly at Portinscale near Keswick. While staying at Portinscale he was seized with a sudden giddiness and fell upon the floor: he afterwards wrote a curious account of the visions which oppressed his brain immediately after the accident. He returned by Solihull, where his son Osmund was residing.—From Oct. 4th to Nov. 8th he was at Playford. While there he drew up a short statement of his general state of health, adverting particularly to the loss of strength in his legs and failure of his walking powers.—His health seems to have failed a good deal in this year: on Feb. 4th he had an accidental fall, and there are several entries in his Journal of mental attacks, pains in his limbs, affection of his eye-sight, &c.—In the early part of the year he was much engaged on the history of the Airy family, particularly on that of his father.—In this year the White House was sold by auction by its owners, and Airy purchased it on May 24th.—He was still in difficulties with his private accounts, but was making efforts to abandon his old and elaborate system.—For his amusement he was chiefly engaged on Theological Notes which he was compiling: and also on early optical investigations, &c.

On June 1st he attended the Visitation of the Royal Observatory, and moved a resolution that a Committee be appointed to consider whether any reduction can be effected in the amount of matter printed in the Volume of Observations of the Royal Observatory. During his tenure of office he had on various occasions brought this subject before the Board of Visitors, and with his usual tenacity of purpose he now as Visitor pressed it upon their notice.—In May he zealously joined with others in an application to get for Dr Huggins a pension on the Civil List.—In January he prepared a short Paper illustrated with diagrams to exhibit the Interference of Solar Light, as used by him in his Lectures at Cambridge in 1836: but it does not appear to have been published.—In April he received a copy of a Paper by Mr Rundell, referring to the complete adoption of his system of compass correction in iron ships, not only in the merchant service, but also in the Navy. This was a matter of peculiar gratification to Airy, who had always maintained that the method of Tables of Errors, which had been so persistently adhered to by the Admiralty, was a mistake, and that sooner or later they would find it necessary to adopt his method of mechanical correction. The passage referred to is as follows: "The name of Sir George Airy, the father of the mechanical compensation of the compass in iron vessels, having just been mentioned, it may not be inappropriate to remind you that the present year is the fiftieth since Sir George Airy presented to the Royal Society his celebrated paper on this subject with the account of his experiments on the 'Rainbow' and 'Ironsides.' Fifty years is a long period in one man's history, and Sir George Airy may well be proud in looking back over this period to see how complete has been the success of his compass investigation. His mode of compensation has been adopted by all the civilized world. Sir William Thomson, one of the latest and perhaps the most successful of modern compass adjusters, when he exhibited his apparatus in 1878 before a distinguished meeting in London, remarked that within the last ten years the application of Sir George Airy's method had become universal, not only in the merchant service, but in the navies of this and other countries, and added—The compass and the binnacles before you are designed to thoroughly carry out in practical navigation the Astronomer Royal's principles."

1890

From May 17th to 24th he was on an expedition to North Wales, stopping at Chester, Conway, Carnarvon, Barmouth, and Shrewsbury.—From June 18th to July 24th he was at Playford; and again from Oct. 11th to Nov. 15th.—In this year his powers greatly failed, and he complained frequently of mental attacks, weakness of limbs, lassitude, and failure of sleep. He occupied himself as usual with his books, papers, and accounts; and read Travels, Biblical History, &c., but nothing very persistently.

On June 7th he attended the Visitation of the Royal Observatory.—From a letter addressed to him by Mr J. Hartnup, of Liverpool Observatory, it appears that there had grown up in the mercantile world an impression that very accurate chronometers were not needed for steam ships, because they were rarely running many days out of sight of land : and Airy's opinion was requested on this matter. He replied as follows on Mar. 3rd : " The question proposed in your letter is purely a practical one. (1) If a ship is *likely* ever to be two days out of sight of land, I think that she ought to be furnished with two *good* chronometers, properly tested. (2) For the proper testing of the rates of the chrono- meters, a rating of the chronometers for three or four days in a meridional observatory is necessary. A longer testing is desirable."—In March he was in correspondence, as one of the Trustees of the Sheepshanks Fund, with the Master of Trinity relative to grants from the Fund for Cambridge Observatory.

1891

From June 16th to July 15th he was at Playford. And again from Oct. 12th to Dec. 2nd (his last visit). Throughout the year his weakness, both of brain power and muscular power, had been gradually increasing, and during this stay at Playford, on Nov. 11th, he fell down in his bed-room (probably from failure of nerve action) and was much prostrated by the shock. For several days he remained in a semi-unconscious condition, and although he rallied, yet he continued very weak, and it was not until Dec. 2nd that he could be removed to the White House. Up to the time of his fall he had been able to take frequent drives and even short walks in the neighbourhood that he was so fond of, but he could take but little exercise afterwards, and on or about Nov. 18th he made the following note: " The saddest expedition that I have ever made. We have not left home for several days."

The rapid failure of his powers during this year is well exemplified by his handwriting in his Journal entries, which, with occasional rallies, becomes broken and in places almost illegible. He makes frequent reference to his decline in strength and brain-power, and to his failing memory, but he continued his ordinary occupations, made frequent drives around Blackheath, and amused himself with his family history researches, arrangement of papers, and miscellaneous reading : and he persisted to the last with his private accounts. His interest in matters around him was still keen. On June 13th he was driving along the Greenwich Marshes in order to track the course of the great sewer; and on August 5th he visited the Crossness Sewage Works and took great interest in the details of the treatment of the sewage.—In March he contributed, with great satisfaction, to the Fund for the Portrait of his old friend Sir G. G. Stokes, with whom he had had so much scientific correspondence.—On July 25th an after-noon party was arranged to celebrate the 90th anniversary of his birthday (the actual anniversary was on July 27th). None of his early friends were there : he had survived them all. But invitations were sent to all his scientific and private friends who could be expected to come, and a large party assembled. The afternoon was very fine, and he sat in the garden and received his friends (many of whom had come from long distances) in good strength and spirits. It was a most successful gathering and was not without its meaning ; for it was felt that, under the circumstances of his failing powers, it was in all probability a final leave-taking.—On July 27th he went down to the Greenwich Parish Church at 9 p.m., to be present at the illumination of the church clock face for the first time—a matter of local interest which had necessitated a good deal of time and money.

On this occasion at the request of the company assembled in and around the Vestry he spoke for about a quarter of an hour on Time—the value of accurate time, the dissemination of Greenwich time throughout the country by time-signals from the Observatory, and the exhibition of it by time-balls, &c., &c.,—the subject to which so large a part of his life had been devoted. It was a pleasant and able speech and gave great satisfaction to the parishioners, amongst whom he had lived for so many years.—He received two illuminated addresses—one from the Astronomer Royal and Staff of the Royal Observatory; the other from the Vorstand of the Astronomische Gesellschaft at Berlin—and various private letters of congratulation. The address from the Staff of the Observatory was worded thus: "We, the present members of the Staff of the Royal Observatory, Greenwich, beg to offer you our most sincere congratulations on the occasion of your 90th birthday. We cannot but feel how closely associated we are with you, in that our whole energies are directed to the maintenance and development of that practical astronomical work, of which you essentially laid the foundation. It affords us great pleasure to think that after the conclusion of your life's work, you have been spared to live so long under the shadow of the noble Observatory with which your name was identified for half a century, and with which it must ever remain associated."

After his return from Playford he seemed to rally a little: but he soon fell ill and was found to be suffering from hernia. This necessitated a surgical operation, which was successfully performed on Dec. 17th. This gave him effectual relief, and after recovering from the immediate effects of the operation, he lay for several days quietly and without active pain reciting the English poetry with which his memory was stored. But the shock was too great for his enfeebled condition, and he died peacefully in the presence of his six surviving children on Jan. 2nd, 1892. He was buried in Playford churchyard on Jan. 7th. The funeral procession was attended at Greenwich by the whole staff of the Royal Observatory, and by other friends, and at his burial there were present two former Fellows of the College to which he had been so deeply attached.

APPENDIX.

LIST OF PRINTED PAPERS BY G. B. AIRY.

LIST OF BOOKS WRITTEN BY G. B. AIRY.

PRINTED PAPERS BY G. B. AIRY.

WITH the instinct of order which formed one of his chief characteristics Airy carefully preserved a copy of every printed Paper of his own composition. These were regularly bound in large quarto volumes, and they are in themselves a striking proof of his wonderful diligence. The bound volumes are 14 in number, and they occupy a space of 2 ft. 6 in. on a shelf. They contain 518 Papers, a list of which is appended, and they form such an important part of his life's work, that his biography would be very incomplete without a reference to them.

He was very careful in selecting the channels for the publication of his Papers. Most of the early Papers were published in the Transactions of the Cambridge Philosophical Society, but several of the most important, such as his Paper " On an inequality of long period in the motions of the Earth and Venus," were published in the Philosophical Transactions of the Royal Society, and others, such as the articles on " The Figure of the Earth," " Gravitation," " Tides and Waves," &c., were published in Encyclopædias. After his removal to Greenwich nearly all his Papers on scientific subjects (except astronomy), such as Tides, Magnetism, Correction of the Compass, &c., &c., were communicated to the Royal Society, and were published in the Philosophical Transactions. But everything astronomical was reserved for the Royal Astronomical Society. His connection with that Society was very close: he had joined it in its earliest days (the date of his election was May 9th, 1828), and regarded it as the proper medium for the discussion of current astronomical questions, and for recording astronomical progress. He was unremitting in his attendance at the Monthly Meetings of the Society, and was several times President. In the Memoirs of the Society 35 of his Papers are printed, and in addition 129 Papers in the Monthly Notices. In fact a meeting of the Society rarely passed without some communication from him, and such was his wealth of matter that sometimes he would communicate as many as 3 Papers on a single evening. For the publication of several short mathematical Papers, and especially for correspondence on disputed points of mathematical investigation, he chose as his vehicle the Philosophical Magazine, to which he contributed 32 Papers. Investigations of a more popular character he published in the Athenæum, which he also used as a vehicle for his replies to attacks on his work, or on the Establishment which he conducted : in all he made 55 communications to that Newspaper. To various Societies, such as the Institution of Civil Engineers, the British Association, the Royal Institution, &c., he presented Papers or made communications on subjects specially suited to each ; and in like manner to various Newspapers : there were 58 Papers in this category. In so long an official life there would naturally be a great number of Official Reports, Parliamentary Returns, &c., and these, with other miscellaneous Papers printed for particular objects and for a limited circulation, amounted in all to 141. Under this head come his Annual Reports to the Board of Visitors, which in themselves contain an extremely full and accurate history of the Observatory during his tenure of office. There are 46 of these Reports, and they would of themselves form a large volume of about 740 pages.

The following summary of his Printed Papers shews the manner in which they were distributed :

SUMMARY OF PRINTED PAPERS BY G. B. AIRY.

	Number of Papers.
In the Transactions of the Cambridge Philosophical Society	30
In the Philosophical Transactions of the Royal Society . .	29
In the Proceedings of the Royal Society	9
In the Memoirs of the Royal Astronomical Society . .	35
In the Monthly Notices of the Royal Astronomical Society .	129
In the Philosophical Magazine and Journal	32
In the Athenæum	55
In Encyclopædias, and in various Newspapers and Transactions	58
In Official Reports, Addresses, Parliamentary Returns, Evidence before Committees, Lectures, Letters, Sundry Treatises, and Papers	141
Total	518

PRINTED PAPERS BY G. B. AIRY.

Date when read or published.	Title of Paper.	Where published.
1822 Nov. 25	On the use of Silvered Glass for the Mirrors of Reflecting Telescopes.	Camb. Phil. Soc.
1824 Mar. 15	On the Figure assumed by a Fluid Homogeneous Mass, whose Particles are acted on by their mutual Attraction, and by small extraneous Forces.	Camb. Phil. Soc.
1824 May 17	On the Principles and Construction of the Achromatic Eye-Pieces of Telescopes, and on the Achromatism of Microscopes.	Camb. Phil. Soc.
1824	Trigonometry.	Encycl. Metrop.
1825 Feb. 21	On a peculiar Defect in the Eye, and a mode of correcting it.	Camb. Phil. Soc.
1825 May 2	On the Forms of the Teeth of Wheels.	Camb. Phil. Soc.
1826 May 8	On Laplace's Investigation of the Attraction of Spheroids differing little from a Sphere.	Camb. Phil. Soc.
1826 June 15	On the Figure of the Earth.	Phil. Trans.
1826 Nov. 26	On the Disturbances of Pendulums and Balances, and on the Theory of Escapements.	Camb. Phil. Soc.
1827 Feb. 15	Remarks on a Correction of the Solar Tables, required by Mr South's observations.	Phil. Trans.
1827 May 9	On some Passages in Mr Ivory's Remarks on a Memoir by M. Poisson relating to the Attraction of Spheroids.	Phil. Mag.
1827 {May 14, May 21}	On the Spherical Aberration of the Eye-pieces of Telescopes.	Camb. Phil. Soc.
1827 Dec. 6	On the corrections in the elements of Delambre's Solar Tables required by the observations made at the Royal Observatory, Greenwich.	Phil. Trans.

Date when read or published.	Title of Paper.	Where published.
1828 Feb. 26	Address to the Members of the Senate, on an Improvement in the Position of the Plumian Professor.	
1828 Nov. 24	On the Longitude of the Cambridge Observatory.	Camb. Phil. Soc.
1829 Nov. 13	On a method of determining the Mass of the Moon from Transit Observations of Venus near her inferior conjunction.	Astr. Soc. (Memoirs.)
1829 Nov. 16	On a Correction requisite to be applied to the Length of a Pendulum consisting of a Ball suspended by a fine Wire.	Camb. Phil. Soc.
1829 Dec. 14	On certain Conditions under which a Perpetual Motion is possible.	Camb. Phil. Soc.
1830 Aug. 17	Figure of the Earth.	Encycl. Metrop.
1831 Feb. 21	On the Nature of the Light in the Two Rays produced by the Double Refraction of Quartz.	Camb. Phil. Soc.
1831 Apr. 18	Addition to the above Paper.	Camb. Phil. Soc.
1831 Nov. 14	On a remarkable Modification of Newton's Rings.	Camb. Phil. Soc.
1831 Nov. 24	On an inequality of long period in the motions of the Earth and Venus.	Phil. Trans.
1832 Jan. 2	Translation of Encke's Dissertation (on Encke's Comet) contained in Nos. 210 and 211 of the Astronomische Nachrichten.	
1832 Mar. 5	On a new Analyzer, and its use in Experiments of Polarization.	Camb. Phil. Soc.
1832 Mar. 19	On the Phænomena of Newton's Rings when formed between two transparent Substances of different refractive Powers.	Camb. Phil. Soc.
1832 May 2	Report on the Progress of Astronomy during the present century.	Trans. Brit. Ass.
1832 Oct.	Report of the Syndicate of the Cambridge Observatory.	
1833 Feb. 2	Remarks on Mr Potter's Experiment on Interference.	Phil. Mag.
1833 Apr. 12	On the Mass of Jupiter, as determined from the Observation of Elongations of the Fourth Satellite.	R. Astr. Soc. (Memoirs.)
1833	Syllabus of a Course of Experimental Lectures.	
1833 May 7	On the Calculation of Newton's Experiments on Diffraction.	Camb. Phil. Soc.

Date when read or published.	Title of Paper.	Where published.
1833 May 7	Remarks on Sir David Brewster's Paper "On the Absorption of Specific Rays" &c.	Phil. Mag.
1833 May 16	Results of the Repetition of Mr Potter's Experiment of interposing a Prism in the Path of Interfering Light.	Phil. Mag.
1833 May	On a supposed black bar formed by Diffraction.	Phil. Mag.
1833 June 20	Report on Mr Barlow's Fluid-Lens Telescope.	R. Soc. (Proc.)
1834 Mar. 14	Continuation of Researches into the Value of the Mass of Jupiter, by observation of the Elongations of the Fourth Satellite.	R. Astr. Soc. (Memoirs.)
1834 Apr. 14	On the Latitude of Cambridge Observatory.	Camb. Phil. Soc.
1834 June	Report of the Syndicate of the Cambridge Observatory.	
1834 June 13	On the Position of the Ecliptic, as inferred from Transit and Circle Observations made at Cambridge Observatory in the year 1833.	R. Astr. Soc. (Memoirs.)
1834 June 13	Observations of the Solar Eclipse of July 16th, 1833, made at Cambridge Observatory, and Calculations of the Observations.	R. Astr. Soc. (Memoirs.)
1834 Nov. 24	On the Diffraction of an Object-Glass with Circular Aperture.	Camb. Phil. Soc.
1834 Dec. 3	On the Calculation of the Perturbations of the Small Planets and the Comets of short period.	Naut. Alm. (1837, App.)
1835 May 8	Continuation of Researches into the Value of Jupiter's Mass.	R. Astr. Soc. (Memoirs.)
1835 June	Report of the Syndicate of the Cambridge Observatory.	
1835 June 12	On the Position of the Ecliptic, as inferred from Observations with the Cambridge Transit and Mural Circle, made in the year 1834.	R. Astr. Soc. (Memoirs.)
1835 June 12	On the Time of Rotation of Jupiter.	R. Astr. Soc. (Memoirs.)
1836 Feb. 12	Speech on delivering the Medal of the R. Astr. Soc. to Sir John Herschel.	R. Astr. Soc. (Proc.)
1836 June 4	Report of the Astronomer Royal to the Board of Visitors.	

Date when read or published.	Title of Paper.	Where published.
1836 June 9	Report upon a Letter (on a Systematic Course of Magnetic Observations) addressed by M. le Baron de Humboldt to His Royal Highness the President of the Royal Society (by S. Hunter Christie and G. B. Airy).	R. Soc. (Proc.)
1837 Jan. 13	Continuation of Researches into the Value of Jupiter's Mass.	R. Astr. Soc. (Memoirs.)
1837 Feb. 10	Speech on delivering the Medal of the R. Astr. Soc. to Professor Rosenberger.	R. Astr. Soc. (Proc.)
1837 Mar. 10	Results of the Observations of the Sun, Moon, and Planets, made at Cambridge Observatory in the years 1833, 1834, and 1835.	R. Astr. Soc. (Memoirs.)
1837 May 12	On the Position of the Ecliptic, as inferred from Observations with the Cambridge Transit and Mural Circle, made in the year 1835.	R. Astr. Soc. (Memoirs.)
1837 June 3	Report of the Astronomer Royal to the Board of Visitors.	
1837 Sept. 9	Address delivered in the Town Hall of Neath.	
1837 Nov. 10	On the Parallax of a Lyræ.	R. Astr. Soc. (Memoirs.)
1838 Feb. 10	Address to the Earl of Burlington on Religious Examination in the University of London.	
1838 Mar. 26	On the Intensity of Light in the neighbourhood of a Caustic.	Camb. Phil. Soc.
1838 June 2	Report of the Astronomer Royal to the Board of Visitors.	
1838 Dec. 14	A Catalogue of 726 Stars, deduced from the Observations made at the Cambridge Observatory, from 1828 to 1835 ; reduced to January 1, 1830.	R. Astr. Soc. (Memoirs.)
1839 Apr. 25	Account of Experiments on Iron-built Ships, instituted for the purpose of discovering a correction for the deviation of the Compass produced by the iron of the Ships.	Phil. Trans.
1839 June 1	Report of the Astronomer Royal to the Board of Visitors.	
1839 Nov. 8	On the Determination of the Orbits of Comets, from Observations.	R. Astr. Soc. (Memoirs.)

Date when read or published.	Title of Paper.	Where published.
1839	Article " Gravitation."	Penny Cyclop.
1839	Article " Greenwich Observatory."	Penny Cyclop.
1840 Mar. 2	On a New Construction of the Going-Fusee.	Camb. Phil. Soc.
1840 Mar. 13	On the Regulator of the Clock-work for effecting uniform Movement of Equatoreals.	R. Astr. Soc.
1840 May 15	On the Correction of the Compass in Iron-built Ships.	Un. Serv. Journ. (Proc.)
1840	Results of Experiments on the Disturbance of the Compass in Iron-built Ships.	J. Weale.
1840 June 6	Report of the Astronomer Royal to the Board of Visitors.	
1840 June 18	On the Theoretical Explanation of an apparent new Polarity in Light.	Phil. Trans.
1840 Nov. 19	Supplement to the above Paper.	Phil. Trans.
1840 Dec. 4	On the Diffraction of an Annular Aperture.	Phil. Mag.
1840 Dec. 9	Remarks on Professor Challis's Investigation of the Motion of a Small Sphere vibrating in a Resisting Medium.	Phil. Mag.
1841 Jan. 20	Correction to the above Paper "On the Diffraction," &c.	Phil. Mag.
1841 Mar. 22	Remarks on Professor Challis's Reply to Mr Airy's Objections to the Investigation of the Resistance of the Atmosphere to an Oscillating Sphere.	Phil. Mag.
1841 June 5	Report of the Astronomer Royal to the Board of Visitors.	
1841 July 10	Reply to Professor Challis, on the Investigation of the Resistance of the Air to an Oscillating Sphere.	Phil. Mag.
1841 Oct. 26	Extraordinary Disturbance of the Magnets.	
1841 Nov. 25	On the Laws of the Rise and Fall of the Tide in the River Thames.	Phil. Trans.
1841 Dec. 21	Report of the Commissioners appointed to consider the steps to be taken for Restoration of the Standards of Weight and Measure.	
1842 Apr. 16	On the Ἰκτὶς of Diodorus.	Athenæum.
1842 May 13	Account of the Ordnance Zenith Sector.	R. Astr. Soc. (Proc.)
1842 June 4	Report of the Astronomer Royal to the Board of Visitors.	
1842 Nov. 11	Observations of the Total Solar Eclipse of 1842 July 7.	R. Astr. Soc. (Memoirs.)

Date when read or published.	Title of Paper.	Where published.
1842 Dec. 1	Remarks on the Present State of Hatcliff's Charity (Greenwich).	Private.
1842	Article on Tides and Waves.	Encyc. Metrop.
1843 Mar. 2	On the Laws of Individual Tides at Southampton and at Ipswich.	Phil. Trans.
1843 Apr. 29	On Monetary and Metrical Systems.	Athenæum.
1843 June 3	Report of the Astronomer Royal to the Board of Visitors.	
1843 Sept. 25	Address to the Individual Members of the Board of Visitors of the Royal Observatory (proposing the Altazimuth).	
1843 Oct. 6	Account of the Northumberland Equatoreal and Dome, attached to the Cambridge Observatory.	
1843 Nov. 10	Address and Explanation of the proposed Altitude and Azimuth Instrument to the Board of Visitors of the Royal Observatory.	
1844 June 1	Report of the Astronomer Royal to the Board of Visitors.	
1844 Dec. 12	On the Laws of the Tides on the Coasts of Ireland, as inferred from an extensive series of observations made in connection with the Ordnance Survey of Ireland.	Phil. Trans.
1845 Jan. 10	On the Flexure of a Uniform Bar supported by a number of equal Pressures applied at equidistant points, &c.	R. Astr. Soc. (Memoirs.)
1845 Feb. 14	Speech on delivering the Medal of the R. Astr. Soc. to Capt. Smyth.	R. Astr. Soc. (Proc.)
1845 May 9	On a New Construction of the Divided Eye-Glass Double-Image Micrometer.	R. Astr. Soc. (Memoirs.)
1845 June 7	Report of the Astronomer Royal to the Board of Visitors.	
1845 July 21	On Wexford Harbour.	
1846	Report of the Gauge Commissioners. And letter to Sir E. Ryan.	
1846 May 7	On the Equations applying to Light under the action of Magnetism.	Phil. Mag.
1846 May 12	Remarks on Dr Faraday's Paper on Ray-vibrations.	Phil. Mag.
1846 May 25	On a Change in the State of an Eye affected with a Mal-formation.	Camb. Phil. Soc.
1846 June 6	Report of the Astronomer Royal to the Board of Visitors.	

Date when read or published.	Title of Paper.	Where published.
1846 June	Account of the Measurement of an Arc of Longitude between the Royal Observatory of Greenwich and the Trigonometrical Station of Feagh Main, in the Island of Valentia.	R. Astr. Soc. (Month. Not.)
1846 July 25	Letter to Sir Robert Harry Inglis, Bart., M.P., in answer to Sir James South's attack on the Observations at the Greenwich Observatory.	Athenæum.
1846 Nov.	On the Bands formed by the partial Interception of the Prismatic Spectrum.	Phil. Mag.
1846 Nov. 13	Account of some circumstances historically connected with the Discovery of the Planet exterior to Uranus.	R. Astr. Soc. (Memoirs.)
1847 Jan. 8	Reduction of the Observations of Halley's Comet made at the Cambridge Observatory in the years 1835 and 1836.	R. Astr. Soc. (Memoirs.)
1847 Jan. 8	On a proposed Alteration of Bessel's Method for the Computation of the Corrections by which the Apparent Places of Stars are derived from the Mean Places.	R. Astr. Soc. (Memoirs.)
1847 Feb.	On Sir David Brewster's New Analysis of Solar Light.	Phil. Mag.
1847 Feb. 20	On the Name of the New Planet.	Athenæum.
1847 Feb. 27	Mr Adams and the New Planet.	Athenæum.
1847	Plan of the Buildings and Grounds of the Royal Observatory, Greenwich, with Explanation and History.	
1847 May 14	Explanation of Hansen's Perturbations of the Moon by Venus.	R. Astr. Soc. (Month. Not.)
1847 June 5	Report of the Astronomer Royal to the Board of Visitors.	
1847 Nov. 30	Address to the Individual Members of the Board of Visitors of the Royal Observatory. (Zenith Tube.)	
1847 Dec. 10	Results deduced from the Occultations of Stars and Planets by the Moon, observed at Cambridge Observatory from 1830 to 1835.	R. Astr. Soc. (Memoirs.)
1848 Feb. 11	Abstract of Struve's " Études d'Astronomie Stellaire."	R. Astr. Soc. (Month. Not.)
1848 Mar. 13	Syllabus of Lectures on Astronomy to be delivered at the Temperance Hall, Ipswich.	

Date when read or published.	Title of Paper.	Where published.
1848 Apr. 10	Remarks on Prof. Challis's Theoretical Determination of the Velocity of Sound.	Phil. Mag.
1848 May 8	Supplement to a Paper on the Intensity of Light in the neighbourhood of a Caustic.	Camb. Phil. Soc.
1848 May 12	Address to Individual Members of the Board of Visitors. (New Transit Circle, Reflex Zenith Tube, &c.)	
1848 June 3	Report of the Astronomer Royal to the Board of Visitors.	
1848 June 9	Corrections of the Elements of the Moon's Orbit, deduced from the Lunar Observations made at the Royal Observatory of Greenwich from 1750 to 1830.	R. Astr. Soc. (Memoirs.)
1848 Aug. 9	Explanation of a proposed construction of Zenith Sector: addressed to the Board of Visitors of the Royal Observatory, Greenwich.	
1848 Oct. 14	On the Construction of Chinese Balls.	Athenæum.
1849	Description of the Instruments and Process used in the Photographic self-registration of the Magnetical and Meteorological Instruments at the Royal Observatory, Greenwich.	
1849	Description of the Altitude and Azimuth Instrument erected at the Royal Observatory, Greenwich, in the year 1847.	
1849	Astronomy. (Tract written for the Scientific Manual.)	
1849 Mar. 9	Substance of the Lecture delivered by the Astronomer Royal on the large Reflecting Telescopes of the Earl of Rosse and Mr Lassell.	R. Astr. Soc. (Month. Not.)
1849 June	On a difficulty in the Problem of Sound.	Phil. Mag.
1849 June 2	Report of the Astronomer Royal to the Board of Visitors.	
1849 June 8	On Instruments adapted to the Measure of small Meridional Zenith Distances.	R. Astr. Soc. (Month. Not.)
1849 Nov. 9	Results of the Observations made by the Rev. Fearon Fallows at the Royal Observatory, Cape of Good Hope, in the years 1829, 1830, 1831.	R. Astr. Soc. (Memoirs.)
1849 Nov. 9	On Bell's Calculating Machine, and on Lord Rosse's Telescope.	R. Astr. Soc. (Month. Not.)
1849 Nov. 10	On the Exodus of the Israelites.	Athenæum.

Date when read or published.	Title of Paper.	Where published.
1849 Dec. 14	On the Method of observing and recording Transits, lately introduced in America, &c.	R. Astr. Soc. (Month. Not.)
1850 Jan. 10	On a problem of Geodesy.	Phil. Mag.
1850 Feb. 8	Address on presenting the Medal of the R. Astr. Soc. to M. Otto von Struve.	R. Astr. Soc. (Month. Not.)
1850 Mar. 15	On the Present State and Prospects of the Science of Terrestrial Magnetism.	R. Inst.
1850 Mar. 16	On the Exodus of the Israelites.	Athenæum.
1850 Mar. 30	On the Exodus of the Israelites.	Athenæum.
1850 May 10	Statement concerning Assistance granted by the Admiralty to Hansen—Also on Henderson's numbers for the teeth of wheels.	R. Astr. Soc. (Month. Not.)
1850 May 10	On the Weights to be given to the separate Results for Terrestrial Longitudes, determined by the Observation of Transits of the Moon and Fixed Stars.	R. Astr. Soc. (Memoirs.)
1850 June 1	Report of the Astronomer Royal to the Board of Visitors.	
1850 June 14	Letter from Hansen on his Lunar Tables.—Valz on an arrangement of double-image micrometer.—On the Computation of Longitude from Lunar Transits.	R. Astr. Soc. (Month. Not.)
1850 Dec. 13	On a Method of regulating the Clock-work for Equatoreals.	R. Astr. Soc. (Month. Not.)
1850 Dec. 13	Supplement to a Paper "On the Regulation of the Clock-work for effecting Uniform Movement of Equatoreals."	R. Astr. Soc. (Memoirs.)
1850 Dec. 27	On the Relation of the Direction of the Wind to the Age of the Moon, as inferred from Observations made at the Royal Observatory, Greenwich, from 1840 Nov. to 1847 Dec.	Phil. Trans.
1851 Jan. 14	Remarks on Mr Wyatt's Paper on the Construction of the Building for the Exhibition of the Works of Industry of all Nations in 1851.	Inst. C. E. (Minutes.)
1851 Feb. 15	Address on presenting the Medal of the R. Astr. Soc. to Dr Annibale de Gasparis.	R. Astr. Soc. (Month. Not.)
1851 Mar. 28	Letter to Professor Challis regarding the Adams Prize.	
1851 Mar. 29	On Cæsar's Place of Landing in Britain.	Athenæum.
1851	Suggestions to Astronomers for the Obser-	Brit. Assoc.

Date when read or published.	Title of Paper.	Where published.
	vation of the Total Eclipse of the Sun on July 28, 1851.	
1851 Apr. 11	On the Determination of the probable Stability of an Azimuthal Circle by Observations of Stars and of a permanent Collimator.	R. Astr. Soc. (Month. Not.)
1851 May 2	On the Total Solar Eclipse of 1851, July 28. (Lecture.)	R. Inst.
1851 May 9	On the Vibration of a Free Pendulum in an Oval differing little from a Straight Line.	R. Astr. Soc. (Memoirs.)
1851 June 7	Report of the Astronomer Royal to the Board of Visitors.	
1851 July 2	The President's Address to the Twenty-first Meeting of the British Association for the Advancement of Science, Ipswich.	Athenæum.
1851 Oct. 17	On Julius Cæsar's Expedition against England, in relation to his places of departure and landing.	Naut. Mag.
1851 Nov. 14	Account of the Total Eclipse of the Sun on 1851, July 28, as observed at Göttenburg, at Christiania, and at Christianstadt.	R. Astr. Soc. (Memoirs.)
1851 Dec. 13	On the Geography of the Exodus.	Athenæum.
1852 Jan. 9	On the Solar Eclipse of July 28, 1851.	{ R. Astr. Soc. (Month. Not.)
1852	On the Place of Cæsar's Departure from Gaul for the Invasion of Britain, and the Place of his Landing in Britain, with an Appendix on the Battle of Hastings.	Soc. of Antiq. (Memoirs.)
1852	On a New Method of computing the Perturbations of Planets, by J. F. Encke—translated and illustrated with notes by G. B. Airy.	Naut. Alm. 1856, App.
1852 June 5	Report of the Astronomer Royal to the Board of Visitors.	
1853 Feb. 3	On the Eclipses of Agathocles, Thales, and Xerxes.	Phil. Trans.
1853 Feb. 4	Lecture on the results of recent calculations on the Eclipse of Thales and Eclipses connected with it.	R. Inst.
1853 May 3	Address to the Individual Members of the Board of Visitors of the Royal Observatory, Greenwich. (Lunar Reductions.)	
1853 May 14	On Decimal Coinage.	Athenæum.

Date when read or published.	Title of Paper.	Where published.
1853 June 4	Report of the Astronomer Royal to the Board of Visitors.	
1853 June	Lecture on the Determination of the Longitude of the Observatory of Cambridge by means of Galvanic Signals.	R. Astr. Soc. (Month. Not.)
1853 Sept. 10	On Decimal Coinage.	Athenæum.
1853 Dec. 14	Description of the Transit Circle of the Royal Observatory, Greenwich. (App. Gr. Observ. 1852.)	
1853 Dec. 14	Regulations of the Royal Observatory, Greenwich. (App. Gr. Observ. 1852.)	
1854 Jan. 14	On the Telegraphic Longitude of Brussels.	Athenæum.
1854 Feb. 10	Address on presenting the Gold Medal of the R. Astr. Soc. to Mr Charles Rümker.	R. Astr. Soc. (Month. Not.)
1854 Feb. 25	On Reforms in the University of Cambridge.	Athenæum.
1854 Apr. 15	Letters relating to "The Late M. Mauvais."	Liter. Gaz.
1854 June 3	Report of the Astronomer Royal to the Board of Visitors.	
1854 Sept.	The Deluge.	Private.
1854 Oct. 28	On the Correction of the Compass in Iron Ships. (Scoresby's Experiments.)	Athenæum.
1854 Nov. 10	On the Difference of Longitude between the Observatories of Brussels and Greenwich, as determined by Galvanic Signals.	R. Astr. Soc. (Memoirs.)
1855 Jan. 1	Lecture at S. Shields on the Pendulum Experiments in the Harton Pit, and Letter on the Results.	
1855 Feb. 2	Lecture on the Pendulum Experiments lately made in the Harton Colliery for ascertaining the mean Density of the Earth.	R. Inst.
1855 Feb. 3	On the Correction of the Compass in Iron Ships. (Remarks on Dr Scoresby's Investigations.)	Athenæum.
1855	Address on presenting the Medal of the R. Astr. Soc. to the Rev. William Rutter Dawes.	R. Astr. Soc. (Month. Not.)
1855 Feb. 15	On the Computation of the Effect of the Attraction of Mountain Masses, as disturbing the Apparent Astronomical Latitude of Stations in Geodetic Surveys.	Phil. Trans.
1855 June 2	Report of the Astronomer Royal to the Board of Visitors.	

Date when read or published.	Title of Paper.	Where published.
1855 Oct. 18	Address to the Individual Members of the Board of Visitors of the Royal Observatory, Greenwich. (Equatoreal.)	
1855 Nov. 21	Remarks upon certain Cases of Personal Equation which appear to have hitherto escaped notice, accompanied with a Table of Results.	R. Astr. Soc. (Month. Not.)
1855 Nov. 22	Discussion of the Observed Deviations of the Compass in several Ships, Wood-built and Iron-built : with a General Table for facilitating the examination of Compass-Deviations.	Phil. Trans.
1855	Description of the Reflex Zenith Tube of the Royal Observatory, Greenwich. (App. to the Greenwich Obs. for 1854.)	
1856 Jan. 9	On Professor Peirce's Criterion for discordant observations.	Astr. Journ. (Cambr.)
1856 Jan. 24	Account of Pendulum Experiments undertaken in the Harton Colliery, for the purpose of determining the Mean Density of the Earth.	Phil. Trans.
1856 June 7	Report of the Astronomer Royal to the Board of Visitors.	
1856 Aug. 25	On Scheutz's Calculating Machine.	Phil. Mag.
1856 Aug. 30	Science and the Government. (Reply to statements in the Morning Chronicle about the instrumental equipment of the Royal Observatory.)	Athenæum.
1857 May 8	On the Means which will be available for correcting the Measure of the Sun's Distance in the next twenty-five years.	R. Astr. Soc. (Month. Not.)
1857 May 12	Knowledge expected in Computers and Assistants in the Royal Observatory.	
1857 June 6	Report of the Astronomer Royal to the Board of Visitors.	
1857 June 12	On the Eclipse of Agathocles, the Eclipse at Larissa, and the Eclipse of Thales. With an Appendix on the Eclipse of Stiklastad.	R. Astr. Soc. (Memoirs.)
1857 June 18	Account of the Construction of the New National Standard of Length, and of its principal copies.	Phil. Trans.
1857 Dec. 5	Letter to the Vice-Chancellor of Cambridge University regarding Smith's Prizes.	

Date when read or published.	Title of Paper.	Where published.
1857 Dec. 7	On the Substitution of Methods founded on Ordinary Geometry for Methods based on the General Doctrine of Proportions, in the treatment of some Geometrical Problems.	Camb. Phil. Soc.
1857	Description of the Galvanic Chronographic Apparatus of the Royal Observatory, Greenwich.	Gr. Obs. 1856, App.
1858 Mar. 8	Suggestions for Observation of the Annular Eclipse of the Sun on 1858, March 14–15.	
1858 Mar. 12	Note on Oltmann's Calculation of the Eclipse of Thales. Also On a Method of very approximately representing the Projection of a Great Circle upon Mercator's Chart.	R. Astr. Soc. (Month. Not.)
1858 May	The Atlantic Cable Problem.	Naut. Mag.
1858 May 20	Report of the Ordnance Survey Commission; together with Minutes of Evidence and Appendix.	
1858 June 5	Report of the Astronomer Royal to the Board of Visitors.	
1858 June 16	On the Mechanical Conditions of the Deposit of a Submarine Cable.	Phil. Mag.
1858 July	Instructions and Chart for Observations of Mars in right ascension at the Opposition of 1860 for obtaining the Measure of the Sun's Distance.	R. Astr. Soc. (Special.)
1858 Aug. 20	On the Advantageous Employment of Stereoscopic Photographs for the representation of Scenery.	Photog. Notes.
1858 Nov. 6	On the " Draft of Proposed New Statutes for Trinity College, Cambridge."	Athenæum.
1858 Nov. 20	Letter to the Vice-Chancellor of the University of Cambridge, offering the Sheepshanks Endowment.	
1858 Dec. 6	Suggestion of a Proof of the Theorem that Every Algebraic Equation has a Root.	Camb. Phil. Soc.
1859	Manual of Astronomy—for the Admiralty.	
1859 Feb. 1	Letter to Lord Monteagle relating to the Standards of Weights and Measures.	Parly. Paper.
1859 Feb. 4	Remarks on Mr Cayley's Trigonometrical Theorem, and on Prof. Challis's Proof that Equations have Roots.	Phil. Mag.

Date when read or published.	Title of Paper.	Where published.
1859 Mar. 11	On the Movement of the Solar System in Space.	R. Astr. Soc. (Memoirs.)
1859 Apr. 8	On the Apparent Projection of Stars upon the Moon's Disc in Occultations. Also Comparison of the Lunar Tables of Burckhardt and Hansen with Observations of the Moon made at the Royal Observatory, Greenwich.	R. Astr. Soc. (Month. Not.)
1859 Apr. 8	On the Apparent Projection of Stars upon the Moon's Disc in Occultations.	R. Astr. Soc. (Memoirs.)
1859 June 4	Report of the Astronomer Royal to the Board of Visitors.	
1859 June 10	Abstract of Maxwell's Paper " On the Stability of the Motion of Saturn's Rings."	R. Astr. Soc. (Month. Not.)
1859 July 8	Corrections of the Elements of the Moon's Orbit, deduced from the Lunar Observations made at the Royal Observatory of Greenwich from 1750 to 1851.	R. Astr. Soc. (Memoirs.)
1859 Sept. 10	On the Invasion of Britain by Julius Cæsar. (Answer to Mr Lewin.)	Athenæum.
1859 Nov. 12	On Iron Ships—The Royal Charter. (Answer to Archibald Smith's Remarks.)	Athenæum.
1859 Nov.	Circular requesting observations of small Planets.	
1859 Dec. 9	Notice of the approaching Total Eclipse of the Sun of July 18, 1860, and suggestions for observation.	R. Astr. Soc. (Month. Not.)
1859 Dec. 12	Supplement to A Proof of the Theorem that Every Algebraic Equation has a Root.	Camb. Phil. Soc.
1860 Jan. 13	Description of the New Equatoreal at the Royal Observatory, Greenwich. Also Abstract of an Essay by Gen. T. F. de Schubert on the Figure of the Earth.	R. Astr. Soc. (Month. Not.)
1860 Jan. 28	On the Claudian or Plautian Invasion of Britain.	Athenæum.
1860 Feb. 2	Examination of Navy 2-foot Telescopes at the Royal Observatory, Greenwich, 1860, Jan. 31 to Feb. 2.	
1860 Feb. 11	Report on the Instrumental Equipments of the Exchequer Office of Weights and Measures, as regards the means for preventing Fraud in the Sale of Gas to the Public; and on the Amendments which	Ho. of Commons. (Parly. Paper.)

Date when read or published.	Title of Paper.	Where published.
	may be required to the existing Legislation on that subject.	
1860 Mar. 9	Address on the approaching Solar Eclipse of July 18, 1860, &c.	R. Astr. Soc. (Month. Not.)
1860 May 10	Correspondence between the Lords Commissioners of Her Majesty's Treasury, &c., and the Astronomer Royal, relating to Gas Measurement, and the Sale of Gas Act.	Ho. of Commons. (Parly. Paper.)
1860 June 2	Report of the Astronomer Royal to the Board of Visitors. And Address to the Members of the Board in reference to Struve's Geodetic suggestions.	
1860 June 7	Correspondence regarding the Grant of £1000 to Prof. Hansen for his Lunar Tables.	
1860 Sept. 13	Remarks on a Paper entitled "On the Polar Distances of the Greenwich Transit Circle, by A. Marth." Addressed to the Members of the Board of Visitors.	
1860 Sept. 22	On Change of Climate, in answer to certain speculations by Sir Henry James.	Athenæum.
1860 Oct. 20	Circular relating to the distribution of Greenwich Observations and other publications of the Royal Observatory.	
1860 Nov. 9	Account of Observations of the Total Solar Eclipse of 1860, July 18, made at Hereña, near Miranda de Ebro; &c. &c.	R. Astr. Soc. (Month. Not.)
1860 Nov. 17	On Change of Climate: further discussion.	Athenæum.
1860	Letters on Lighthouses, to the Commission on Lighthouses.	
1860 Dec. 14	Note on the translation of a passage in a letter of Hansen's relating to coefficients.	R. Astr. Soc. (Month. Not.)
1861 Feb. 9	On the Temperature-correction of Syphon Barometers.	Athenæum.
1861 March	Results of Observations of the Solar Eclipse of 1860 July 18 made at the Royal Observatory, Greenwich, for determination of the Errors of the Tabular Elements of the Eclipse. Also Suggestion of a new Astronomical Instrument, for which the name "Orbit-Sweeper" is proposed. Also Theory of the Regulation of a Clock by Galvanic Currents acting on the Pendulum.	R. Astr. Soc. (Month. Not.)

Date when read or published.	Title of Paper.	Where published.
1861 June 1	Report of the Astronomer Royal to the Board of Visitors.	
1861 June 5	On a supposed Failure of the Calculus of Variations.	Phil. Mag.
1861 July	Report of a Committee of the R. Soc. on the advisability of re-measuring the Indian Arc of Meridian.	R. Soc. (Proc.)
1861 Sept. 21	Lecture at Manchester on the Great Solar Eclipse of July 18, 1860.	Athenæum.
1861 Sept. 21	The same Lecture.	London Review.
1861 Oct.	Examination Paper for the Sheepshanks Exhibition.	
1861 Nov. 1	Translation of Dr Lamont's Paper "On the most Advantageous Form of Magnets."	Phil. Mag.
1861 Nov. 8	Note on a Letter received from Hansen on the Lunar Theory. Also Discussion of a Result deduced by Mr D'Abbadie from Observations of the Total Solar Eclipse of 1860, July 18.	R. Astr. Soc. (Month. Not.)
1861 Nov. 16	Instructions for observing the Total Eclipse of the Sun on December 31.	
1861 Dec.	On a Projection by Balance of Errors for Maps.	Phil. Mag.
1861 Dec. 28	On the Circularity of the Sun's Disk. Also Table of Comparative Number of Observations of Small Planets.	R. Astr. Soc. (Month. Not.)
1862 Jan.	On the Direction of the Joints in the Faces of Oblique Arches.	Phil. Mag.
1862 Mar. 15	Review of "An Historical Survey of the Astronomy of the Ancients" by the Rt Hon. Sir G. Cornewall Lewis.	Athenæum.
1862 Apr. 24	Notes for the Committee on Weights and Measures, 1862.	
1862 May 15	On the Magnetic Properties of Hot-Rolled and Cold-Rolled Malleable Iron.	Phil. Trans.
1862 June 7	Report of the Astronomer Royal to the Board of Visitors.	
1862 June 24	Evidence given before the Select Committee on Weights and Measures.	
1862 Oct. 4	Biography of G. B. Airy (probably in part based upon data supplied by himself).	London Review.
1862 Oct. 11	Abstract of Paper "On the Strains in the Interior of Beams and Tubular Bridges."	Athenæum.
1862 Oct. 11	Translation of a Letter from Prof. Lamont on Dalton's Theory of Vapour, &c.	Phil. Mag.

Date when read or published.	Title of Paper.	Where published.
1862 Nov. 6	On the Strains in the Interior of Beams.	Phil. Trans.
1862 Nov.	Correspondence with Sabine concerning his attack on the Greenwich Magnetic Observations. (Confidentially communicated to the Board of Visitors.)	
1862 Nov. 21	Evidence given before the Public Schools Commission.	
1862 Nov.	Abstract of M. Auwers's Paper on the proper motion of Procyon, and Note on same.	R. Astr. Soc. (Month. Not.)
1862 Dec.	Abstract of Mr Safford's Paper on the Proper Motion of Sirius. Also on the Forms of Lenses proper for the Negative Eye-pieces of Telescopes. Also on the measurements of the Earth, and the dimensions of the Solar System. Also on Fringes of Light in Solar Eclipses.	R. Astr. Soc. (Month. Not.)
1863 Jan.	Address to the Board of Visitors on a further attack by Sabine on the Greenwich Magnetic Observations (confidential).	
1863 Jan. 9	On the Observations of Saturn made at Pulkowa and Greenwich.	R. Astr. Soc. (Month. Not.)
1863 Feb. 24	Report to the Board of Trade on the Proposed Lines of Railway through Greenwich Park.	
1863 Mar. 2	Determination of the Longitude of Valencia in Ireland by Galvanic Signals in the summer of 1862 (App. III. to the Gr. Astr. Obsns. 1862).	
1863 Mar. 13	On the Movement of the Solar System in Space, deduced from the Proper Motions of 1167 Stars. By Edwin Dunkin (for G. B. A.).	R. Astr. Soc. (Memoirs.)
1863 Mar. 13	On the Visibility of Stars in the Pleiades to the unarmed eye.	R. Astr. Soc. (Month. Not.)
1863 Mar. 21	On Marriage Odes.	Athenæum.
1863 Apr. 9	Further Report as to the Probable Effects of the London, Chatham and Dover Railway on the Royal Observatory in Greenwich Park.	
1863 Apr. 10	Determination of the Sun's Parallax from observations of Mars during the Opposition of 1862. By E. J. Stone (for G. B. A.). Also Remarks on Struve's	R. Astr. Soc. (Month. Not.)

Date when read or published.	Title of Paper.	Where published.
	account of a Local deviation in the direction of Gravity, near Moscow. Also an Account of an apparatus for the observation of the spectra of stars, and results obtained.	
1863 Apr. 23	On the Diurnal Inequalities of Terrestrial Magnetism, as deduced from observations made at the Royal Observatory, Greenwich, from 1841 to 1857.	Phil. Trans.
1863 May 8	On the Discordance between the Results for Zenith-Distances obtained by Direct Observation, and those obtained by Observation by Reflection from the Surface of Quicksilver.	R. Astr. Soc. (Memoirs.)
1863 June 6	Report of the Astronomer Royal to the Board of Visitors.	
1863 July 2	On the Amount of Light given by the Moon at the greatest stage in the Excentrically-total Eclipse, 1863, June 1.	R. Astr. Soc. (Month. Not.)
1863 Aug.	Plan of the Buildings and Grounds of the Royal Observatory, Greenwich, with Explanation and History.	
1863 Sept. 5	On the origin of the apparent luminous band which, in partial eclipses of the Sun, has been seen to surround the visible portion of the Moon's limb.	R. Astr. Soc. (Month. Not.)
{ 1863 Sept. 5 { 1863 Oct. 3	On the Invasions of Britain by Julius Cæsar.	Athenæum.
1863 Oct. 17	The Earthquake as observed from Greenwich.	Athenæum.
1863 Nov.	On the Numerical Expression of the Destructive Energy in the Explosions of Steam-Boilers, &c.	Phil. Mag.
1863 Nov. 13	Convention arranged between M. Le Verrier and the Astronomer Royal for meridional observations of the small Planets, &c.	R. Astr. Soc. (Month. Not.)
1863 Nov. 13	Translation of Hansen's Paper "Calculation of the Sun's Parallax from the Lunar Theory," with Notes by G. B. A.	R. Astr. Soc. (Month. Not.)
1863 Dec. 17	First Analysis of 177 Magnetic Storms, registered by the Magnetic Instruments in the Royal Observatory, Greenwich, from 1841 to 1857.	Phil. Trans.

Date when read or published.	Title of Paper.	Where published.
1864 Jan. 8	Pontécoulant's Paper "Sur le Coefficiant de l'Équation Parallactique déduit de la Théorie," with Notes by G. B. A.	R. Astr. Soc. (Month. Not.)
1864 Jan. 26	Remarks on Redman's Paper on the East Coast (Chesil Bank, &c.).	Inst. C. E. (Minutes.)
1864 Mar. 10	Note on a Passage in Capt. Jacob's "Measures of Jupiter," &c.	R. Astr. Soc. (Month. Not.)
1864 Mar. 11	Notes for the Committee on Weights and Measures, 1862.	Ho. of Comm. (Parly. Paper.)
1864 Mar. 17	On a Method of Slewing a Ship without the aid of the Rudder.	Inst. Nav. Arch.
1864 Apr. 5	Comparison of the Chinese Record of Solar Eclipses in the Chun Tsew with the Computations of Modern Theory.	R. Astr. Soc. (Month. Not.)
1864 June 4	Report of the Astronomer Royal to the Board of Visitors.	
1864 June 10	On the Transit of Venus, 1882, Dec. 6.	R. Astr. Soc. (Month. Not.)
1864 June 10	On the bright band bordering the Moon's Limb in Photographs of Eclipses.	R. Astr. Soc. (Month. Not.)
1864	Notes on Methods of Reduction applicable to the Indian Survey.	
1864 Sept. 3	A Visit to the Corryvreckan.	Athenæum.
1864 Sept. 29	Examination Paper for the Sheepshanks Scholarship.	
1865 Jan. 13	Comparison of the Transit-Instrument in its ordinary or reversible form with the Transit-Instrument in its non-reversible form, as adopted at Greenwich, the Cape of Good Hope, and other Observatories.	R. Astr. Soc. (Month. Not.)
1865 Mar. 9	Syllabus of a course of three Lectures on "Magnetical Errors, &c., with special reference to Iron Ships and their Compasses," delivered at the South Kensington Museum.	
1865 Apr. 1	Remarks on Mr Ellis's Lecture on the Greenwich System of Time Signals.	Horolog. Journ.
1865 Apr. 1	Free Translation of some lines of Virgil, "Citharâ crinitus Iopas," &c.	Athenæum.
1865 June 3	Report of the Astronomer Royal to the Board of Visitors.	
1865 June 17	Note on my Recommendation (in 1839) of Government Superintendence of the	Athenæum.

Date when read or published.	Title of Paper.	Where published.
	Compasses of Iron Ships. Also Note on the birthplace of Thomas Clarkson.	
1865 July	On Hemiopsy.	Phil. Mag.
1865 Aug. 22	On the Value of the Moon's Semidiameter as obtained by the Investigations of Hugh Breen, Esq., from Occultations observed at Cambridge and Greenwich.	R. Astr. Soc. (Month. Not.)
1865 Sept. 16	On " The Land of Goshen"—Reply to " A Suffolk Incumbent."	Athenæum.
1865 Oct. 21	Address of the Astronomer Royal to the individual members of the Board of Visitors. (On improved Collimators.)	
1865 Oct. 23	Note on an Error of Expression in two previous Memoirs. Also Description and History of a Quadrant made by Abraham Sharp.	R. Astr. Soc. (Month. Not.)
1865 Nov. 11	On the Possible Derivation of the National Name " Welsh."	Athenæum.
1865	Essays on the Invasion of Britain by Julius Cæsar; The Invasion of Britain by Plautius, and by Claudius Cæsar; The Early Military Policy of the Romans in Britain; The Battle of Hastings. (With corr.)	Private.
1866 Mar. 10	On " The Compass in Iron Ships." Objections to passages in a Lecture by Archibald Smith.	Athenæum.
1866 Apr. 13	On the Supposed Possible Effect of Friction in the Tides, in influencing the Apparent Acceleration of the Moon's Mean Motion in Longitude. Also on a Method of Computing Interpolations to the Second Order without Changes of Algebraic Sign.	R. Astr. Soc. (Month. Not.)
1866 June 2	Report of the Astronomer Royal to the Board of Visitors.	
1866 July 17	Papers relating to Time Signals on the Start Point.	Ho. of Comm. (Parly. Paper.)
1866 Sept. 1	On the Campaign of Aulus Plautius in Britain. (Reply to Dr Guest.)	Athenæum.
1866 Nov. 19	On the Continued Change in an Eye affected with a peculiar malformation.	Camb. Phil. Soc.
1866 Dec.	On the Simultaneous Disappearance of Jupiter's Satellites in the year 1867. Also Inference from the observed Movement	R. Astr. Soc. (Month. Not.)

Date when read or published.	Title of Paper.	Where published.
	of the Meteors in the appearance of 1866, Nov. 13–14.	
1867 Jan. 1	Memorandum for the consideration of the Commission on Standards. (Policy of introducing Metrical Standards.)	
1867 Jan. 12	On Decimal Weights and Measures.	Athenæum.
1867 Feb. 19	On the use of the Suspension Bridge with Stiffened Roadway for Railway and other Bridges of Great Span.	Inst. C. E. (Minutes.)
1867 Mar. 21	Computation of the Lengths of the Waves of Light corresponding to the Lines in the Dispersion Spectrum measured by Kirchhoff.	Phil. Trans.
1867 Mar.	Corresponding Numbers of Elevation in English Feet, and of Readings of Aneroid or Corrected Barometer in English Inches.	R. Obs. (Also Meteor. Soc. Apr. 17, 1867.)
1867 Apr. 16	Remarks on Sir W. Denison's Paper on "The Suez Canal."	Inst. C. E. (Minutes.)
1867 May 3	Statement of the History and Position of the Blue-coat Girls' School, Greenwich.	Private.
1867 June 1	Report of the Astronomer Royal to the Board of Visitors.	
1867 June 14	On Certain Appearances of the Telescopic Images of Stars described by the Rev. W. R. Dawes.	R. Astr. Soc. (Month. Not.)
1867 Dec. 13	Note on the Total Solar Eclipse of 1868, Aug. 17–18.	R. Astr. Soc. (Month. Not.)
1868	Biography of G. B. Airy. (Probably corrected by himself.)	
1868 Jan. 4	Biography (with portrait) of G. B. Airy. (Probably corrected by himself.)	Ill. Lond. News.
1868 Feb. 6	Comparison of Magnetic Disturbances recorded by the Self-registering Magnetometers at the Royal Observatory, Greenwich, with Magnetic Disturbances deduced from the corresponding Terrestrial Galvanic Currents recorded by the Self-registering Galvanometers of the Royal Observatory.	Phil. Trans.
1868 Mar. 13	Address of the Astronomer Royal to the Individual Members of the Board of Visitors. (Number of Copies of Observations.)	

Date when read or published.	Title of Paper.	Where published.
1868 June 6	Report of the Astronomer Royal to the Board of Visitors.	
1868 July 24	First Report of the Commissioners appointed to enquire into The Condition of the Exchequer Standards.	Parly. Paper.
1868 Sept. 19	The Inundation at Visp.	Athenæum.
1868 Nov. 9	On the Factorial Resolution of the Trinomial $x^n - 2 \cos n.a. + \dfrac{1}{x^n}$.	Camb. Phil. Soc.
1868 Dec. 10	On the Diurnal and Annual Inequalities of Terrestrial Magnetism, as deduced from Observations made at the Royal Observatory from 1858 to 1863, &c.	Phil. Trans.
1868 Dec. 11	On the Preparatory Arrangements for the Observation of The Transits of Venus 1874 and 1882.	R. Astr. Soc. (Month. Not.)
1868 Dec. 12	On the Migrations of the Welsh Nations.	Athenæum.
1869 Mar. 8	Memorandum by the Chairman (on the use of the Troy Weight) for the consideration of the Members of the Standards Commission.	
1869 Apr. 3	Second Report of the Commissioners appointed to enquire into the condition of the Exchequer (now Board of Trade) Standards.—The Metric System.	Parly. Paper.
1869 April	Syllabus of Lectures on Magnetism to be delivered in the University of Cambridge.	
1869 Apr. 27	Remarks on Shelford's Paper "On the Outfall of the River Humber."	Inst. C. E. (Minutes.)
1869 June 1	Memorandum for the consideration of the Standards Commission, on the state of the Question now before them regarding the suggested Abolition of Troy Weight.	
1869 June 5	Report of the Astronomer Royal to the Board of Visitors.	
1869	Supplementary Memorandum by the Astronomer Royal on the proposed Abolition of Troy Weight.	
1869 July 6	Correspondence between the Treasury, the Admiralty, and the Astronomer Royal, respecting the arrangements to be made for Observing the Transits of Venus, which will take place in the years 1874 and 1882.	Ho. of Comm. (Parly. Paper.)

Date when read or published.	Title of Paper.	Where published.
1869 Aug. 7	Note on Atmospheric Chromatic Dispersion as affecting Telescopic Observation, and on the Mode of Correcting it.	R. Astr. Soc. (Month. Not.)
1869 Oct. 19	Description of the Great Equatoreal of the Royal Observatory, Greenwich. Greenwich Observations, 1868. App.	
1870 Feb. 3	Note on an Extension of the Comparison of Magnetic Disturbances with Magnetic Effects inferred from observed Terrestrial Galvanic Currents ; &c. &c.	Phil. Trans.
1870 Apr. 8	On the question of a Royal Commission for Science.	Journ. Soc. Arts.
1870 May 2	Letters to the First Lord of the Admiralty enclosing Application of the Assistants for an increase of Salaries.	
1870 May 13	On Decimal and Metrical Systems.	Journ. Soc. Arts.
1870 June 4	Report of the Astronomer Royal to the Board of Visitors.	
1870 Aug. 27	On the meaning of the word " Whippultree."	Athenæum.
1870 Oct. 22	On the Locality of " Paradise."	Athenæum.
1870 Nov. 12	On the Locality of the Roman Gesoriacum.	Athenæum.
1870 Nov. 30	Recommendation of Prof. Miller for a Royal Medal of the Royal Society. (Quoted by the President.)	R. Soc. (Proc.)
1870	Revised Edition of " Astronomy."	Man. Naut. Sci.
1871 Jan. 21	The Burial of Sir John Moore.	Athenæum.
1871 Mar. 14	Letter to the Hydrographer of the Admiralty on the qualifications and claims of the Assistants of the Royal Observatory.	
1871 Apr. 5	Remarks on the Determination of a Ship's Place at Sea.	R. Soc. (Proc.)
1871 May 2	Remarks on Samuelson's Paper " Description of two Blast Furnaces," &c.	Inst. C. E. (Minutes.)
1871 May 3	Note on Barometric Compensation of the Pendulum.	Phil. Mag.
1871 June 3	Report of the Astronomer Royal to the Board of Visitors.	
1871 June 9	Remarks on Mr Abbott's observations on η Argûs. Also on A. S. Herschel's and J. Herschel's Mechanism for measuring Time automatically in taking Transits.	R. Astr. Soc. (Month. Not.)
1871	Erratum in Results of Greenwich Observations of the Solar Eclipse of 1860, July 18. Also Observations of the Solar	R. Astr. Soc. (Month. Not.)

Date when read or published.	Title of Paper.	Where published.
	Eclipse of 1870, Dec. 21–22, made at the Royal Observatory, Greenwich.	
1871 Aug.	Investigation of the Law of the Progress of Accuracy in the usual process for Forming a Plane Surface.	Phil. Mag.
1871 Nov. 16	Corrections to the Computed Lengths of Waves of Light for Kirchhoff's Spectral Lines.	Phil. Trans.
1871	On a supposed alteration in the amount of Astronomical Aberration of Light, produced by the passage of the Light through a considerable thickness of Refracting Medium.	R. Soc. (Proc.)
1871 Nov. 29	Biography of G. B. Airy. (Probably corrected by himself.)	Daily Telegraph.
1871 Dec. 8	Note on a special point in the determination of the Elements of the Moon's Orbit from Meridional Observations of the Moon.	R. Astr. Soc. (Month. Not.)
1871 Dec. 26	Proposed devotion of an Observatory to observation of the phenomena of Jupiter's Satellites.	R. Astr. Soc. (Month. Not.)
1872 Jan.	Address to the Council of the Royal Society on the propriety of continuing the Grant to the Kew Observatory for meteorological observations.	
1872 Feb. 8	Experiments on the Directive Power of large Steel Magnets, of Bars of magnetized Soft Iron, and of Galvanic Coils, in their Action on external small Magnets —with Appendix by James Stuart.	Phil. Trans.
1872 Feb. 12	Further Observations on the state of an Eye affected with a peculiar malformation.	Camb. Phil. Soc.
1872 Mar. 20	Notes on Scientific Education, submitted to the Royal Commission on Scientific Instruction and the Advancement of Science.	
1872 May 9	On a Supposed Periodicity in the Elements of Terrestrial Magnetism, with a period of 26¼ days.	R. Soc. (Proc.)
1872 Nov. 30	Address (as President) delivered at the Anniversary Meeting of the Royal Society.	

Date when read or published.	Title of Paper.	Where published.
1872 Dec. 19	Magnetical Observations in the Britannia and Conway Tubular Iron Bridges.	Phil. Trans.
1873 Feb. 25	Remarks on Mr Thornton's Paper on "The State Railways of India"—chiefly in reference to the proposed break of gauge.	Inst. C. E. (Minutes.)
1873 Mar. 12	Note on the want of Observations of Eclipses of Jupiter's First Satellite from 1868 to 1872.	R. Astr. Soc. (Month. Not.)
1873 Mar. 14	Letter to the Secretary of the Admiralty on certain Articles which had appeared in the Public Newspapers in regard to the approaching Transit of Venus.	R. Astr. Soc. (Month. Not.)
1873	Additional Note to the Paper on a supposed Alteration in the Amount of Astronomical Aberration of Light produced by the passage of the Light through a considerable thickness of Refracting Medium.	R. Soc. (Proc.)
1873 Apr. 10	List of Candidates for election into the Royal Society—classified.	
1873	On the Topography of the "Lady of the Lake."	Private.
1873 June 7	Report of the Astronomer Royal to the Board of Visitors.	
1873 Nov. 14	On the rejection, in the Lunar Theory, of the term of Longitude depending for argument on eight times the mean longitude of Venus minus thirteen times the mean longitude of the Earth, introduced by Prof. Hansen; &c.	R. Astr. Soc. (Month. Not.)
1873 Dec. 1	Address (as President) delivered at the Anniversary Meeting of the Royal Society.	
1874 Jan.	On a Proposed New Method of treating the Lunar Theory.	R. Astr. Soc. (Month. Not.)
1874 May 4	British Expeditions for the Observation of the Transit of Venus, 1874, December 8. Instructions to Observers.	
1874 June 6	Report of the Astronomer Royal to the Board of Visitors.	
1874 Aug. 6	Regulations of the Royal Observatory, Greenwich. Appendix to the Greenwich Observations, 1873.	
1874 Oct. 3	Science and Art. The Moon as carved on Lee church.	Athenæum.

Date when read or published.	Title of Paper.	Where published.
1874 Nov. 13	Preparations for the Observation of the Transit of Venus 1874, December 8-9.	R. Astr. Soc. (Month. Not.)
1874 Nov. 17	Remarks on the Paper "On the Nagpur Waterworks."	Inst. C. E. (Minutes.)
1874 Dec.	Telegrams relating to the Observations of the Transit of Venus 1874, Dec. 9.	R. Astr. Soc. (Month. Not.)
1875 Feb. 2	Remarks on Mr Prestwich's Paper on the Origin of the Chesil Bank.	Inst. C. E. (Minutes.)
1875 Feb. 25	Letter to the Rev. N. M. Ferrers, on the subject of the Smith's Prizes.	
1875 Mar. 12	On the Method to be used in Reducing the Observations of the Transit of Venus 1874, Dec. 8.	R. Astr. Soc. (Month. Not.)
1875 Mar.	Report on the Progress made in the Calculations for a New Method of treating the Lunar Theory.	R. Astr. Soc. (Month. Not.)
1875 June 5	Report of the Astronomer Royal to the Board of Visitors.	
1875 June 7	Apparatus for Final Adjustment of the Thermal Compensation of Chronometers, by the Astronomer Royal.	Horolog. Journ.
1875 Nov.	Chart of the Apparent Path of Mars, 1877, with neighbouring Stars. Also Spectroscopic Observations made at the Royal Observatory, Greenwich. Also Observations of the Solar Eclipse of 1875, September 28-29, made at the Royal Observatory, Greenwich.	R. Astr. Soc. (Month. Not.)
1876 Jan.	Report by the Astronomer Royal on the present state of the Calculations in his New Lunar Theory.	R. Astr. Soc. (Month. Not.)
1876 Jan. 27	Note on a point in the life of Sir William Herschel.	Athenæum.
1876 Mar. 15	Evidence given before the Government Committee on the Meteorological Committee.	
1876 May 20	On Toasting at Public Dinners.	Public Opinion.
1876 June 3	Report of the Astronomer Royal to the Board of Visitors.	
1876 Aug. 7	On a Speech attributed to Nelson.	Athenæum.
1876 Dec.	Spectroscopic Results for the Rotation of Jupiter and of the Sun, obtained at the Royal Observatory, Greenwich.	R. Astr. Soc. (Month. Not.)
1877 Jan.	Stars to be compared in R.A. with Mars,	R. Astr. Soc.

Date when read or published.	Title of Paper.	Where published.
	1877, for Determination of the Parallax of Mars.	(Month. Not.)
1877 Mar.	Note by the Astronomer Royal on the Numerical Lunar Theory. Also Remarks on Le Verrier's intra-Mercurial Planet. Also on Observations for the Parallax of Mars.	R. Astr. Soc. (Month. Not.)
1877 Mar. 27	Remarks on a Paper on "The River Thames."	Inst. C. E. (Minutes.)
1877 Apr.	On observing for Le Verrier's intra-Mercurial Planet. Also on the Parallax of Mars, and Mr Gill's proposed expedition.	R. Astr. Soc. (Month. Not.)
1877 May	On the vulgar notion that the Sun or Moon is smallest when overhead.	The Observatory (No. 2).
1877 June 2	Report of the Astronomer Royal to the Board of Visitors.	
1877 July 16	Report on the Telescopic Observations of the Transit of Venus 1874, made in the Expedition of the British Government, and on the Conclusion derived from those Observations.	Ho. of Commons Parly. Paper.
1877 Sept. 13	On Spurious Discs of Stars produced by oval object-glasses.	The Observatory (No. 7).
1877 Sept. 24	Obituary Notice of the work of Le Verrier —died Sept. 23, 1877.	Daily News.
1877 Nov. 20	On the Value of the Mean Solar Parallax &c. from the British telescopic Observations of the Transit of Venus 1874. Also Remarks on Prof. Adams's Lunar Theory.	The Observatory (No. 8).
1877 Nov.	On the Inferences for the Value of Mean Solar Parallax &c. from the Telescopic Observations of the Transit of Venus 1874, which were made in the British Expedition for the Observation of that Transit.	R. Astr. Soc. (Month. Not.)
1877	Numerical Lunar Theory: Appendix to Greenwich Astronomical Observations 1875.	
1877 Dec. 6	On the Tides at Malta.	Phil. Trans.
1878	Correspondence with Le Verrier on his Planetary Tables in 1876.	The Observatory (No. 10).
1878	On the Proposal of the French Committee to erect a Statue to Le Verrier. Also	The Observatory (No. 13).

Date when read or published.	Title of Paper.	Where published.
	on the Observation of the approaching Transit of Mercury.	
1878 Mar. 11	On the Correction of the Compass in Iron Ships without use of a Fixed Mark.	Phil. Mag.
1878 Mar. 30	On the Standards of Length in the Guildhall, London.	The Times.
1878 Apr. 27	Report of Lecture on " The probable condition of the Interior of the Earth."	W. Cumberland Times.
	On the probable condition of the Interior of the Earth—Revised Edition of above Lecture.	Trans. of the Cumberland Assoc., &c.
1878 June 1	Discussion of the Observations of the Transit of Mercury on May 6.	The Observatory (No. 14).
1878	Abstract of Lecture delivered at Cockermouth on " The Interior of the Earth."	The Observatory (No. 14).
1878 June 1	Report of the Astronomer Royal to the Board of Visitors.	
1878 July 1	Remarks on the measurement of the photographs taken in the Transit of Venus Observations.	The Observatory (No. 15).
1878 July 13	On the Variable Star R. Scuti: distortion in the Photo-heliograph.	The Observatory (No. 16).
1878	Remarks on Mr Gill's Heliometric Observations of Mars.	The Observatory (No. 20).
1878 Dec.	Note on a Determination of the Mass of Mars, and reference to his own determination in 1828. Also Note on the Conjunction of Mars and Saturn, 1879, June 30.	R. Astr. Soc. (Month. Not.)
1879 Jan. 1	On the remarkable conjunction of the Planets Mars and Saturn which will occur on 1879, June 30.	The Observatory (No. 21).
1879 Feb. 15	On the names " Cabul" and " Malek."	Athenæum.
1879 Feb. 25	On Faggot Votes in Cornwall in 1828.	Athenæum.
1879 Mar. 13	Letter on the Examination Papers for the Smith's Prizes.	
1879 Apr. 7	Drafts of Resolutions proposed concerning Sadler's Notes on the late Admiral Smyth's " Cycle of Celestial Objects."	
1879 June 1	Letter to Le Verrier, dated 1875, Feb. 5, in support of the Method of Least Squares.	The Observatory (No. 26).
1879 June 1	Remarks in debate on Sadler's " Notes" above-mentioned.	The Observatory (No. 26).
1879 June 7	Report of the Astronomer Royal to the Board of Visitors.	

Date when read or published.	Title of Paper.	Where published.
1879 July 29	Index to the Records of occasional Observations and Calculations made at the Royal Observatory, Greenwich, and to other miscellaneous Papers connected with that Institution.	R. Astr. Soc. (Month. Not. supplementary.)
1879	Biography of G. B. Airy (perhaps corrected by himself) in French, published at Geneva.	
1879 Sept.	On the Construction and Use of a Scale for Gauging Cylindrical Measures of Capacity.	Phil. Mag.
1880	On the Theoretical Value of the Acceleration of the Moon's Mean Motion.	The Observatory (No. 37).
1880	On the Secular Acceleration of the Moon —additional note.	The Observatory (No. 37).
1880 Apr. 27	Memoranda for the Commission appointed to consider the Tay Bridge casualty.	
1880 Apr.	On the Theoretical Value of the Acceleration of the Moon's Mean Motion in Longitude produced by the Change of Eccentricity of the Earth's Orbit.	R. Astr. Soc. (Month. Not.)
1880 May	On the Preparations to be made for Observation of the Transit of Venus 1882, Dec. 6.	R. Astr. Soc. (Month. Not.)
1880	On the present Proximity of Jupiter to the Earth, and on the Intervals of Recurrence of the same Phænomena.	The Observatory (No. 42).
1880 June 5	Report of the Astronomer Royal to the Board of Visitors.	
1880 Sept. 4	On the *e muet* in French.	Athenæum.
1880 Sept. 4	Excursions in the Keswick District.	Keswick Guardian.
1880 Dec. 1	Description of Flamsteed's Equatoreal Sextant, and Remarks on Graham.	The Observatory (No. 44).
1880	Addition to a Paper entitled "On the Theoretical Value of the Moon's Mean Motion in Longitude," &c.	R. Astr. Soc. (Month. Not. supplementary.)
1881 Mar.	Effect on the Moon's Movement in Latitude, produced by the slow change of Position of the Plane of the Ecliptic.	R. Astr. Soc. (Month. Not.)
1881 June 4	Report of the Astronomer Royal to the Board of Visitors.	
1881	Logarithms of the Values of all Vulgar Fractions with Numerator and Denomi-	Inst. C. E. (Minutes.)

A. B,

26

Date when read or published.	Title of Paper.	Where published.
	nator not exceeding 100 : arranged in order of magnitude.	
1881 July 6	A New Method of Clearing the Lunar Distance.—Admiralty.	
1881 Aug. 4	On a Systematic Interruption in the order of numerical values of Vulgar Fractions, when arranged in a series of consecutive magnitudes.	Phil. Mag.
1882 Sept. 15	Monthly Means of the Highest and Lowest Diurnal Temperatures of the Water of the Thames, and Comparison with the corresponding Temperatures of the Air at the Royal Observatory, Greenwich.	R. Soc. (Proc.)
1882 Oct. 19	On the Proposed Forth Bridge.	Nature.
1882 Dec. 7	On the Proposed Forth Bridge.	Nature.
1883 Jan. 21	On the Ossianic Poems.	Athenæum.
1883 Mar. 12	On the proposed Braithwaite and Buttermere Railway.	Daily News. Times. Standard.
1883 Apr. 28	Memorandum on the progress of the Numerical Lunar Theory, addressed to the Board of Visitors of the Royal Observatory, Greenwich.	
1883	Letter on The Apparent Inequality in the Mean Motion of the Moon.	The Observatory (No. 74).
1883 Aug. 18	On a Singular Morning Dream.	Nature.
1883 Sept. 10	Power of organization of the common mouse.	Nature.
1883 Nov. 17	On Chepstow Railway Bridge, with general remarks suggested by that Structure.	Nature.
1884 Mar. 8	On the Erroneous Usage of the term "arterial drainage."	Athenæum.
1884	On the Comparison of Reversible and Non-reversible Transit Instruments.	The Observatory (No. 85).
1884 Nov. 10	On an obscure passage in the Koran.	Nature. (?)
1885 May 28	An Incident in the History of Trinity College, Cambridge.	Athenæum.
1885 June 8	Incident No. 2 in the History of Trinity College, Cambridge.	Athenæum.
1885 Nov. 26	Results deduced from the Measure of Terrestrial Magnetic Force in the Horizontal Plane, at the Royal Observatory, Greenwich, from 1841 to 1876.	Phil. Trans.

Date when read or published.	Title of Paper.	Where published.
1886 Apr. 6	Integer Members of the First Centenary satisfying the Equation $A^2 = B^2 + C^2$.	Nature.
1887 Feb. 12	On the earlier Tripos of the University of Cambridge : in MSS.	Nature. (?)
1887 Apr. 14	On the Establishment of the Roman Dominion in South-East Britain.	Nature.
1887 July 23	On a special Algebraic function, and its application to the solution of some Equations: in MSS.	Camb. Phil Soc. (?)

BOOKS WRITTEN BY G. B. AIRY.

Mathematical Tracts on Physical Astronomy, the Figure of the Earth, Precession and Nutation, and The Calculus of Variations. This was published in 1826. In a 2nd Edition published in 1831 the Undulatory Theory of Optics was added to the above list. Four Editions of this work have been published, the last in 1858. The Undulatory Theory of Optics was published separately in 1877.

Gravitation: an Elementary Explanation of the Principal Perturbations in the Solar System. Written for the Penny Cyclopædia, and published previously as a book in 1834. There was a 2nd Edition in 1884.

Trigonometry. This was written for the Encyclopædia Metropolitana about 1825, and was published as a separate book in 1855 under the Title of "A Treatise on Trigonometry."

Six Lectures on Astronomy delivered at the meetings of the friends of the Ipswich Museum at the Temperance Hall, Ipswich, in the month of March 1848. These Lectures under the above Title, and that of "Popular Astronomy, a series of Lectures," have run through twelve editions.

On the Algebraical and Numerical Theory of Errors of Observations and the Combination of Observations. 1st Edition in 1861, 2nd in 1875, 3rd in 1879.

Essays on the Invasion of Britain by Julius Cæsar; The Invasion of Britain by Plautius, and by Claudius Cæsar; The Early Military Policy of the Romans in Britain; The Battle of Hastings; with Correspondence. Collected and printed for private distribution in 1865.

An Elementary Treatise on Partial Differential Equations. 1866.

On Sound and Atmospheric Vibrations, with the Mathematical Elements of Music. The 1st Edition in 1868, the 2nd in 1871.

A Treatise on Magnetism, published in 1870.

Notes on the Earlier Hebrew Scriptures, published in 1876.

Numerical Lunar Theory, published in 1886.

INDEX.

Printed in the United States
By Bookmasters